AF361656

Machine Learning in Tribology

Machine Learning in Tribology

Editors

Stephan Tremmel
Max Marian

MDPI • Basel • Beijing • Wuhan • Barcelona • Belgrade • Manchester • Tokyo • Cluj • Tianjin

Editors
Stephan Tremmel
University of Bayreuth
Germany

Max Marian
Pontificia Universidad Católica de Chile
Chile

Editorial Office
MDPI
St. Alban-Anlage 66
4052 Basel, Switzerland

This is a reprint of articles from the Special Issue published online in the open access journal *Lubricants* (ISSN 2075-4442) (available at: https://www.mdpi.com/journal/lubricants/special_issues/machine_learning_tribology).

For citation purposes, cite each article independently as indicated on the article page online and as indicated below:

LastName, A.A.; LastName, B.B.; LastName, C.C. Article Title. *Journal Name* **Year**, *Volume Number*, Page Range.

ISBN 978-3-0365-3981-2 (Hbk)
ISBN 978-3-0365-3982-9 (PDF)

© 2022 by the authors. Articles in this book are Open Access and distributed under the Creative Commons Attribution (CC BY) license, which allows users to download, copy and build upon published articles, as long as the author and publisher are properly credited, which ensures maximum dissemination and a wider impact of our publications.
The book as a whole is distributed by MDPI under the terms and conditions of the Creative Commons license CC BY-NC-ND.

Contents

About the Editors ix

Stephan Tremmel and Max Marian
Machine Learning in Tribology—More than Buzzwords?
Reprinted from: *Lubricants* **2022**, *10*, 68, doi:10.3390/lubricants10040068 1

Andreas Rosenkranz, Max Marian, Francisco J. Profito, Nathan Aragon and Raj Shah
The Use of Artificial Intelligence in Tribology—A Perspective
Reprinted from: *Lubricants* **2021**, *9*, 2, doi:10.3390/lubricants9010002 5

Andreas Almqvist
Fundamentals of Physics-Informed Neural Networks Applied to Solve the Reynolds Boundary
Value Problem
Reprinted from: *Lubricants* **2021**, *9*, 82, doi:10.3390/lubricants9080082 17

**Josef Prost, Ulrike Cihak-Bayr, Ioana Adina Neacșu, Reinhard Grundtner, Franz Pirker and
Georg Vorlaufer**
Semi-Supervised Classification of the State of Operation in Self-Lubricating Journal Bearings
Using a Random Forest Classifier
Reprinted from: *Lubricants* **2021**, *9*, 50, doi:10.3390/lubricants9050050 27

**Valentina Zambrano, Markus Brase, Belén Hernández-Gascón, Matthias Wangenheim,
Leticia A. Gracia, Ismael Viejo, Salvador Izquierdo and José Ramón Valdés**
A Digital Twin for Friction Prediction in Dynamic Rubber Applications with Surface Textures
Reprinted from: *Lubricants* **2021**, *9*, 57, doi:10.3390/lubricants9050057 45

Diwang Ruan, Xinzhou Song, Clemens Gühmann and Jianping Yan
Collaborative Optimization of CNN and GAN for Bearing Fault Diagnosis under Unbalanced
Datasets
Reprinted from: *Lubricants* **2021**, *9*, 105, doi:10.3390/lubricants9100105 69

Patricia Kügler, Max Marian, Rene Dorsch, Benjamin Schleich and Sandro Wartzack
A Semantic Annotation Pipeline towards the Generation of Knowledge Graphs in Tribology
Reprinted from: *Lubricants* **2022**, *10*, 18, doi:10.3390/lubricants10020018 87

**Christopher Sauer, Benedict Rothammer, Nicolai Pottin, Marcel Bartz, Benjamin Schleich
and Sandro Wartzack**
Design of Amorphous Carbon Coatings Using Gaussian Processes and Advanced Data
Visualization
Reprinted from: *Lubricants* **2022**, *10*, 22, doi:10.3390/lubricants10020022 113

**Sebastian Schwarz, Hannes Grillenberger, Oliver Graf-Goller, Marcel Bartz,
Stephan Tremmel and Sandro Wartzack**
Using Machine Learning Methods for Predicting Cage Performance Criteria in an Angular
Contact Ball Bearing
Reprinted from: *Lubricants* **2022**, *10*, 25, doi:10.3390/lubricants10020025 129

Christoph Bienefeld, Eckhard Kirchner, Andreas Vogt and Marian Kacmar
On the Importance of Temporal Information for Remaining Useful Life Prediction of Rolling
Bearings Using a Random Forest Regressor
Reprinted from: *Lubricants* **2022**, *10*, 67, doi:10.3390/lubricants10040067 153

Max Marian and Stephan Tremmel
Current Trends and Applications of Machine Learning in Tribology—A Review
Reprinted from: *Lubricants* **2021**, *9*, 86, doi:10.3390/lubricants9090086 . . . . **165**

About the Editors

Stephan Tremmel is Chair of Engineering Design and CAD at the Faculty of Engineering at the University of Bayreuth. His research focuses on finite element analysis (development, modeling, simulation) and on design of functionally integrated machine elements with main application in drive technology and energy engineering. Furthermore, he has special experience in surface engineering (texturing, PVD coatings) and tribology. He has 90 peer-reviewed journal publications. He is on the editorial board of *Lubricants* and is a member of the German Research Association for Drive Technology and the German Society for Tribology.

Max Marian is an Assistant Professor for Multiscale Engineering Mechanics at the Department of Mechanical and Metallurgical Engineering of Pontificia Universidad Católica de Chile. His research focuses on energy efficiency and sustainability through tribology, with an emphasis on the modification of surfaces through micro-texturing and coatings (DLC, MoS_2, MXenes). Besides machine elements and engine components, he expanded his fields towards biotribology and artificial joints as well as triboelectric nanogenerators. His research is particularly related to the development of numerical multiscale tribo-simulation and machine learning approaches. He has published more than 30 peer-reviewed journal publications, as well as a book chapter, and is on the editorial boards of *Frontiers in Chemistry Nanoscience*, *Industrial Lubrication* and *Tribology* as well as *Lubricants*. Furthermore, he is a member of the Society of Tribologists and Lubrication Engineers (STLE) and the German Society for Tribology (GfT).

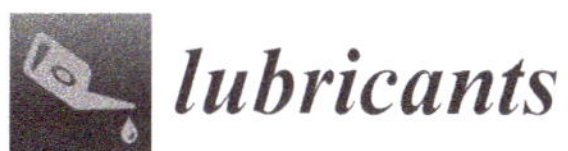

Editorial

Machine Learning in Tribology—More than Buzzwords?

Stephan Tremmel [1,*] and Max Marian [2,*]

[1] Engineering Design and CAD, University of Bayreuth, Universitätsstr. 30, 95447 Bayreuth, Germany
[2] Department of Mechanical and Metallurgical Engineering, School of Engineering, Pontificia Universidad Católica de Chile, Macul, Santiago 6904411, Chile
* Correspondence: stephan.tremmel@uni-bayreuth.de (S.T.); max.marian@ing.puc.cl (M.M.)

Citation: Tremmel, S.; Marian, M. Machine Learning in Tribology—More than Buzzwords? *Lubricants* **2022**, *10*, 68. https://doi.org/10.3390/lubricants10040068

Received: 22 March 2022
Accepted: 12 April 2022
Published: 15 April 2022

Publisher's Note: MDPI stays neutral with regard to jurisdictional claims in published maps and institutional affiliations.

Copyright: © 2022 by the authors. Licensee MDPI, Basel, Switzerland. This article is an open access article distributed under the terms and conditions of the Creative Commons Attribution (CC BY) license (https://creativecommons.org/licenses/by/4.0/).

Tribology has been and continues to be one of the most relevant fields, being present in almost all aspects of our lives. The understanding of tribology provides us with solutions for future technical challenges. At the root of all advances made so far are multitudes of precise experiments and an increasing number of advanced computer simulations across different scales and multiple physical disciplines. Based upon this sound and data-rich foundation, advanced data handling, analysis and learning methods can be developed and employed to expand existing knowledge. Therefore, modern machine learning (ML) or artificial intelligence (AI) methods provide opportunities to explore the complex processes in tribological systems and to classify or quantify their behavior in an efficient or even real-time way. Thus, their potential also goes beyond purely academic aspects into actual industrial applications.

To help pave the way, this Special Issue (SI) aimed to present the latest research on ML or AI approaches for solving tribology-related issues. The focus was less on presenting new ML or AI methods but rather on demonstrating the possible applications of existing methods and their adaptation to problems in tribology. We are pleased that the SI has collected ten articles including a perspective [1], a technical note [2], seven original research articles [3–9], and a review [10]. The contributions came from both academia and industry all around the globe and presented cutting-edge research in the field and provided deep insights into the development or the application of sophisticated ML or AI approaches to resolve problems broadly related to friction, lubrication and wear.

Rosenkranz et al. [1] opened the SI by highlighting successful case studies using AI methods in a tribological context, e.g., online condition monitoring, designing material compositions, lubricant formulations, or lubrication and fluid film formation.

Almqvist [2] derived a physics-informed neural network (PINN) applicable to solve initial and boundary value problems described by linear ordinary differential equations in the context of hydrodynamic lubrication. In contrast to finite-element- or finite-difference-based methods, the fully explicit mathematical description of the PINN is a meshless method, and the training did not require large amounts of data as are typically employed for other AI/ML training procedures.

Prost et al. [3] trained a semi-supervised Random Forest (RF) online classifier for the operational state of a self-lubricating steel shaft/bronze pairing using experimental data. Thereby, automatically generated labels or full manual labelling by an expert user can be employed. They reported that the labelling of the individual cycles from the lateral force tribometer data was crucial for a high prediction accuracy.

Zambrano et al. [4] utilized Reduced Order Modeling (ROM) to predict the friction behavior of dynamic rubber applications under different operating conditions and to find optimized micro-texture parameters such as depth, diameter, or distance. The approach was also used to evaluate the influence manufacturing deviations of the surface textures on friction. With respect to an industrial context, it is believed that the product performance of rubber products could be optimized by tailoring micro-textures and controlling nominal texture tolerances prior to production.

Ruan et al. [5] combined a Convolutional Neural Network (CNN) with a Generative Adversarial Network (GAN) for bearing fault diagnosis with unbalanced datasets. Thereby, the GAN provided a more balanced dataset for the CNN, and the CNN gave the fault diagnosis as a correction term in the GAN generator's loss function. The envelope spectrum error between the generated data and the original measurement of the fault characteristic frequencies was taken as another correction in the GAN generator's loss function. Thus, it was reported that the CNN's fault classification accuracy was substantially improved.

Kügler et al. [6] employed semantic annotation and natural language processing (NLP) techniques for generating knowledge graphs in the domain of tribology. The pipeline was built on Bidirectional Encoder Representations from Transformers (BERT) and involved some NLP tasks such as information extraction, named entity recognition and question answering. The authors verified a satisfactory performance compared to a manual annotation of publications on tribological model testing. It is believed that the approach will decrease manual effort involving time-consuming literature review by providing a semi-automatic support in knowledge acquisition.

Schwarz et al. [7] utilized ML regression methods trained by multibody simulations to predict the dynamic behavior of various cages in angular-contact ball bearings. Thereby, the hyperparameters of RF, extreme gradient boosting (XGBoost), and ANN models were optimized by an evolutionary algorithm. It was reported that all regression algorithms predicted the highly non-linear interplay of operational conditions and cage geometry with satisfactory accuracy. The authors emphasized that the ML approaches will allow to analyze a new dataset in the shortest time without the need to perform new dynamics simulations.

Sauer et al. [8] compared various supervised ML approaches for predicting the elastic and hardness characteristics of diamond-like carbon (DLC) coatings on polymeric medical materials in dependency of the sputter process parameters. It was reported that Gaussian Process Regression (GPR) featured the highest accuracy compared to polynomial regression, support vector machines (SVM), or ANN. Slicing-based data visualization and process maps can further provide support to experts when designing coating systems and processes.

Bienefeld et al. [9] used RF regression for predicting the remaining useful life (RUL) of deep-groove ball bearings from temporal information, such as the means of structure-borne sound signals. The authors reported that by taking temporal past information into account, the prediction quality could be increased by 37% compared to conventional lifetime prediction.

Finally, we [10] systematically reviewed the trends and applications of ML in tribology. We demonstrated that ML has already been employed in many fields of tribology, from composite materials and drive technology to manufacturing, surface engineering, and lubricants. It was emphasized that the intent of ML might not necessarily be to create conclusive predictive models but can be seen as complementary tool to efficiently achieve optimum designs for problems, which elude other physically motivated mathematical and numerical formulations. Therefore, ML and AI might change the landscape of what is possible, going beyond the mere understanding of mechanisms towards designing novel and/or potentially smart tribological systems. One of the challenges is that ML approaches do not necessarily guide towards specific solutions and the selection/optimization of ML algorithms is crucial.

This SI shows that there already is a wide variety of approaches that have been successfully applied to tackle tribological challenges generating true added value beyond just buzzwords. In this sense, the SI can support researchers in identifying initial selections and best practice solutions for ML in tribology.

Ultimately, the Guest Editors would like to express their sincere gratitude to all authors and reviewers contributing to this SI for their exceptional efforts and to the editorial staff of MDPI *Lubricants* for their valuable support and professional guidance.

Funding: This research received no external funding.

Institutional Review Board Statement: Not applicable.

Lubricants **2022**, *10*, 68

Informed Consent Statement: Not applicable.

Data Availability Statement: Not applicable.

Conflicts of Interest: The authors declare no conflict of interest.

References

1. Rosenkranz, A.; Marian, M.; Profito, F.J.; Aragon, N.; Shah, R. The Use of Artificial Intelligence in Tribology—A Perspective. *Lubricants* **2021**, *9*, 2. [CrossRef]
2. Almqvist, A. Fundamentals of Physics-Informed Neural Networks Applied to Solve the Reynolds Boundary Value Problem. *Lubricants* **2021**, *9*, 82. [CrossRef]
3. Prost, J.; Cihak-Bayr, U.; Neacşu, I.A.; Grundtner, R.; Pirker, F.; Vorlaufer, G. Semi-Supervised Classification of the State of Operation in Self-Lubricating Journal Bearings Using a Random Forest Classifier. *Lubricants* **2021**, *9*, 50. [CrossRef]
4. Zambrano, V.; Brase, M.; Hernández-Gascón, B.; Wangenheim, M.; Gracia, L.A.; Viejo, I.; Izquierdo, S.; Valdés, J.R. A Digital Twin for Friction Prediction in Dynamic Rubber Applications with Surface Textures. *Lubricants* **2021**, *9*, 57. [CrossRef]
5. Ruan, D.; Song, X.; Gühmann, C.; Yan, J. Collaborative Optimization of CNN and GAN for Bearing Fault Diagnosis under Unbalanced Datasets. *Lubricants* **2021**, *9*, 105. [CrossRef]
6. Kügler, P.; Marian, M.; Dorsch, R.; Schleich, B.; Wartzack, S. A Semantic Annotation Pipeline towards the Generation of Knowledge Graphs in Tribology. *Lubricants* **2022**, *10*, 18. [CrossRef]
7. Schwarz, S.; Grillenberger, H.; Graf-Goller, O.; Bartz, M.; Tremmel, S.; Wartzack, S. Using Machine Learning Methods for Predicting Cage Performance Criteria in an Angular Contact Ball Bearing. *Lubricants* **2022**, *10*, 25. [CrossRef]
8. Sauer, C.; Rothammer, B.; Pottin, N.; Bartz, M.; Schleich, B.; Wartzack, S. Design of Amorphous Carbon Coatings Using Gaussian Processes and Advanced Data Visualization. *Lubricants* **2022**, *10*, 22. [CrossRef]
9. Bienefeld, C.; Kirchner, E.; Vogt, A.; Kacmar, M. On the Importance of Temporal Information for Remaining Useful Life Prediction of Rolling Bearings Using a Random Forest Regressor. *Lubricants* **2022**, *10*, 48. [CrossRef]
10. Marian, M.; Tremmel, S. Current Trends and Applications of Machine Learning in Tribology—A Review. *Lubricants* **2021**, *9*, 86. [CrossRef]

 *lubricants*

Perspective

The Use of Artificial Intelligence in Tribology—A Perspective

Andreas Rosenkranz [1,*], Max Marian [2,*], Francisco J. Profito [3], Nathan Aragon [4] and Raj Shah [4]

[1] Department of Chemical Engineering, Biotechnology and Materials, University of Chile, Santiago 7820436, Chile

[2] Engineering Design, Friedrich-Alexander-University Erlangen-Nuremberg (FAU), 91058 Erlangen, Germany

[3] Department of Mechanical Engineering, Polytechnic School, University of São Paulo, São Paulo 17033360, Brazil; fprofito@usp.br

[4] Koehler Instrument Company, Holtsville, NY 11742, USA; nathan.aragon@stonybrook.edu (N.A.); rshah@koehlerinstrument.com (R.S.)

* Correspondence: arosenkranz@ing.uchile.cl (A.R.); marian@mfk.fau.de (M.M.)

Abstract: Artificial intelligence and, in particular, machine learning methods have gained notable attention in the tribological community due to their ability to predict tribologically relevant parameters such as, for instance, the coefficient of friction or the oil film thickness. This perspective aims at highlighting some of the recent advances achieved by implementing artificial intelligence, specifically artificial neutral networks, towards tribological research. The presentation and discussion of successful case studies using these approaches in a tribological context clearly demonstrates their ability to accurately and efficiently predict these tribological characteristics. Regarding future research directions and trends, we emphasis on the extended use of artificial intelligence and machine learning concepts in the field of tribology including the characterization of the resulting surface topography and the design of lubricated systems.

Keywords: artificial intelligence; machine learning; artificial neural networks; tribology

Citation: Rosenkranz, A.; Marian, M.; Profito, F.J.; Aragon, N.; Shah, R. The Use of Artificial Intelligence in Tribology—A Perspective. *Lubricants* 2021, 9, 2. https://dx.doi.org/ 10.3390/lubricants9010002

Received: 29 November 2020
Accepted: 23 December 2020
Published: 26 December 2020

Publisher's Note: MDPI stays neutral with regard to jurisdictional claims in published maps and institutional affiliations.

Copyright: © 2020 by the authors. Licensee MDPI, Basel, Switzerland. This article is an open access article distributed under the terms and conditions of the Creative Commons Attribution (CC BY) license (https://creativecommons.org/ licenses/by/4.0/).

1. Introduction and Background

There have been very recent advances in applying methods of deep or machine learning (ML) to improve tribological characteristics of materials by means of artificial intelligence (AI). AI is generally concerned with the design and construction of intelligent agents, which is anything that acts in the best way possible in any situation [1]. ML refers to a vast set of data-driven methods and computational tools for modelling and understanding complex datasets. These methods can be used to detect automatically patterns in datasets thus creating models to predict future data or other outcomes of interest under uncertainty [2–4]. Generally, ML methods can be divided into supervised learning and unsupervised learning [3,5], see Figure 1. Regarding predictive or supervised learning approaches, the aim is to learn a mapping from input vectors (training data) to their corresponding output vectors (target data). Depending on the nature of the target data, supervised approaches can be subdivided into classification or regression methods. When the output is a categorical or nominal variable from a finite set of discrete categories (e.g., type of surface finish, oil grade, lubricant additive, etc.), the problem is known as classification or pattern recognition. In contrast, when the output consists of one or more real-valued continuous variables (e.g., coefficient of friction, film thickness, temperature rise, etc.), the problem is defined as regression. The second type of machine learning approaches is denoted as descriptive or unsupervised learning. In this case, only inputs are provided without any corresponding output vectors. The goal is to find meaningful patterns and groups of similar features within the dataset (clustering), to determine the distribution of data in the input space (density estimation), or to reduce high-dimensional data space to two or three dimensions for visualization purposes (dimensionality reduction) [5]. Unlike supervised learning, for which comparisons can be made between the predictions to the observed values, problems involving unsupervised learning are not well-defined since no additional information or obvious error metric is provided about the patterns to be 'discovered' in the dataset [5].

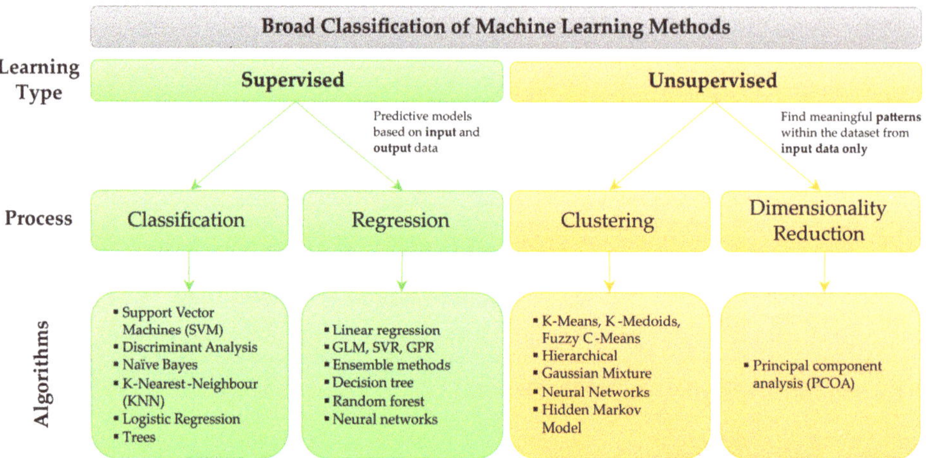

Figure 1. Diagram generally classifying existing machine learning methods and algorithms.

A prominent method that machines employ to learn is by using artificial neural networks (ANNs). These networks are based upon the network of neurons in the human brain and have the ability to "learn" in a fashion similar to the way humans do. An ANN is made of a network of model neurons, which can use algorithms to make them function like biological neurons. In this context, each model neuron has a threshold. The model neurons will receive many different inputs, which are summed up and sent an output equal to 1, if the sum is larger than the threshold. Otherwise, the output is 0. Machines are able to learn by modifying the thresholds of each model neuron, when a new example is introduced, until the thresholds reach a point to where they don't change much [6].

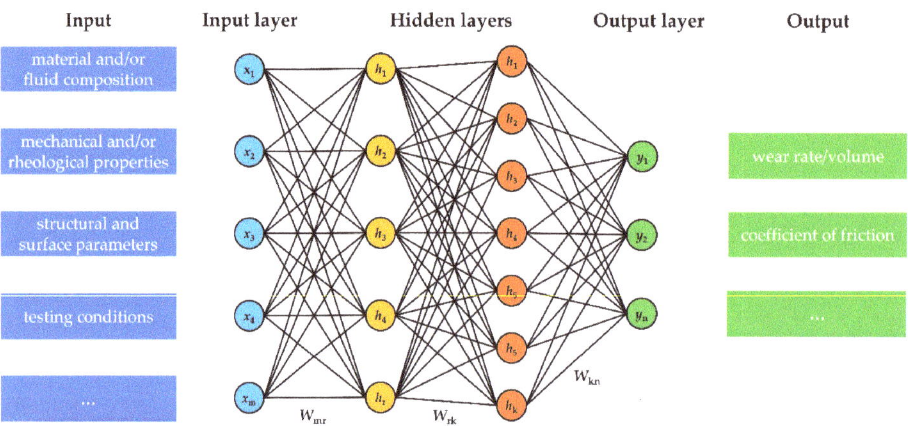

Figure 2. Correlation between material properties and testing conditions using artificial neural networks. Redrawn from [7,8].

In addition to ANNs, fuzzy systems are another type of models used in AI. These systems are based on fuzzy logic and represent a more human way of thinking in their application of inference. They are characterized by displaying a range of truth from 0 to 1 instead of displaying Boolean true/false results [9]. In the field of tribology, many tests on materials are typically performed, which define a set of tribological properties. This dataset can for example be incorporated to develop an ANN (Figure 2), which can be used for further optimization [8,10]. This perspective attempts to display some of the recent advances done in implementing AI, specifically but not limited to ANNs. Furthermore, we intend to address current challenges and future research directions towards tribological-related problems.

2. Application Fields of AI in Tribology

As briefly evidenced by several examples below, the use of AI and ML approaches already covers various fields of tribology, ranging from condition monitoring over the design of material compositions to lubricant formulations or film thickness predictions.

2.1. Online Condition Monitoring

As early as in 1998, Umeda et al. [11] trained a multilayer and a self-organizing feature map ANN with microscopy data from lubricated ball-on-disk sliding experiments to classify wear particles by means of various descriptors (width, length, projection area, perimeter, representative diameter, elongation, reflectivity etc.). When trained with representative data, the multilayer ANN successfully predicted the relation between the experimental conditions and the obtained particle features. The self-organizing feature map ANN was found to be capable of classifying the data without any supervised data. Thus, on the one hand, characteristic particle features can be identified, and, on the other hand, the authors suggested that these approaches could be used for condition monitoring, while the (partly unknown) sliding conditions can be derived from the automated particle analysis. Shortly after, Subrahmanyam and Sujatha [12] applied two ANN approaches, namely a multilayered feed forward neural network trained with supervised error back propagation (EBP) technique and an unsupervised adaptive resonance theory-2 (ART2) based neural network for the detection and diagnosis of localized defects in ball bearings. These networks were trained with vibration acceleration signals from a rolling bearing test-rig under various load and speed conditions. Thereby, the EBP and the ART2 model were found to be accurate in distinguishing a defective bearing from a normal one (100% reliability), with the ART2 being 100 times faster. Moreover, the EBP network was capable to classify the ball bearings into different states, i.e., ball or raceway defect, with a success rate over 95%. A more recent ANN-based approach for monitoring and classifying the multi-variant wear behavior of lubricated journal bearings was presented by König et al. [13]. As illustrated in Figure 3, an autoencoder was used for anomaly detection. Moreover, acoustic emission signals with continuous wavelet transformation were utilized to train a convolutional neural network to classify the modes of running-in, insufficient lubrication and particle-contamination of the oil. While the first and second were sometimes mistaken, the contaminated lubricant was detected with an accuracy and a sensitivity of 97 and 100%, respectively.

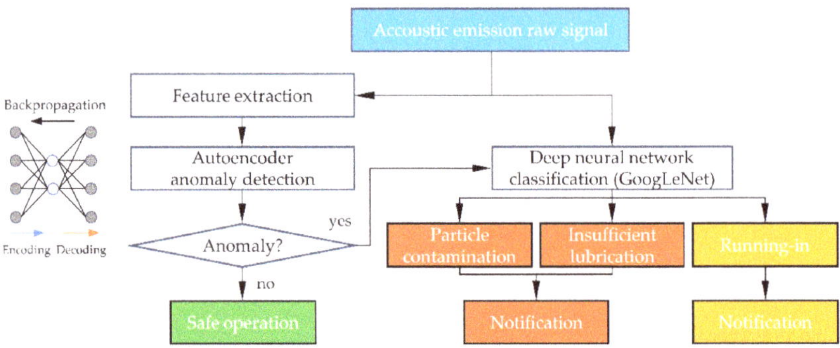

Figure 3. Framework of the machine learning approach from [13]. Redrawn and adopted with permission.

2.2. Design of Material Composition

In addition, various researchers have applied ML and AI approaches to predict and optimize the tribological behavior of different materials and operating conditions with manifold applications in mind [14–20]. For instance, Alambeigi et al. [21] investigated the accuracy of predictive AI models by comparing them with experimental results obtained by testing the dry sliding contact of sintered steels [21]. The steel for this study was manufactured by powder metallurgy, which is known to have many industrial applications in engine parts and transmission systems with problems related to friction and wear. Three different types of predictive models were used in this study. The first approach made use of an ANN. ANNs have a significant advantage in their learning ability and their handling of nonlinear functions, which the wear tests of these steel materials are mainly characterized by. The second model used is known as a "fuzzy system," specifically known as a fuzzy C-means clustering algorithm (FCM). Fuzzy systems also work well in decision making with nonlinear functions and for systems with many time-dependent parameters. The third model was based upon a fuzzy-neuro system known as adaptive neuro fuzzy inference system (ANFIS), which combines the qualities of both ANNs and regular fuzzy networks. Their tests were conducted using input parameters such as cooling rates, applied loads, sliding distances, and the type of powders. Therefore, tests were done at three different cooling rates and three different applied loads. In this study, all three methods displayed high accuracy in predicting the behavior of the wear tests with the ANN model performing the best, generating an R^2 value above 0.9911 and a root mean square error of 3.98×10^{-4} for testing data sets [21].

Apart from steels, composite materials have also been the subject of investigations based upon AI and ML approaches. Senatore et al. [22] developed an EBP model to study the tribological behavior of different brakes and clutch materials thus elucidating the influence of different materials, loading, sliding and acceleration conditions. Therefore, the three-layer ANNs were optimized by an evolutionary genetic algorithm to maximize the prediction quality and the data base was extracted from experimental pin-on-disc tests. By means of a sensitivity analysis, it was demonstrated, for instance, that the sliding velocity particularly contributed to the friction coefficient in the experiments carried out. Moreover, it was verified that the behavior within the data limits was predicted well, whereas an extrapolation rather serves merely as a first indication for future research directions. Besides, Busse and Schlarb [23] developed an ANN architecture with the Levenberg–Marquardt (LM) training algorithm and mean squared error with regularization (MSEREG) as performance function to predict the tribological properties of polyphenylene sulfide (PPS) reinforced on different scales. Using this approach, the coefficient of friction (COF) was predicted with two times and the wear rate with six times higher accuracy than with conventional ANN pruned by the optimal brain surgeon (OBS) method. In addition, the predicted error scales for both friction and wear were ten times smaller than the standard deviations from the tribological pin-on-disk experiments of the database. Both tribological properties were predicted well by using the material composition, sliding speed and contact pressure as input variables. It was also demonstrated that additional input variables such as tensile or compression properties only slightly improve the predictions for friction and wear.

2.3. Lubricant Formulations

In addition to material composition, AI/ML approaches can be used for designing lubricant formulations. Bhaumik et al. [24] used the ANN approach to create a biolubricant with optimized properties and characteristics. They used a genetic algorithm to optimize the properties and the ANN acted as the objective function for the genetic algorithm. Their goal was to create a blend of different vegetable oils, including castor, coconut, and palm oils using multiple friction modifiers, including carbon nanotubes and graphene, which were intended to be optimized using an ANN. The ANN was used to verify the effect of these inputs on the COF. Two different ANN models were used to optimize and design two different lubricants. The first lubricant (Lube A) had a composition of 40% palm oil, 40% castor oil, and 20% coconut oil with 0.7 wt.-% carbon nanotubes and no graphene. The second lubricant (Lube B) contained equal percentages of all oils with 1 wt.-% each of carbon nanotubes and graphene. Based upon the experimental results, it was shown that the COF was reduced by the addition of friction modifiers in both lubricants. Additionally, experimental work demonstrated a sensitivity regarding the respective testing conditions (four-ball versus pin-on-disk tester). A similar study was performed by Bhaumik et al. [25], in which a genetic algorithm and ANN were used to optimize and design a castor oil lubricant with graphite, graphene, multi-walled carbon nanotubes, and zinc oxide nanoparticles. Pin-on-disk tests were used to gather tribological data for the castor oil with different concentrations of these modifiers. The ANN used concentration, load, and speed as input parameters and the COF as output parameter. The composition of the designed lubricants using the ANN came out to have a total concentration of friction modifiers of 2 wt.-% with a distribution of 0.66 wt.-% each of graphite, carbon nanotubes, and zinc oxide nanoparticles. There was no graphene in the designed lubricant since the amount of graphene was shown to have a negligible effect. Afterward, it was experimentally verified that the designed lubricant induced a friction reduction by about 50% compared to most conventional mineral oils.

Other researchers [26] also analyzed the use of ANNs to design lubricants with significantly lower COFs (Figure 4). They considered the optimization of mixtures of vegetable oil (sunflower and rapeseed oils) with diesel oil for use in diesel engines. The ANN predicted a lower COF for a mixture of 4 wt.-% sunflower oil and 0 wt.-% rapeseed oil compared to a mixture of 0 wt.-% sunflower oil and 20 wt.-% rapeseed oil. The ANN also predicted a lower COF for a mixture of 6.5 wt.-% sunflower oil and 0 wt.-% rapeseed oil compared to a mixture of 0 wt.-% sunflower oil and 0 wt.-% rapeseed oil. Both predictions aligned well with experimental results.

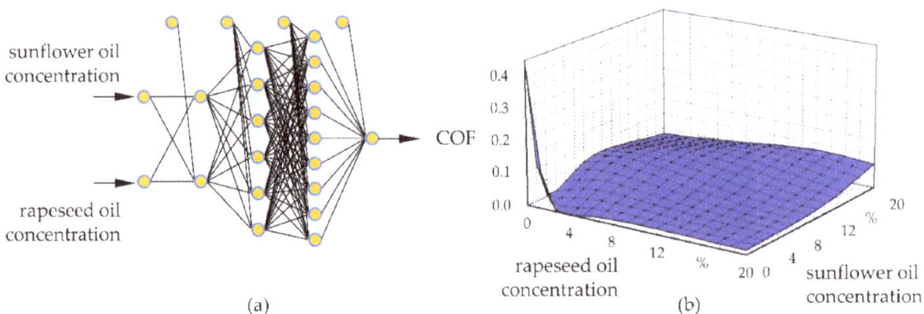

Figure 4. Artificial neural network (ANN) architecture (**a**) and coefficient of friction for different vegetable oil concentrations (**b**) according to [26] (CC BY 4.0).

2.4. Lubrication and Fluid Film Formation

In addition, AI/ML algorithms can be used to predict lubricant film formation and friction behavior in thermo-hydrodynamically (THL) and thermo-elastohydrodynamically lubricated (TEHL) contacts. For example, Moder et al. [27] used highspeed data signals of a torque sensor obtained from a journal bearing test-rig to train ML models for predicting lubrication regimes (Figure 5). Main results showed that deep and shallow neural networks performed equally well, reaching high accuracies. Furthermore, logistic regression yielded the same level of accuracy as neural networks. It was also emphasized the potential use of the proposed methodology for further investigations of ML applications on other tribological experiments.

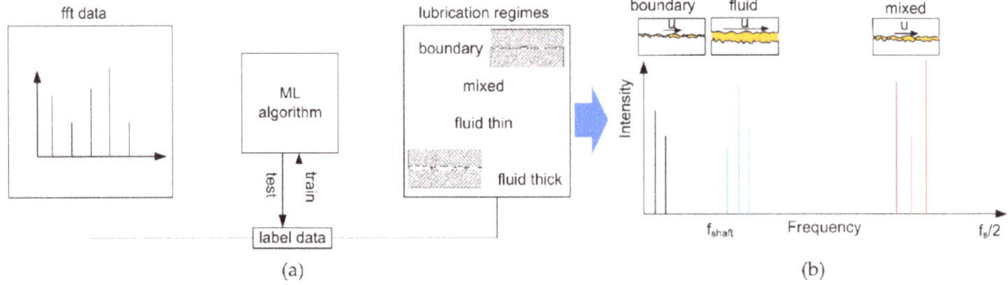

Figure 5. Lubricant regime classification approach (**a**) and schematic illustration of lubricant regimes and frequency ranges (**b**) according to [27] (CC BY 4.0).

Senatore and Ciortan [28] trained an ANN with excellent prediction quality to optimize the frictional performance of the piston-liner using data obtained from numerical HL simulations. Wang and Tsai [29] proposed a surrogate model using ANN for fast prediction of the THL lubrication performance of a slider bearing. The goal of creating the meta-model was to reduce the computational efforts of conventional THL analysis without compromising the solution accuracy. The dataset used to train and validate the ANN were obtained from numerical simulations. Results showed that when using an appropriate, results can be predicted with reasonable accuracy. Furthermore, it was also verified that the training algorithm and the sample size affect the prediction accuracy significantly. Gorasso and Wang [30] proposed a journal bearing optimization process, in which the performance functions were an ANN trained with a dataset obtained from numerical solutions of the Reynolds equation and Computational Fluid Dynamics (CFD) simulations. The optimization strategies adopted for the calculations were non-sorted genetic algorithm and artificial bee colony algorithm. Otero et al. [31] investigated the use of ANNs for predicting the friction coefficient in EHL point contacts. The model was fed with friction data obtained from tribological tests carried out for different lubricants and a range of operating conditions. It was shown that properly trained networks are capable to offer excellent predictions with a high level of correlation with the corresponding experimental data. Nonetheless, it was highlighted that special care is needed when using ANNs models as predictive since they are accompanied with the loss of relevant information, or intermediate results of interest (e.g., the complex rheological response of the lubricant at high pressure, temperature and shear-strain rate conditions), associated with the physical phenomena taking place in TEHL contacts. However, the supervised ML approach could be extended by using data from mixed-TEHL simulations for a wide range of materials and lubricant properties, contact geometries and working conditions. This would enable a fast and powerful design tool to predict the lubrication performance (e.g., film thickness, friction, temperature rise, leakage, among others) of different types of bearings and other lubricated systems [32]. For example, Marian et al. [33] applied a metamodel of optimal prognosis (MOP) to predict the influence of surface micro-textures on the frictional behavior of EHL point-contacts, see Figure 6. Thereby, the database was generated by numerical simulations considering mixed lubrication, whereby geometrical micro-texture parameters such as dimple depth and diameter were varied. Non-significant variables were then filtered and various metamodels, such as polynomial regression, moving least squares and kriging were trained. The most suitable approach was then automatically selected using a coefficient of prognosis and used for optimization by a genetic algorithm. Thus, tailored and load-case dependent surface textures can be determined.

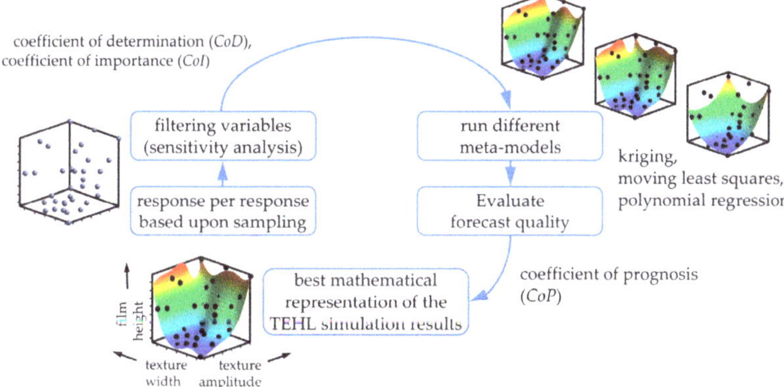

Figure 6. Framework of the metamodel of optimal prognosis to predict the tribological behavior of micro-textured EHL contacts utilized in [33].

Furthermore, Boidi et al. [34] employed the radial basis function (RBF) method for predicting the friction coefficient in lubricated contacts with textured and porous surfaces. The RBF model was trained with friction data obtained from tribological tests conducted on surfaces with different features and for a range of entrainment velocity and slide-roll ratio. The main results show that the hardy multiquadric radial basis function provided satisfactory overall correlation with the experimental data. It was also pointed out that the application of the suggested methodology could be extended to other experimental results to train more robust ML models for predicting tribological performances of textured and structured surfaces. In this respect, unsupervised ML methods could be used to construct design charts and to identify patterns of optimum performance. Furthermore, the reliability and accuracy of these ML-based tools are expected to improve continuously as more data is available. Regarding the analysis of the involved surface topography, unsupervised ML methods could be applied to achieve robust segmentation procedures for the characterization of real surface topographies. In this regard, clustering methods can help to identify and separate different surface features with tribological functionalities, such as plateau regions, bumps, honing grooves, textures, pores, wear zones, among others. Once the surface features are detected and separated in clusters, specific characterization methods and statistical assessments can be individually applied to each cluster. Furthermore, the analysis of the surface topography could further benefit from the use of supervised ML methods, in which ML algorithms are trained using post-processed data (e.g., roughness parameters, features statistics, etc.) of a variety of surfaces to create classification learning models. These trained models are intended to be used to assess surface finishing processes or correlated with tribological results to reveal specific characteristics of the surfaces that most affect the tribological performance.

3. Current Challenges, Future Research Directions and Concluding Remarks

Summarizing, the application of AI and ML has been shown to be a powerful and efficient way in predicting tribological characteristics and performance of materials with respect to valuable resources and time. Thereby, these techniques combine statistics and machine learning, imitating human intelligence at a more unconscious or untransparent level. ANNs are suitable for highly complex, non-linear fundamental and applied problems, which makes them particularly interesting for various fields of tribology. In addition, however, there are some other approaches that should certainly receive attention from tribologists. With this perspective, we intended to highlight some successful examples that show the potential for further research and future applications. In the future, AI methods could be applied in a lot more fields of tribology, e.g., the additivation of base oils with viscosity and friction modifies (for instance, nano-particles) to predict the results of experiments done on various materials compositions and test conditions.

Besides the prediction of optimal concentration, we hypothesize that AI and ML approaches will be useful to predict the size of the nano-particles (x-, y- and z-dimension) to effectively reduce friction and wear. Moreover, these approaches may be used to predict the likelihood of a tribo-layer formation, which largely depends on a complex interplay between different operational conditions. Moreover, AI and ML methods may be useful to support the characterization and classification of the involved surface topography in case of a stochastic surface roughness or even deterministic textured surfaces. Thereby, the change of the surface topography during running-in or wearing may also be addressed. From a more applied point of view, it can be also expected that AI and ML methods will be greatly involved in the design process of dry or lubricated components, see Figure 7. Apart from the prediction of optimum oil film thicknesses in machine components depending on the operational conditions such as sliding speed or load, surface topographies and textures, which are designed for specific conditions resulting in certain oil film thicknesses, are likely to be predictable by AI and ML algorithms.

With the fast-paced developments in the area of algorithms and computing power as well as the increasing availability and reusability of data [35], the utilization of AI in tribology will certainly increase in the upcoming years. To increase the range of applications and enhance the accuracy of the AI models, an online open platform could be created, on which the tribology community could share data of numerical simulations, surface characterizations and experiments. It is also conceivable that the database will not only be based on numerical simulations or experimental work on a research/laboratory scale but will also include actual operating data from real applications such as machine elements or engine components. This would enable controllers with ML/AI algorithms to be incorporated directly into these applications, e.g., rolling/sliding bearings, gears, brakes, clutches or the piston assembly, and used for performance prediction and adaptation to discontinuous and critical operating conditions.

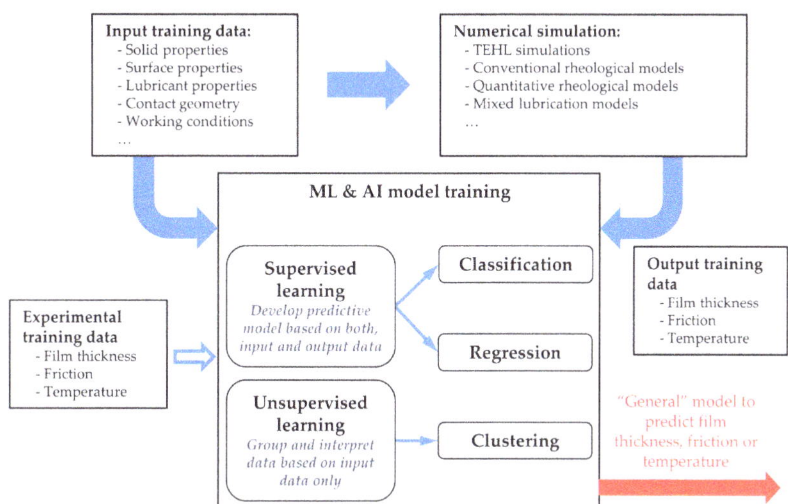

Figure 7. Schematic showing the main steps and parameters associated with the training process of ML/AI models to predict the lubrication performance from experimental and simulation data-sets.

The recently emerging big data trend won't bypass tribology and AI techniques have already be shown to be effective for many tribological questions. However, one of the biggest obstacles remaining is the handling of uncertainties in experimental and practical data sets due to differences in setups and sensor systems as well as the inherent multi-scale and statistical character of tribology with partially pronounced scattering of targeted parameters. These are not hard data but correspond to time-dependent and very specific loss variables, also resulting in frequently difficult transferability to other conditions or even tribosystems. Therefore, further fundamental research is essential for the application of new AI methods to ensure suitability and reliability in solving tribological issues. In particular, strongly domain-specific expert knowledge is crucial. The interdisciplinary character of tribology represents a great opportunity, but also a great challenge for the intense collaborations between different disciplines including physics, chemistry, materials science, mechanical engineering and computational science. Therefore, we would like to encourage tribologists all over the world to be open to new approaches/methods and interdisciplinary collaborations. Together with the aid of AI/ML algorithms, this can enable deeper insights in the incredibly important domain of tribology thus guiding us towards a new, greener and more energy-efficient era.

Author Contributions: Conceptualization, A.R., M.M. and R.S.; methodology, A.R., M.M. and F.J.P.; investigation, A.R., M.M., F.J.P., N.A. and R.S.; writing—original draft preparation, A.R., M.M., F.J.P. and R.S.; writing—review and editing, A.R., M.M., F.J.P., N.A. and R.S. All authors have read and agreed to the published version of the manuscript.

Funding: A. Rosenkranz greatly acknowledges the financial support given by CONICYT-ANID (Fondecyt Iniciación 11180121) as well as by the project "U-Inicia UI 013/2018" sponsored by the VID of the University of Chile.

Institutional Review Board Statement: Not applicable.

Informed Consent Statement: Not applicable.

Data Availability Statement: No new data were created or analyzed in this study. Data sharing is not applicable to this article.

Acknowledgments: A. Rosenkranz greatly acknowledges the continuous support of the Department of Chemical Engineering, Biotechnology and Materials of the University of Chile including VID. M. Marian would like to thank Friedrich-Alexander-University Erlangen-Nuremberg (FAU), Germany for the continuous support.

Conflicts of Interest: The authors declare no conflict of interest.

References

1. Russell, S.J.; Norvig, P.; Canny, J.F.; Edwards, D.D.; Malik, J.M.; Thrun, S. *Artificial Intelligence: A Modern Approach*, 2nd ed.; Prentice-Hall: Upper Saddle River, NJ, USA, 2003; ISBN 0-13-790395-2.
2. James, G.; Witten, D.; Hastie, T.; Tibshirani, R. *An Introduction to Statistical Learning: With Applications in R*; Springer: Basel, Switzerland, 2017; ISBN 978-1-4614-7138-7.
3. Murphy, K.P. *Machine Learning: A Probabilistic Perspective*; MIT Press: Cambridge, UK, 2013; ISBN 978-0262018029.
4. Barber, D. *Bayesian Reasoning and Machine Learning*; Cambridge University Press: Cambridge, UK, 2012; ISBN 978-0521518147.
5. Bishop, C.M. *Pattern Recognition and Machine Learning*; Springer: New York, NY, USA, 2016; ISBN 978-0-387-31073-2.
6. Krogh, A. What are artificial neural networks? *Nat. Biotechnol.* **2008**, *26*, 195–197. [CrossRef] [PubMed]
7. Friedrich, K.; Reinicke, R.; Zhang, Z. Wear of polymer composites. *Proc. Inst. Mech. Eng.* **2002**, *216*, 415–426. [CrossRef]
8. Argatov, I. Artificial Neural Networks (ANNs) as a Novel Modeling Technique in Tribology. *Front. Mech. Eng.* **2019**, *5*, 1074. [CrossRef]
9. Vlamou, E.; Papadopoulos, B. Fuzzy logic systems and medical applications. *AIMS Neurosci.* **2019**, *6*, 266–272. [CrossRef] [PubMed]
10. El Kadi, H. Modeling the mechanical behavior of fiber-reinforced polymeric composite materials using artificial neural networks—A review. *Compos. Struct.* **2006**, *73*, 1–23. [CrossRef]
11. Umeda, A.; Sugimura, J.; Yamamoto, Y. Characterization of wear particles and their relations with sliding conditions. *Wear* **1998**, *216*, 220–228. [CrossRef]
12. Subrahmanyam, M.; Sujatha, C. Using neural networks for the diagnosis of localized defects in ball bearings. *Tribol. Int.* **1997**, *30*, 739–752. [CrossRef]
13. König, F.; Sous, C.; Ouald Chaib, A.; Jacobs, G. Machine learning based anomaly detection and classification of acoustic emission events for wear monitoring in sliding bearing systems. *Tribol. Int.* **2021**, *155*, 106811. [CrossRef]
14. Senthil Kumar, P.; Manisekar, K.; Narayanasamy, R. Experimental and Prediction of Abrasive Wear Behavior of Sintered Cu-SiC Composites Containing Graphite by Using Artificial Neural Networks. *Tribol. Trans.* **2014**, *57*, 455–471. [CrossRef]
15. Zhu, J.; Shi, Y.; Feng, X.; Wang, H.; Lu, X. Prediction on tribological properties of carbon fiber and TiO$_2$ synergistic reinforced polytetrafluoroethylene composites with artificial neural networks. *Mater. Des.* **2009**, *30*, 1042–1049. [CrossRef]
16. Cavaleri, L.; Asteris, P.G.; Psyllaki, P.P.; Douvika, M.G.; Skentou, A.D.; Vaxevanidis, N.M. Prediction of Surface Treatment Effects on the Tribological Performance of Tool Steels Using Artificial Neural Networks. *Appl. Sci.* **2019**, *9*, 2788. [CrossRef]
17. Li, D.; Lv, R.; Si, G.; You, Y. Hybrid neural network-based prediction model for tribological properties of polyamide6-based friction materials. *Polym. Compos.* **2017**, *38*, 1705–1711. [CrossRef]
18. Sardar, S.; Dey, S.; Das, D. Modelling of tribological responses of composites using integrated ANN-GA technique. *J. Compos. Mater.* **2020**, *75*. [CrossRef]
19. Vinoth, A.; Datta, S. Design of the ultrahigh molecular weight polyethylene composites with multiple nanoparticles: An artificial intelligence approach. *J. Compos. Mater.* **2020**, *54*, 179–192. [CrossRef]
20. Aleksendrić, D.; Barton, D.C. Neural network prediction of disc brake performance. *Tribol. Int.* **2009**, *42*, 1074–1080. [CrossRef]
21. Alambeigi, F.; Khadem, S.M.; Khorsand, H.; Mirza Seied Hasan, E. A comparison of performance of artificial intelligence methods in prediction of dry sliding wear behavior. *Int. J. Adv. Manuf. Technol.* **2016**, *84*, 1981–1994. [CrossRef]
22. Senatore, A.; D'Agostino, V.; Di Giuda, R.; Petrone, V. Experimental investigation and neural network prediction of brakes and clutch material frictional behaviour considering the sliding acceleration influence. *Tribol. Int.* **2011**, *44*, 1199–1207. [CrossRef]
23. Busse, M.; Schlarb, A.K. A novel neural network approach for modeling tribological properties of polyphenylene sulfide reinforced on different scales. In *Tribology of Polymeric Nanocomposites*; Elsevier: Oxford, UK, 2013; pp. 779–793, ISBN 9780444594556.
24. Bhaumik, S.; Mathew, B.R.; Datta, S. Computational intelligence-based design of lubricant with vegetable oil blend and various nano friction modifiers. *Fuel* **2019**, *241*, 733–743. [CrossRef]
25. Bhaumik, S.; Pathak, S.D.; Dey, S.; Datta, S. Artificial intelligence based design of multiple friction modifiers dispersed castor oil and evaluating its tribological properties. *Tribol. Int.* **2019**, *140*, 105813. [CrossRef]
26. Humelnicu, C.; Ciortan, S.; Amortila, V. Artificial Neural Network-Based Analysis of the Tribological Behavior of Vegetable Oil-Diesel Fuel Mixtures. *Lubricants* **2019**, *7*, 32. [CrossRef]
27. Moder, J.; Bergmann, P.; Grün, F. Lubrication Regime Classification of Hydrodynamic Journal Bearings by Machine Learning Using Torque Data. *Lubricants* **2018**, *6*, 108. [CrossRef]
28. Senatore, A.; Ciortan, S. An application of artificial neural networks to piston ring friction losses prediction. *Mech. Test. Diagn.* **2011**, *1*, 7–14.
29. Wang, N.; Tsai, C.-M. Assessment of artificial neural network for thermohydrodynamic lubrication analysis. *Ind. Lubr. Tribol.* **2020**, *72*, 1233–1238. [CrossRef]

30. Gorasso, L.; Wang, L. Journal Bearing Optimization Using Nonsorted Genetic Algorithm and Artificial Bee Colony Algorithm. *Adv. Mech. Eng.* **2014**, *6*, 213548. [CrossRef]
31. Echávarri Otero, J.; De La Guerra Ochoa, E.; Chacón Tanarro, E.; Lafont Morgado, P.; Díaz Lantada, A.; Munoz-Guijosa, J.M.; Muñoz Sanz, J.L. Artificial neural network approach to predict the lubricated friction coefficient. *Lubr. Sci.* **2014**, *26*, 141–162. [CrossRef]
32. Marian, M.; Bartz, M.; Wartzack, S.; Rosenkranz, A. Non-Dimensional Groups, Film Thickness Equations and Correction Factors for Elastohydrodynamic Lubrication: A Review. *Lubricants* **2020**, *8*, 95. [CrossRef]
33. Marian, M.; Grützmacher, P.; Rosenkranz, A.; Tremmel, S.; Mücklich, F.; Wartzack, S. Designing surface textures for EHL point-contacts—Transient 3D simulations, meta-modeling and experimental validation. *Tribol. Int.* **2019**, *137*, 152–163. [CrossRef]
34. Boidi, G.; Rodrigues da Silva, M.; Profito, F.J.J.; Machado, I.F. Using Machine Learning Radial Basis Function (RBF) Method for Predicting Lubricated Friction on Textured and Porous Surfaces. *Surf. Topogr. Metrol. Prop.* **2020**, *8*, 044002. [CrossRef]
35. Kügler, P.; Marian, M.; Schleich, B.; Tremmel, S.; Wartzack, S. tribAIn—Towards an Explicit Specification of Shared Tribological Understanding. *Appl. Sci.* **2020**, *10*, 4421. [CrossRef]

Technical Note

Fundamentals of Physics-Informed Neural Networks Applied to Solve the Reynolds Boundary Value Problem

Andreas Almqvist

Division of Machine Elements, Luleå University of Technology, SE-971 87 Luleå, Sweden; andreas.almqvist@ltu.se

Abstract: This paper presents a complete derivation and design of a physics-informed neural network (PINN) applicable to solve initial and boundary value problems described by linear ordinary differential equations. The objective with this technical note is not to develop a numerical solution procedure which is more accurate and efficient than standard finite element- or finite difference-based methods, but to give a fully explicit mathematical description of a PINN and to present an application example in the context of hydrodynamic lubrication. It is, however, worth noticing that the PINN developed herein, contrary to FEM and FDM, is a meshless method and that training does not require big data which is typical in machine learning.

Keywords: PINN; machine learning; reynolds equation

Citation: Almqvist, A. Fundamentals of Physics-Informed Neural Networks Applied to Solve the Reynolds Boundary Value Problem. *Lubricants* **2021**, *9*, 82. https://doi.org/10.3390/lubricants9080082

Received: 31 July 2021
Accepted: 14 August 2021
Published: 19 August 2021

Publisher's Note: MDPI stays neutral with regard to jurisdictional claims in published maps and institutional affiliations.

Copyright: © 2021 by the authors. Licensee MDPI, Basel, Switzerland. This article is an open access article distributed under the terms and conditions of the Creative Commons Attribution (CC BY) license (https://creativecommons.org/licenses/by/ 4.0/).

1. Introduction

There are various categories of artificial neural networks (ANN) and a physics-informed neural network (PINN), see [1] for a recent review on the matter, is a neural network trained to solve both supervised and unsupervised learning tasks while satisfying some given laws of physics, which may be described in terms of nonlinear partial differential equations (PDE). For example, the balance of momentum and conservation laws in solid- and fluid mechanics and various types of initial value problems (IVP) and boundary value problems (BVP), see e.g., [2,3]. The application of a PINN (of this type) to solve differential equations, renders meshless numerical solution procedures [4], and an important feature from a machine learning perspective, is that it is not a data-driven approach requiring a large set of training data to learn the solution.

In fluid mechanics, under certain assumptions, i.e., that the fluid is incompressible, iso-viscous, the balance of linear momentum and the continuity equation, for flows in narrow interfaces reduces to the classical Reynolds equation [5]. For more recent work establishing lower-dimensional models in a similar manner, see e.g., [6–8]. The present work describes how a PINN can be adapted and trained to solve both initial and boundary value problems, described by ordinary differential equations, numerically. The theoretical description starts by presenting the neural network's architecture and it is first applied to solve an initial value problem, which is described by a first order ODE, which can be solved analytically so that the validity of the solution can be thoroughly assessed. Thereafter, it is used to obtain a PINN for the classical one-dimensional Reynolds equation, which is a boundary value problem governing, e.g., the flow of lubricant between the runner and the stator in a 1D slider bearing. The novelty and originality of the present work lays in the explicit mathematical description of the cost function, which constitutes the *physics-informed* feature of the ANN, and the associated gradient with respect to the networks weights and bias. Important features of this particular numerical solution procedure, that is publicly available here: [9], are that it is not data driven, i.e., no training data need to be provided and that it is a meshless method [4].

2. PINN Architecture

Knowing the characteristics of the solution to the differential equation under consideration is very helpful when designing the PINN architecture, including structure, number of hidden layers, activation function, etc. For this reason, the PINN developed here has one input node x (the independent variable representing the spatial coordinate), one hidden layer consisting of N nodes and one output node y (the dependent variable representing pressure). Figure 1 depicts a graphical illustration of the present architecture, which when trained solves both the IVP example and the Reynolds BVP considered here.

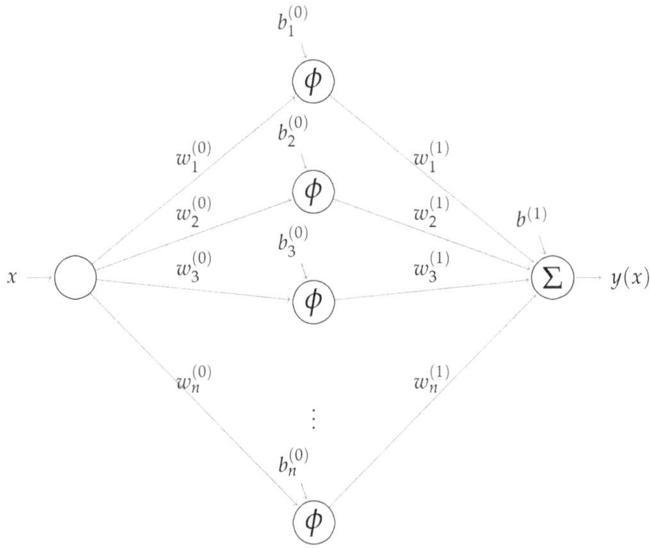

Figure 1. Architecture of the PINN employed to solve the IVP and BVP considered here.

The Sigmoid function, i.e.,

$$\phi(\xi) = \frac{1}{1 + e^{-\xi}}, \tag{1}$$

which is mapping $\mathbb{R}$ to $[0, 1]$ and exhibits the property

$$\phi'(\xi) = \phi(\xi)(1 - \phi(\xi)). \tag{2}$$

is employed as activation function for the hidden layer. This means that the neural network has $3N + 1$ trainable parameters. That is, the weights $w_i^{(0)}$ and bias $b_i^{(0)}$ for the nodes in the hidden layer and the weights $w_i^{(1)}$, $i = 1 \ldots N$, for each synapses connecting them with the output node, plus the bias $b^{(1)}$ applied there.

Based on this particular architecture, the output z_i of each node in the first hidden layer is,

$$z_i(x) = \phi\left(w_i^{(0)} x + b_i^{(0)}\right). \tag{3}$$

The output value is then given by applying the Sigmoid activation function scaled by the weight from the node in the second layer and yields

$$y(x) = b^{(1)} + \sum_{i=1}^{N} w_i^{(1)} z_i(x) = b^{(1)} + \sum_{i=1}^{N} w_i^{(1)} \phi\left(w_i^{(0)} x + b_i^{(0)}\right). \tag{4}$$

Let us now construct the cost function which the network will be trained to minimise. While the cost function appearing in a typical machine learning procedure is just the quadratic difference between the predicted and the target values, it will here be defined by means of the operators $\mathcal{L}$ and $\mathcal{B}$. The cost function applied here reads

$$l = \left\langle (\mathcal{L}y - f)^2 \right\rangle + ((\mathcal{B}y - \mathbf{b}) \cdot \mathbf{e_1})^2 + ((\mathcal{B}y - \mathbf{b}) \cdot \mathbf{e_2})^2, \tag{5}$$

where $\langle f \rangle$ defines the average value of f, and this is exactly the feature that makes an ANN "physics informed", i.e., a PINN.

Since $\mathcal{L}y$ is a differential operator the cost function contains derivatives of the network output (4). In order to obtain an expression of the cost function, in terms of the input x, the weights w and bias b, the network output (4), must be differentiated twice with respect to (w.r.t.) x. This can be accomplished by some kind of automatic differentiation (AD) (also referred to as algorithmic differentiation, computer differentiation, auto-differentiation or simply autodiff), which is a computerised methodology based on the chain rule, which can be applied to efficiently and accurately evaluate derivatives of numeric functions, see e.g., [10,11]. The present work instead applies symbolic differentiation to clearly explain all the essential details of the PINN. Indeed, differentiating one yield

$$y'(x) = \frac{\partial}{\partial x}\left(\left(\sum_{i=1}^{N} w_i^{(1)} z_i(x)\right) + b^{(1)}\right) = \frac{\partial}{\partial x}\left(\left(\sum_{i=1}^{N} w_i^{(1)} \phi\left(w_i^{(0)} x + b_i^{(0)}\right)\right) + b^{(1)}\right) =$$

$$= \sum_{i=1}^{N} w_i^{(1)} w_i^{(0)} \phi'\left(w_i^{(0)} x + b_i^{(0)}\right) = \sum_{i=1}^{N} w_i^{(1)} w_i^{(0)} \phi\left(w_i^{(0)} x + b_i^{(0)}\right)\left(1 - \phi\left(w_i^{(0)} x + b_i^{(0)}\right)\right), \tag{6}$$

and, because of (2), a consecutive differentiation then yields

$$y''(x) = \frac{\partial}{\partial x}\left(\sum_{i=1}^{N} w_i^{(1)} w_i^{(0)} \phi'\left(w_i^{(0)} x + b_i^{(0)}\right)\right) = \sum_{i=1}^{N} w_i^{(1)} \left(w_i^{(0)}\right)^2 \phi''\left(w_i^{(0)} x + b_i^{(0)}\right) =$$

$$= \sum_{i=1}^{N} w_i^{(1)} \left(w_i^{(0)}\right)^2 \phi'\left(w_i^{(0)} x + b_i^{(0)}\right)\left(1 - 2\phi\left(w_i^{(0)} x + b_i^{(0)}\right)\right) = \tag{7}$$

$$= \sum_{i=1}^{N} w_i^{(1)} \left(w_i^{(0)}\right)^2 \phi\left(w_i^{(0)} x + b_i^{(0)}\right)\left(1 - \phi\left(w_i^{(0)} x + b_i^{(0)}\right)\right)\left(1 - 2\phi\left(w_i^{(0)} x + b_i^{(0)}\right)\right).$$

Moreover, finding the set of weights and bias minimising the cost function requires its partial derivatives w.r.t. to each weight and bias defining the PINN. In the subsections below, we will present how to achieve this, by first considering a first order differential equation with an analytical solution, and, thereafter, we will consider the classical Reynolds equation which is a second order (linear) ODE that describes laminar flow of incompressible and iso-viscous fluids in narrow interfaces.

3. A First Order ODE Example

Let us consider the first order ODE, describing the initial value problem (IVP) given by

$$\mathcal{L}y - f = y' + 2xy = 0, \quad x > 0 \tag{8a}$$
$$\mathcal{B}y - \mathbf{b} = y(0) - 1 = 0, \tag{8b}$$

with the exact solution $y = e^{-x^2}$. By means of (6), a cost function suitable for solving (8) may be generated by

$$l = \left\langle \left[\sum_{i=1}^{N} w_i^{(1)} w_i^{(0)} \phi\left(w_i^{(0)} x + b_i^{(0)}\right)\left(1 - \phi\left(w_i^{(0)} x + b_i^{(0)}\right)\right) + \right.\right.$$

$$\left.\left. 2x\left(\left(\sum_{i=1}^{N} w_i^{(1)} \phi\left(w_i^{(0)} x + b_i^{(0)}\right)\right) + b^{(1)}\right)\right]^2 \right\rangle + [y(0) - 1]^2 \tag{9}$$

The solution of (8) can be obtained by implementing a training routine which iteratively finds the set of weights w and bias b that minimises (9) (and similarly for (19) minimising (17)). The most well-known of these is the Gradient Decent method attributed to Cauchy, who first suggested it in 1847 [12]. For an overview, see, e.g., [13].

As mentioned in the previous section, the derivatives of (4) w.r.t. to the weights w and bias b are required to find them, and *automatic differentiation* is, normally, employed to perform the differentiation. However, here we carry out symbolic differentiation to demonstrate exactly the explicit expressions that constitute the gradient of the cost function. Indeed, by taking the partial derivatives we obtain

$$\frac{\partial y}{\partial w_i^{(0)}} = \frac{\partial}{\partial w_i^{(0)}} \left(\left(\sum_{i=1}^{N} w_i^{(1)} \phi\left(w_i^{(0)} x + b_i^{(0)} \right) \right) + b^{(1)} \right) = w_i^{(1)} \phi'\left(w_i^{(0)} x + b_i^{(0)} \right) x, \quad (10a)$$

$$\frac{\partial y}{\partial w_i^{(1)}} = \frac{\partial}{\partial w_i^{(1)}} \left(\left(\sum_{i=1}^{N} w_i^{(1)} \phi\left(w_i^{(0)} x + b_i^{(0)} \right) \right) + b^{(1)} \right) = \phi\left(w_i^{(0)} x + b_i^{(0)} \right), \quad (10b)$$

$$\frac{\partial y}{\partial b_i^{(0)}} = \frac{\partial}{\partial b_i^{(0)}} \left(\left(\sum_{i=1}^{N} w_i^{(1)} \phi\left(w_i^{(0)} x + b_i^{(0)} \right) \right) + b^{(1)} \right) = w_i^{(1)} \phi'\left(w_i^{(0)} x + b_i^{(0)} \right), \quad (10c)$$

$$\frac{\partial y}{\partial b^{(1)}} = 1. \quad (10d)$$

Moreover, the derivatives of the cost function (5) w.r.t. to the weights and bias are also required. For the derivative w.r.t. $w_i^{(0)}$ for the first order ODE (8), this means that

$$\left\langle 2\left(y' + 2xy\right) \left(\frac{\partial y'}{\partial w_i^{(0)}} + 2x \frac{\partial y}{\partial w_i^{(0)}} \right) \right\rangle + 2(y(0) - 1) \frac{\partial y(0)}{\partial w_i^{(0)}}. \quad (11)$$

To complete the analysis, we also need expressions for the derivatives of y' w.r.t. $w_i^{(0)}$, $w_i^{(1)}$, $b_i^{(0)}$ and $b^{(1)}$. By the chain rule, the following expressions can be obtained, viz.

$$\frac{\partial y'}{\partial w_i^{(0)}} = \frac{\partial}{\partial w_i^{(0)}} \sum_{i=1}^{N} w_i^{(1)} w_i^{(0)} \phi'\left(w_i^{(0)} x + b_i^{(0)} \right) =$$

$$= w_i^{(1)} \phi'\left(w_i^{(0)} x + b_i^{(0)} \right) + x w_i^{(1)} \left(w_i^{(0)} \right)^2 \phi''\left(w_i^{(0)} x + b_i^{(0)} \right), \quad (12a)$$

$$\frac{\partial y'}{\partial w_i^{(1)}} = \frac{\partial}{\partial w_i^{(1)}} w_i^{(1)} w_i^{(0)} \phi'\left(w_i^{(0)} x + b_i^{(0)} \right) = w_i^{(0)} \phi'\left(w_i^{(0)} x + b_i^{(0)} \right), \quad (12b)$$

$$\frac{\partial y'}{\partial b_i^{(0)}} = \frac{\partial}{\partial b_i^{(0)}} \sum_{i=1}^{N} w_i^{(1)} w_i^{(0)} \phi'\left(w_i^{(0)} x + b_i^{(0)} \right) = w_i^{(1)} w_i^{(0)} \phi''\left(w_i^{(0)} x + b_i^{(0)} \right), \quad (12c)$$

$$\frac{\partial y'}{\partial b^{(1)}} = 0. \quad (12d)$$

What remains now is to obtain expressions for $y(0)$ and the partial derivatives of $y(0)$, w.r.t. to the weights and bias. Let us start with $y(0)$. With $y(x)$ given by (4) we directly have

$$y(0) = \left(\sum_{i=1}^{N} w_i^{(1)} \phi\left(b_i^{(0)} \right) \right) + b^{(1)}, \quad (13)$$

which, in turn, means that

$$\frac{\partial y(0)}{\partial w_i^{(0)}} = 0, \tag{14a}$$

$$\frac{\partial y(0)}{\partial w_i^{(1)}} = \frac{\partial}{\partial w_i^{(1)}}\left(\left(\sum_{i=1}^{N} w_i^{(1)}\phi\left(b_i^{(0)}\right)\right) + b^{(1)}\right) = \phi\left(b_i^{(0)}\right), \tag{14b}$$

$$\frac{\partial y(0)}{\partial b_i^{(0)}} = \left(\left(\sum_{i=1}^{N} w_i^{(1)}\phi\left(b_i^{(0)}\right)\right) + b^{(1)}\right) = w_i^{(1)}\phi'\left(b_i^{(0)}\right) \tag{14c}$$

$$\frac{\partial y(0)}{\partial b^{(1)}} = 1. \tag{14d}$$

The PINN (following the architecture presented above) was implemented as computer program in MATLAB. The program was employed to obtain a numerical solution to the IVP in (8), using the parameters in Table 1.

Table 1. Parameters used to defined the PINN to for the IVP in (8).

Parameter	Description	Value
N_i	# of grid points for the solution domain $[0,2]$	41
N_e	# of training batches (# or corrections during 1 Epoch)	1000
T_b	# of Epochs (1 Epoch contains T_b training batches)	100
L_r	Learning rate coefficient (relaxation for the update)	0.01
N	# of nodes/neurons in the hidden layer	10

The weights $w_i^{(0)}$ and bias $b_i^{(0)}$ were initialised using randomly generated and uniformly distributed numbers in the interval $[-2,2]$, while the weights $w_i^{(1)}$ were initially set to zero and the bias $b^{(1)}$ to one, to ensure fulfilment of the initial condition ($y(0) = 1$).

Table 2 lists the weights an bias corresponding to the solution presented in Figure 2. We note that, with the weights and bias given by Table 2, the trained network's prediction exhibits the overall error

$$\frac{1}{N_i}\sqrt{\sum_{k=1}^{N_i}\left(e^{-x_k^2} - y(x_k)\right)^2} = 5.8 \times 10^{-4}, \tag{15}$$

and $1 - y(0) = 2.2 \times 10^{-4}$, when comparing against the initial condition.

Table 2. Parameters used to defined the PINN for the IVP (8).

Node	$w^{(0)}$	$b^{(0)}$	$w^{(1)}$	$b^{(1)}$
1	1.8500	−0.5946	−3.5805	0.3055
2	1.8588	1.5974	0.9712	
3	0.3025	1.9241	0.8921	
4	1.4546	0.3742	−0.9955	
5	0.5065	1.2535	−0.1430	
6	−1.0898	−1.0199	−1.1067	
7	−0.8302	0.3519	−1.1668	
8	0.3789	1.6502	0.1754	
9	2.5012	0.7657	1.2955	
10	2.2743	1.4172	1.2787	

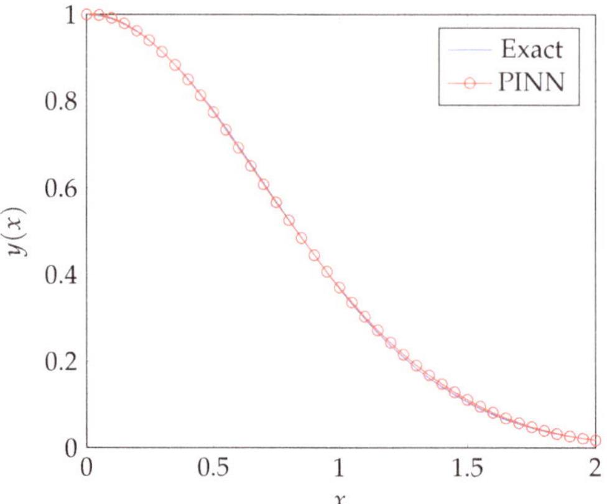

Figure 2. The solution to the IVP (8), predicted by the PINN (red line with circle markers) and the exact solution obtained by integration (blue continuous line).

4. A PINN for the Classical Reynolds Equation

The Reynolds equation for a one-dimensional flow situation, where the lubricant is assumed to be incompressible and iso-viscous, is a second order Boundary Value Problem (BVP), which in dimensionless form can be formulated as

$$\frac{d}{dx}\left(c(x)\frac{dy}{dx}\right) = f(x), \quad 0 < x < 1, \tag{16a}$$

$$y(0) = 0, \quad y(1) = 0, \tag{16b}$$

where $c(x) = H^3$, $f(x) = dH/dX$ and H is the dimensionless film thickness, if it is assumed that the pressure y at the boundaries is zero. For the subsequent analysis it is, however, more suitable work with a condensed form which can be obtained by defining the operators $\mathcal{L}$ and $\mathcal{B}$ as

$$\mathcal{L}y = c(x)y'' + c'(x)y', \tag{17a}$$

$$\mathcal{B}y = \begin{bmatrix} y(0) \\ y(1) \end{bmatrix}. \tag{17b}$$

The Reynolds BVP given by (16) can then be presented as

$$\mathcal{L}y - f = 0, \quad 0 < x < 1, \tag{18a}$$

$$\mathcal{B}y - \mathbf{b} = \mathbf{0}, \tag{18b}$$

where $\mathbf{b} = \mathbf{0}$.

For the Reynolds BVP, the cost function (5) becomes

$$l = \left\langle \left(c(x)y'' + c'(x)y' - f\right)^2 \right\rangle + y^2(0) + y^2(1), \tag{19}$$

and from the analysis presented for the IVP in Section 3 above, we have all the "ingredients" except for the partial derivatives of y'' and $y(1)$ w.r.t. to the weights and bias. For y'', based on (7) and (12), we obtain

$$\frac{\partial y''}{\partial w_i^{(0)}} = \frac{\partial}{\partial w_i^{(0)}} \sum_{i=1}^{N} w_i^{(1)} \left(w_i^{(0)}\right)^2 \phi''\left(w_i^{(0)}x + b_i^{(0)}\right) =$$

$$= 2w_i^{(1)} w_i^{(0)} \phi''\left(w_i^{(0)}x + b_i^{(0)}\right) + x w_i^{(1)} \left(w_i^{(0)}\right)^2 \phi'''\left(w_i^{(0)}x + b_i^{(0)}\right), \tag{20a}$$

$$\frac{\partial y''}{\partial w_i^{(1)}} = \frac{\partial}{\partial w_i^{(1)}} \sum_{i=1}^{N} w_i^{(1)} \left(w_i^{(0)}\right)^2 \phi''\left(w_i^{(0)}x + b_i^{(0)}\right) = \left(w_i^{(0)}\right)^2 \phi''\left(w_i^{(0)}x + b_i^{(0)}\right), \tag{20b}$$

$$\frac{\partial y''}{\partial b_i^{(0)}} = \frac{\partial}{\partial b_i^{(0)}} \sum_{i=1}^{N} w_i^{(1)} \left(w_i^{(0)}\right)^2 \phi''\left(w_i^{(0)}x + b_i^{(0)}\right) = w_i^{(1)} \left(w_i^{(0)}\right)^2 \phi'''\left(w_i^{(0)}x + b_i^{(0)}\right), \tag{20c}$$

$$\frac{\partial y''}{\partial b^{(1)}} = 0, \tag{20d}$$

where the third derivative of the Sigmoid function (1) is required. It yields

$$\frac{d}{d\xi}\left(\phi''(\xi)\right) = \frac{d}{d\xi}\left(\phi'(\xi)(1 - 2\phi(\xi))\right) = \phi''(\xi)(1 - 2\phi(\xi)) - 2\left(\phi'(\xi)\right)^2 =$$

$$= \phi(\xi)(1 - \phi(\xi))(1 - 2\phi(\xi))^2 - 2(\phi(\xi)(1 - \phi(\xi)))^2 =$$

$$= \phi(\xi)(1 - \phi(\xi))^2(1 - 3\phi(\xi)).$$

For $y(1)$ we obtain

$$\frac{\partial y(1)}{\partial w_i^{(0)}} = \frac{\partial}{\partial w_i^{(0)}} \left(\left(\sum_{i=1}^{N} w_i^{(1)} \phi\left(w_i^{(0)} + b_i^{(0)}\right)\right) + b^{(1)}\right) = w_i^{(1)} \phi'\left(w_i^{(0)} + b_i^{(0)}\right), \tag{22a}$$

$$\frac{\partial y(1)}{\partial w_i^{(1)}} = \frac{\partial}{\partial w_i^{(1)}} \left(\left(\sum_{i=1}^{N} w_i^{(1)} \phi\left(w_i^{(0)} + b_i^{(0)}\right)\right) + b^{(1)}\right) = \phi\left(w_i^{(0)} + b_i^{(0)}\right), \tag{22b}$$

$$\frac{\partial y(1)}{\partial b_i^{(0)}} = \frac{\partial}{\partial b_i^{(0)}} \left(\left(\sum_{i=1}^{N} w_i^{(1)} \phi\left(w_i^{(0)} + b_i^{(0)}\right)\right) + b^{(1)}\right) = w_i^{(1)} \phi'\left(w_i^{(0)} + b_i^{(0)}\right), \tag{22c}$$

$$\frac{\partial y(1)}{\partial b^{(1)}} = \frac{\partial}{\partial b^{(1)}} \left(\left(\sum_{i=1}^{N} w_i^{(1)} \phi\left(w_i^{(0)} + b_i^{(0)}\right)\right) + b^{(1)}\right) = 1, \tag{22d}$$

and we now have all the "ingredients" required to fully specify (19). To test the performance of the PINN, a Reynolds BVP was specified for a linear slider with dimensionless film thickness defined by

$$H(x) = 1 + K - Kx. \tag{23}$$

This means that $c(x) = (1 + K - Kx)^3$ and $f(x) = dH/dx = -K$ and that the exact solution is

$$y_{exact}(x) = \left[\frac{1}{K}\left(\frac{1}{1 + K - Kx} - \frac{1 + K}{2 + K}\frac{1}{(1 + K - Kx)^2} - \frac{1}{2 + K}\right)\right], \tag{24}$$

see, e.g., [14].

The PINN (following the architecture suggested herein) was implemented in MATLAB and a numerical solution to (16) was obtained using the parameters in Table 3. As for the IVP, addressed in the previous section, the weights $w_i^{(0)}$ and bias $b_i^{(0)}$ were, again, initialised using randomly generated numbers, uniformly distributed in $[-2, 2]$, while the weights $w_i^{(1)}$ and the bias $b^{(1)}$ was initially set to zero, to ensure fulfilment of the boundary conditions.

Figure 3 depicts solution predicted by the PINN (red line with circle markers) and the exact solution obtained by integration (blue continuous line).

Table 3. Parameters used to defined the ANN to for the Reynolds equation.

Parameter	Description	Value
N_i	# of grid points for the solution domain $[0,1]$	21
K	Slope parameter for the Reynolds equation	1
N_e	# of training batches (# or corrections during 1 epoch)	2000
T_b	# of Epochs (1 epoch contains T_b training batches)	600
L_r	Learning rate coefficient (relaxation for the update)	0.005
N	# of nodes/neurons in the hidden layer	10

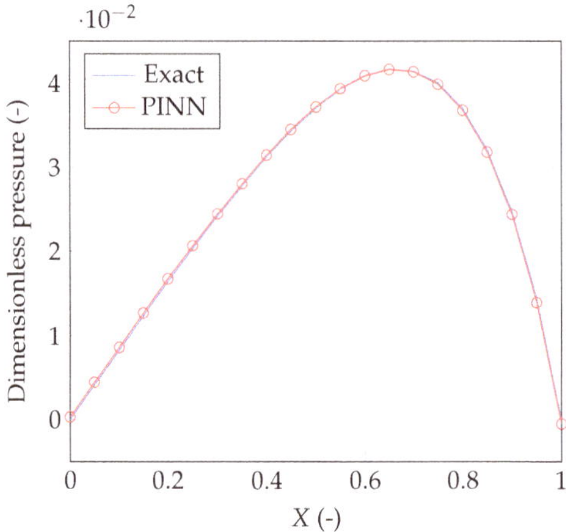

Figure 3. The solution achieved by the ANN (red line with circle markers) and the exact solution obtained by integration (blue continuous line).

Table 4 lists the weights an bias corresponding to the solution presented in Figure 3.

Table 4. Parameters used to defined the ANN.

Node	$w^{(0)}$	$b^{(0)}$	$w^{(1)}$	$b^{(1)}$
1	0.0557	1.9808	-0.2186	-0.0641
2	-6.3047	6.1664	0.1220	
3	-9.3674	11.4571	0.3843	
4	-4.5473	3.3266	0.0305	
5	-2.4464	-1.9884	0.1188	
6	-0.1365	-0.1674	0.4155	
7	0.8581	0.5253	0.5089	
8	1.0901	2.0858	0.3348	
9	0.2085	0.2523	-0.2024	
10	-3.2168	5.9722	-0.9899	

We note that, with these weights and bias, the trained network's prediction of the solution to the Reynolds BVP exhibits the overall error

$$\frac{1}{N_i}\sqrt{\sum_{k=1}^{N_i}\left(y_{exact}(x_k) - y(x_k)\right)^2} = 6.2 \times 10^{-5}, \tag{25}$$

while $y(0) = 4.1 \times 10^{-4}$ and $y(1) = -4.0 \times 10^{-4}$.

Remark 1. *The formulation of the PINN presented here is applicable as a numerical solution procedure for the Reynolds BVP (16) and it does not consider the effect of cavitation. Exactly how the effect of cavitation can be included is, however, out of the scope of this paper.*

5. Concluding Remarks

A physics-informed neural network (PINN) applicable to solve initial and boundary value problems has been established. The PINN was applied to solve an initial value problem described by a first order ordinary differential equation and to solve the Reynolds boundary value problem, described by a second order ordinary differential equation. Both these problems were selected since they can be solved analytically, and the error analysis showed that the predictions returned by the PINN was in good agreement with the analytical solutions for the specifications given. The advantage of the present approach is, however, neither accuracy nor efficiency when solving these linear equations, but that it presents a meshless method and that it is not data driven. This concept may, of course, be generalised, and it is hypothesised that future research in this direction may lead to more accurate and efficient in solving related but nonlinear problems than currently available routines.

Funding: The author acknowledges support from VR (The Swedish Research Council): DNR 2019-04293.

Institutional Review Board Statement: Not applicable.

Informed Consent Statement: Not applicable.

Data Availability Statement: Not applicable.

Conflicts of Interest: The author declares no conflict of interest.

References

1. Karniadakis, G.E.; Kevrekidis, I.G.; Perdikaris, L.L.P.; Wang, S.; Yang, L. Physics-informed machine learning. *Nat. Rev. Phys.* **2021**, *3*, 422–440 . [CrossRef]
2. Bai, X.D.; Wang, Y.; Zhang, W. Applying physics informed neural network for flow data assimilation. *J. Hydrodyn.* **2020**, *32*, 1050–1058. [CrossRef]
3. Lu, L.; Meng, X.; Mao, Z.; Karniadakis, G.E. DeepXDE: A deep learning library for solving differential equations. *SIAM Rev.* **2021**, *63*, 208–228. [CrossRef]
4. Liu, G.R. *Mesh Free Methods: Moving beyond the Finite Element Method*; Taylor & Francis: Boca Raton, FL, USA, 2003.
5. Reynolds, O. On the theory of lubrication and its application to Mr. Beauchamps tower's experiments, including an experimental determination of the viscosity of olive oil. *Philos. Trans. R. Soc. Lond. A* **1886**, *177*, 157 234.
6. Almqvist, A.; Burtseva, E.; Pérez-Rà fols, F.; Wall, P. New insights on lubrication theory for compressible fluids. *Int. J. Eng. Sci.* **2019**, *145*, 103170. [CrossRef]
7. Almqvist, A.; Burtseva, E.; Rajagopal, K.; Wall, P. On lower-dimensional models in lubrication, part a: Common misinterpretations and incorrect usage of the reynolds equation. *Proc. Inst. Mech. Eng. Part J J. Eng. Tribol.* **2020**, *235*, 1692–1702. [CrossRef]
8. Almqvist, A.; Burtseva, E.; Rajagopal, K.; Wall, P. On lower-dimensional models in lubrication, part b: Derivation of a reynolds type of equation for incompressible piezo-viscous fluids. *Proc. Inst. Mech. Eng. Part J J. Eng. Tribol.* **2020**, *235*, 1703–1718. [CrossRef]
9. Almqvist, A. Physics-Informed Neural Network Solution of 2nd Order Ode:s. MATLAB Central File Exchange. Retrieved 31 July 2021. Available online: https://www.mathworks.com/matlabcentral/fileexchange/96852-physics-informed-neural-network-solution-of-2nd-order-ode-s (accessed on 7 July 2020).
10. Neidinger, R.D. Introduction to automatic differentiation and MATLAB object-oriented programming. *SIAM Rev.* **2010**, *52*, 545–563. [CrossRef]
11. Baydin, A.G.; Pearlmutter, B.A.; Radul, A.A.; Siskind, J.M. Automatic differentiation in machine learning: A survey. *J. Mach. Learn. Res.* **2018**, *18*, 1–43.
12. Cauchy, A.M. Méthode générale pour la résolution des systèmes d'équations simultanées. *Comp. Rend. Sci. Paris* **1847**, *25*, 536–538.
13. Ruder, S. An overview of gradient descent optimization algorithms. *arXiv* **2016**, arXiv:1609.04747.
14. Almqvist, A.; Pérez-Ràfols, F. Scientific Computing with Applications in Tribology: A Course Compendium. 2019. Available online: http://urn.kb.se/resolve?urn=urn:nbn:se:ltu:diva-72934 (accessed on 26 July 2021).

Article

Semi-Supervised Classification of the State of Operation in Self-Lubricating Journal Bearings Using a Random Forest Classifier

Josef Prost *, Ulrike Cihak-Bayr, Ioana Adina Neacșu, Reinhard Grundtner, Franz Pirker and Georg Vorlaufer

AC2T research GmbH, Viktor-Kaplan-Straße 2/C, 2700 Wiener Neustadt, Austria;
ulrike.cihak-bayr@ac2t.at (U.C.-B.); adina.neacsu@ac2t.at (I.A.N.); reinhard.grundtner@ac2t.at (R.G.);
franz.pirker@ac2t.at (F.P.); georg.vorlaufer@ac2t.at (G.V.)
* Correspondence: josef.prost@ac2t.at

Abstract: For a tribological experiment involving a steel shaft sliding in a self-lubricating bronze bearing, a semi-supervised machine learning method for the classification of the state of operation is proposed. During the translatory oscillating motion, the system may undergo different states of operation from normal to critical, showing self-recovering behaviour. A Random Forest classifier was trained on individual cycles from the lateral force data from four distinct experimental runs in order to distinguish between four states of operation. The labelling of the individual cycles proved to be crucial for a high prediction accuracy of the trained RF classifier. The proposed semi-supervised approach allows choosing within a range between automatically generated labels and full manual labelling by an expert user. The algorithm was at the current state used for ex post classification of the state of operation. Considering the results from the ex post analysis and providing a sufficiently sized training dataset, online classification of the state of operation of a system will be possible. This will allow taking active countermeasures to stabilise the system or to terminate the experiment before major damage occurs.

Keywords: condition monitoring; semi-supervised learning; random forest classifier; self-lubricating journal bearings

Citation: Prost, J.; Cihak-Bayr, U.; Neacșu, I.A.; Grundtner, R.; Pirker, F.; Vorlaufer, G. Semi-Supervised Classification of the State of Operation in Self-Lubricating Journal Bearings Using a Random Forest Classifier. *Lubricants* **2021**, *9*, 50. https://doi.org/10.3390/lubricants9050050

Received: 26 March 2021
Accepted: 30 April 2021
Published: 4 May 2021

Publisher's Note: MDPI stays neutral with regard to jurisdictional claims in published maps and institutional affiliations.

Copyright: © 2021 by the authors. Licensee MDPI, Basel, Switzerland. This article is an open access article distributed under the terms and conditions of the Creative Commons Attribution (CC BY) license (https://creativecommons.org/licenses/by/4.0/).

1. Introduction

Predictive maintenance has been a topic of increasing interest in research and industry over the past few years [1]. As part of predictive maintenance techniques, condition monitoring [2–4] is used to detect anomalies and to predict the health of machinery in real time. It uses both sensor data and monitoring software to establish whether a component failure is likely. While some types of failure occur gradually and can be prevented by routine examinations, sudden failures are of course very difficult to forecast. This is the reason why artificial intelligence (AI), especially machine learning (ML) techniques, has gained increasing popularity in the recent years. ML algorithms are trained to learn from the available data and help identify certain behaviours or parameters that contribute to failure with high accuracy. ML algorithms can be divided into two main groups, namely supervised and unsupervised learning [5], differing in whether prior knowledge on the expected output is considered or not. The prerequisite for supervised learning is a set of labelled training data, while unsupervised learning aims at uncovering features on its own.

In tribology research, AI has already been applied to various fields, including in-process tool condition monitoring [3], anomaly detection [6–8], failure prediction [9], classification of the lubrication regime [10], optimisation of tribological performance of copper composites [11], as well as AI-based lubricant design [12]. Deshpande et al. [13] give a good summary of the most common machine learning algorithms used in the classification of tribological states of operation and prediction of wear, depending on the application. Classical ML techniques, such as Support Vector Machine (SVM) [3,6,14], Random Forest (RF) [9,15] and Radial Base Function (RBF) methods [16] are widely used. An approach for fast bearing fault diagnosis in rolling elements, combining traditional pattern recognition methods with meta-heuristic search and ML, was presented by Sun et al. [17]. Additionally, deep-learning techniques based on Artificial Neural Networks (ANN) have gained increased popularity over the past few years [10,18,19]. The recently published article by Rosenkranz et al. [20] gives a comprehensive overview of the various application fields and methods in tribology and shows the extended use of AI and ML techniques in the field of tribology as a future perspective.

Acoustic emission (AE), both airborne [8] and structure-borne [9,14], have proven to provide well-suited datasets for training ML algorithms. Other datasets used in tribology-related applications include torque [10] and force [2] data, accelerometer signals [21], as well as images of worn tool surfaces [22]. Thermal imaging has also been applied successfully to fault diagnosis [23].

RF classifiers may have a slightly lower prediction accuracy compared to ANN-based classifiers. However, ANN algorithms require careful parameter tuning and large training datasets. RF classifiers already give good prediction accuracy without or with little fine tuning of their hyperparameters. This makes RF models very suitable for industrial use, as they are easier to adopt for specific applications [24].

In general, self-lubricating sliding elements are composed of porous sintered materials filled with a lubricant [25]. Often, the bearing itself is made out of a porous material, such as sintered metal compounds [26–28] or oil-bearing self-lubricating layers [29], as well as polymer composites [30]. Another variant of self-lubricating elements is the use of solid lubricants as coatings, e.g., PTFE in roller bearings [31].

In contrast, the bearings used in this study consist of a base material equipped with a grid of bores, which are filled with a porous polymer compound infiltrated with lubricant. This kind of bearing is common in industrial applications. However, scientific literature on these specific systems is not very widely available; e.g., [32,33], information on this topic is often restricted to company-owned empirical know-how. Consequently, precise knowledge of the main acting mechanisms has not been reported publicly. It is assumed that the variety of commercially available products is based on proprietary know-how and engineering experience.

In 2007, Jisa [34] performed a fundamental review and studied sliding elements in the shape of plates and bearings with different copper-based alloys forming the supporting structure. Jisa has shown that the thermal expansion of the liquid lubricant in the gap between the two sliding components, assisted by capillary effects of the pore and surface topography structures, determine friction levels and lifetime. Generally, the lubricating effect is assisted by a moderate rise of temperature, as the bearing is most likely to operate in boundary or mixed friction conditions. This made a stepwise increase of loading necessary during the run-in phase of the experiment, as a too-high temperature would lead to inferior lubrication due to lower oil viscosity, resulting in adhesive wear and finally end of lifetime by increase of the friction force up to the limit of the specific machine.

For axial sliding operation conditions as studied in the current work, wear is predominantly taking place at the bearing edges and at the edges of the lubricant macrodepots. The wear debris generated at these positions causes abrasive wear in the whole contact zone, leading to gradually growing grooves. As long as the lubricant macrodepots are in contact with the counterbody, these grooves are no lifetime-limiting feature, and most of the wear debris particles are quickly transported out of the contact zone. The re-disposition of wear debris particles into the lubricant macrodepots may lead to a temporary strong increase of the friction force. These events occur rather statistically, accompanied by friction peaks, but they do not result in permanent damage of the lubricant macrodepots and eventually the removal of the loosened wear debris from the sliding contact. Due to these mechanisms, this type of bearings shows self-recovery effects [34].

The judgment of critical operation and useful remaining lifetime in industrial applications relies on specific experience and empirical data exhibiting large variance. The system studied in this work is interesting for application of ML techniques to explore the opportunities of ML for a self-recovering complex tribological system.

2. Materials and Methods

2.1. Experimental Setup

The experiments were performed on a laboratory-built tribometer setup for bidirectional, translatory movements with high normal loads, and large sliding amplitudes (Figure 1a). In this setup, a self-lubricating journal bearing made of a bronze alloy with polymer lubricant macrodepots is horizontally mounted on the tribometer and held in a fixed position. The counterpart, a shaft made of hardened and polished Cr-steel, slides inside this bearing in a translatory oscillating movement driven by a pneumatic cylinder. Two adjustable electronic switches define the reversal points of the oscillating movement. The normal load is applied by a second pneumatic cylinder. The executed force is transmitted via a parallelogram structure, which ensures that the horizontal position of the bearing is always maintained even in the event of wear-induced lowering (Figure 1b). The pressure applied on the bearing was calculated as the normal force acting on the nominal cross-section, according to engineering standards for journal bearings.

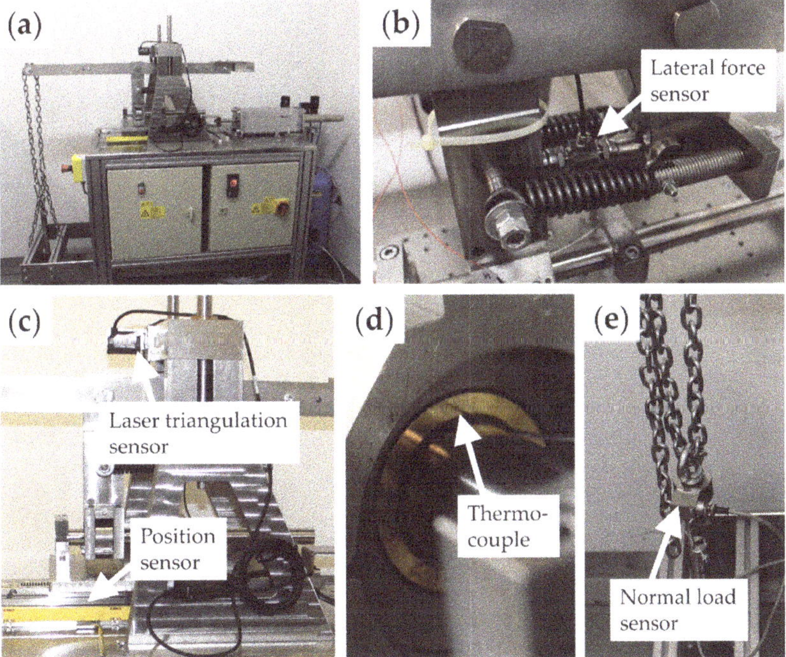

Figure 1. (**a**) Translatory oscillating tribometer setup, (**b**) lateral force measurement, (**c**) position measurement, (**d**) bearing temperature measurement, and (**e**) normal force measurement.

The tribometer is equipped with several sensors to monitor and document the defined experimental parameters, the environmental situation, and the reactions of the tribometer to the different friction conditions. The instrumentation of the setup is described in detail below. In the current study, the focus lies on the data generated by the lateral force sensor.

A commercially available linear inductive position sensor (Turck Li300P0-Q17LM0-LiU5X2) measures the oscillating movement of the shaft. The applied normal load and the lateral force, i.e., the force in sliding direction, are recorded by two load sensors

(HBM Type U2B and HBM U9C, respectively). The wear of the journal bearing can be qualitatively monitored by the vertical movement of the cantilever, which is measured by a laser triangulation sensor (Keyence IL-030). The temperature of the bearing is measured by a thermocouple (type K, diameter 0.5 mm), which is mounted inside a drilled hole at the top point of the bearing's front face, where the highest contact pressure and therefore the highest temperature is to be expected. Figure 1 indicates the mounting positions of the force, position, and laser triangulation sensors as well as the thermocouple.

In addition, several sensing techniques are used to detect friction-induced vibrations. Two acoustic emission sensors (NF-Corporation AE-900M-WB), one mounted at the shaft and one mounted at the bearing holder, measure high-frequency structure-borne noise in the range between 100 kHz and 5 MHz. Three MEMS (micro-electro-mechanical system)-acceleration sensors (Analog Devices ADXL1002) are mounted at the bearing holder and detect low-frequency vibrations up to 11 kHz in all spatial directions. The emitted airborne noise is measured in the frequency range between 20 Hz and 20 kHz using a high-precision microphone (Brüel & Kjaer 4189-A-021).

Furthermore, the ambient air temperature and humidity is monitored in the vicinity of the experiment by a TE Connectivity HTM 2500 LF sensing module and the supply air pressure of the pneumatic drive by a Telemecanique XMLP016BC71V pressure transducer.

The oscillation frequency of the shaft was set to a nominal value of 1 Hz and a stroke amplitude of 30 mm, which ensured that each contact point of the shaft was moved out of the contact completely in each stroke. It has to be noted that the oscillation frequency was not constant during the experiment but varied with the resistance the pneumatic cylinder had to overcome to move the shaft. During the first 1.5 h of the experiment, the normal load was gradually increased until a nominal bearing pressure of 8 N/mm^2, corresponding to a normal load of 6 kN, was reached. The experiments were performed until at least one of two thresholds was exceeded. The first threshold was set for the bearing temperature at 150 °C and the second one was set for the uncorrected lateral force at ±3.5 kN. However, it should be noted that the temperature threshold was never exceeded, and all experiments were stopped after exceeding the lateral force threshold.

In total, data from 9 experiments performed under the described conditions were used for this study.

2.2. Data Preprocessing

As mentioned above, the sensor measuring the lateral force F_L is part of the lever system. In order to obtain the coefficient of friction (μ), the geometry of the lever system has to be taken into account, resulting in the following relation:

$$\mu = \frac{100 \, F_L - 2 \, F_N}{175 \, F_N}. \tag{1}$$

Before feeding the algorithm, several data preprocessing steps were necessary. This was done using the programming language Python in the form of interactive Jupyter notebooks [35], using NumPy arrays [36] and pandas DataFrame objects [37] for efficient computing.

The time-series signals acquired by the force, acceleration, and supplementary sensors, sampled at rates of up to 5 kHz, were stored in the hdf5 file format [38] on a file server dedicated to the storage of large amounts of raw measurement data. Since the amount of raw data was too large for efficient processing on a conventional workstation, data were directly read from the server and downsampled to 100 Hz, thereby carefully retaining the main characteristics of the sensor data.

In a second step, noise was removed from the lateral force and position signals by smoothing with a third-degree Savitzky–Golay filter [39] with a window length of 25 samples.

Due to the oscillating nature of the setup, periodic patterns repeating with the oscillation frequency of the system are present in the lateral force data. Each one of these patterns describes the evolution of the lateral force during one cycle. Deviations from the normal state of operation can be seen as distortions of the individual cycle shapes, which are discussed in more detail in Section 3.1. This leads to an increase in the cycle periods as well as the lateral force levels and maxima.

The zero position of single-cycle curves were triggered using the zero-crossings of the normalised position signal in negative stroke direction. Thus, the length of the extracted curves was normalised to 100 data points per curve using linear interpolation. In the end, an $m \times 100$ matrix, with m being the number of individual cycles of the respective experiment, was obtained as input for the Random Forest classifier.

2.3. Random Forest Classifiers

The RF algorithm was first described in detail by Breiman [40]. RF is an ensemble learning algorithm and is based on the aggregation of a large number of independent decision trees. When used for classification, the class votes of each tree determine the classification by majority vote [5], resulting in enhanced classification accuracy and reduced overfitting. Each tree within this RF is grown using random feature selection; each new training set being drawn with replacement from the original training set. This method is known as bootstrap aggregation or *bagging* [40,41].

In RFs, bagging is combined with a randomised selection of the p input features to be considered for splitting an internal node. At each node, a random subset of k features is selected, from which only the best split is determined [42]. For classification, the default value for k is typically set as the square root of p. At each split, the total reduction in the split criterion, usually measured by the Gini index [43], can be used as an importance measure for the corresponding splitting feature. The feature importance is obtained by accumulating this importance measure over all trees separately for each feature [5]. The size of an individual tree is typically controlled by predefined parameters, such as the terminal node size and tree depth. As a consequence, for every tree in the RF ensemble, a set of observations exists that are not used for growing the tree. These so-called out-of-bag observations (OOB) can be used to estimate the prediction accuracy of the individual decision trees [43].

Generally speaking, the larger the number of estimators, the better the prediction accuracy becomes. However, beyond a critical number of trees, there is no significant performance gain in adding more trees, at the cost of increasing computing demand. Numbers available in the literature include 128 [44], 200 [5], or 250 [45] trees.

In order to assess the prediction quality of the trained RF algorithm, a series of classification metrics is used [46,47].

The most straightforward metric is the *accuracy* (q_a), which is defined as the ratio between the number of correct predictions (N_T) and the total number of samples (N), i.e.,

$$q_a = \frac{N_T}{N}.$$ (2)

If a sample that has been labelled as positive is also predicted as positive, the classification is counted as a *True Positive* (N_{TP}). If it is predicted as negative, the classification is a *False Negative* (N_{FN}). *True Negatives* (N_{TN}) and *False Positives* (N_{FP}) are defined analogously. These four numbers can be displayed as a 2×2 *confusion matrix* C. In the present study, we follow scikit-learn's implementation [48]; other sources may use the transposed version, e.g., [46].

$$C = \begin{pmatrix} N_{TP} & N_{FN} \\ N_{FP} & N_{TN} \end{pmatrix}.$$ (3)

Using above four definitions, the number of correct predictions is given by

$$N_T = N_{TP} + N_{TN}.$$ (4)

The *precision* (q_p) or *confidence* is defined as the fraction of all positively predicted samples (N_{PP}), which are actually labelled as positive (N_{TP}), i.e.,

$$q_p = \frac{N_{TP}}{N_{PP}} = \frac{N_{TP}}{N_{TP} + N_{FP}}. \tag{5}$$

Conversely, the *recall* (q_r) or *sensitivity* gives the fraction of all positively labelled samples (N_{PL}), which are correctly identified as positive, i.e.,

$$q_r = \frac{N_{TP}}{N_{PL}} = \frac{N_{TP}}{N_{TP} + N_{FN}}. \tag{6}$$

In the case of multi-label classification, precision and recall values are calculated separately for each class, with 'positives' meaning samples belonging to the respective class. Each row in the confusion matrix represents a 'true' class, with the 'predicted' class labels as columns. In this case, the confusion matrix contains the number of correct predictions of each class in the diagonal, and false predictions are contained in the respective off-diagonal elements. Given a classification with N labels, the precision and recall can be calculated separately for each class (denoted by index $i, i = 1 \ldots N$) from the coefficients of the $N \times N$ confusion matrix as follows:

$$q_p^{(i)} = \frac{C_{ii}}{\sum_{j=1}^{N} C_{ji}} \tag{7}$$

and

$$q_r^{(i)} = \frac{C_{ii}}{\sum_{j=1}^{N} C_{ij}}. \tag{8}$$

2.4. Labelling of Datasets and RF Model

In this and the following sections, the term 'state' refers to the current state of operation on the basis of individual cycles. The term 'phase' denotes a longer period of time, in which the system is in a certain state of operation. The term 'class' describes a specific categorical label in a set of labels that is assigned to the individual cycles of the dataset during the training period of the RF algorithm, based on their state. Thus, each class consists of a set of individual cycles belonging to one state of operation.

As manual labelling of tens of thousands of cycles would be a very tedious and time-consuming task, a pre-labelling of the cycles via clustering methods was performed. Given the large size of the data, Principal Components Analysis (PCA) was applied to reduce the dimensionality of the input and visualise the general shape of the data. After selecting an appropriate number of principal components, based on the amount of variance covered, a k-means clustering algorithm was applied to the reduced dataset. Each cycle was assigned to a cluster such that the squared Euclidean distances within each cluster were minimised. The implementations of these two steps were performed in R using the *prcomp* and *kmeans* functions [49].

The results of this pre-labelling stage are given in Table 1. Before applying PCA, the datasets were centred and scaled to have unit variance. For the k-means clustering, the first two principal components were selected, showing cumulative proportions of variance between 0.79 and 0.93. The number of clusters for the k-means algorithm was set to 5, based on the expected tribological regimes of the studied tribological experiment. The resulting cluster sizes for each experiment are unevenly distributed, as can be seen in Table 1.

Table 1. Overview of k-means clustering results.

Experiment No.	Cumulative Variance	Total Number of Cycles	Number of Cycles in Each Cluster (in Ascending Order)				
Experiment 1	0.79	46,485	66	1447	10,029	16,271	18,672
Experiment 2	0.83	44,265	458	1992	4075	15,684	22,074
Experiment 3	0.89	38,605	364	2690	5553	12,467	17,531
Experiment 4	0.86	57,516	4532	8075	13,281	14,484	17,144
Experiment 5	0.87	44,944	2043	3279	9904	10,905	18,813
Experiment 6	0.80	35,388	38	2569	5214	13,443	14,124
Experiment 7	0.80	39,822	245	1918	6411	12,718	18,530
Experiment 8	0.84	54,782	1368	3359	9603	19,197	21,255
Experiment 9	0.85	35,734	1193	5678	8896	9920	10,047

The clusters were subsequently assigned to tribological states of operation: 'Steady1', 'Steady2', 'Pre-critical', and 'Critical'. The first 5000 cycles were defined 'Run-in' and discarded due to the high variability of the data. The preliminary labelling was refined in a second step by closer inspection of the data, taking into account distinctive features in the other sensor signals, e.g., sudden temperature increases or distortions of the position signal. This resulted in the inclusion of additional 'Pre-critical' areas–typically before and after short-term ('Critical') anomalies or before critical operation at the end of the experiments as well as physically meaningful merging of regions fragmented into various states of operation by the clustering algorithm. Figure 2 shows the comparison of the classification obtained by k-means clustering and the final labelling for one of the experiments used for training the RF algorithm. Here, single cycles or groups of cycles that did not differ significantly from their surroundings, which were marked as 'Pre-critical' (cluster 4) by the k-means clustering, were assigned to the respective steady state. Furthermore, the area preceding the final critical state was labelled as 'Pre-critical' in its entirety, while the k-means result switched between 'Pre-critical' and 'Steady2' in this region. This led to an overall increase of cycles labelled as 'Pre-critical' after manual adaptation (see Table 2). The 'Steady1' state is reduced in size after manual adaptation, as the first 5000 cycles were discarded.

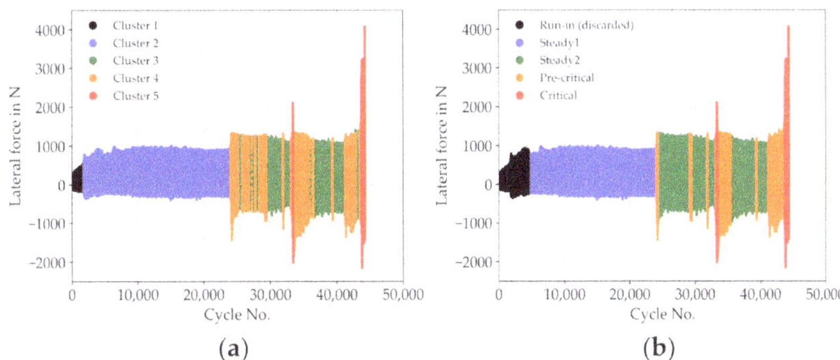

Figure 2. Labelling of the lateral force signal of one experiment (Experiment 2 in Table 1). (**a**) Classification obtained using k-means clustering, (**b**) Labels used for training the RF model after manual adaptation. The clusters were assigned to tribological states of operation: 'Run-in' (Cluster 1), 'Steady1' (Cluster 2), 'Steady2' (Cluster 3), 'Pre-critical' (Cluster 4), and 'Critical' (Cluster 5).

Table 2. Number of cycles for one experiment (Experiment 2 in Table 1) as classified by k-means clustering before and after manual adaptation.

State	No. of Cycles after k-Means	No. of Cycles after Manual Adaptation
Steady1	22,074	19,120
Steady2	15,684	15,331
Pre-critical	4075	4263
Critical	458	551

In the end, four classes representing the individual states of operation were distinguished; see Figure 3. 'Steady1' was used for steady operation, typically right after the run-in period, with little fluctuation and few distortions in the data. 'Steady2' typically occurred after major events. The system stabilises, but higher lateral forces are measured, and the curve shapes of the cycles are more distorted and variable. After sufficient running time without major events, the system may reach the 'Steady1' state again. 'Pre-critical' cycles are typically found before and after cycles labelled as 'Critical'. The 'Pre-critical' label is also associated with short-time events, typically lasting less than 100 cycles. During these short-time events, maximum lateral force values of 1.5 times the maxima of the surrounding steady-state cycles or lager were measured. 'Critical' cycles show heavily distorted curves with the lateral force increasing considerably at one or both turning points. This indicates that the bearing was stuck in its turning position and could only be brought back into motion when a sufficiently high lateral force was applied. One has to note that the *x*-axis in the graphs of Figure 3 corresponds to a relative position in time within each cycle rather than the actual physical encoder position. The length of the half-cycle, in which the deadlock occurred (the case for the positive half-cycle is depicted in Figure 3d), is extended, leading to an overall asymmetric cycle shape. As all cycles were normalised to a length of 100 data points, the steepness of the lateral force curve in the turning points is related to the cycle duration, which itself depends on the friction in the system at that moment.

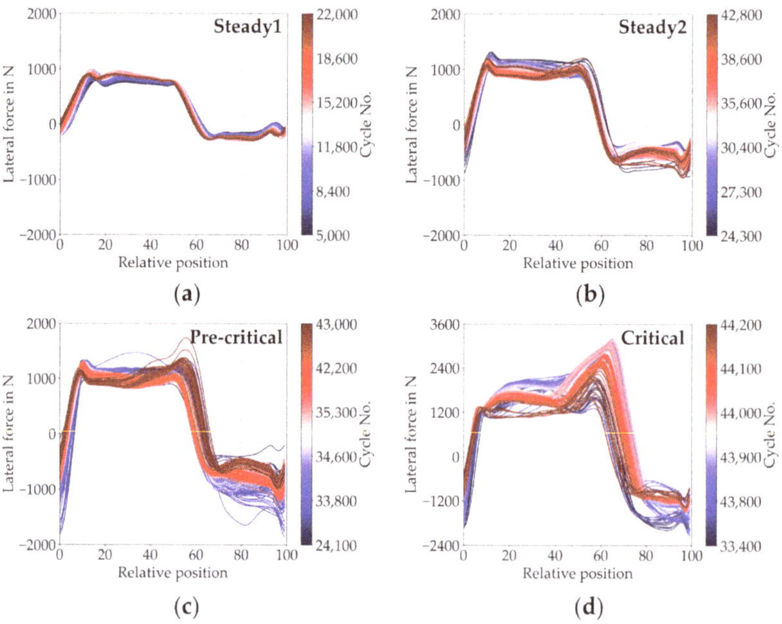

Figure 3. Characteristic cycle shapes of the four operation states: (**a**) Steady1, (**b**) Steady2, (**c**) Pre-critical, and (**d**) Critical. Please note the different scaling of the y-axis for the critical state in (**d**).

The RF algorithm was developed using the Python ML package scikit-learn [48]. The workflow for training and application of the algorithm is described in detail below, and the corresponding flowchart is shown in Figure 4.

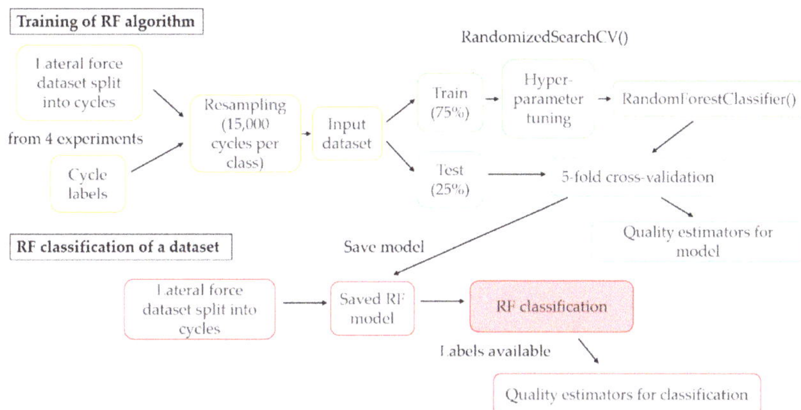

Figure 4. Flowchart of the training and classification of lateral force datasets with an RF classifier.

The dataset for training the algorithm was created using labelled cycles from four experiments, namely the numbers 2, 4, 7, and 9 in Table 1. As mentioned above, the first 5000 cycles from each experiment were considered as run-in and discarded from the dataset. Data from multiple experiments were chosen in order to cover the diversity of cycle shapes within each state and to equalise bias towards a certain state introduced by manual labelling. This includes, above all, the distortions introduced by pre-critical and critical operation, which can happen in either positive, negative, or both stroke directions.

As the distribution of the cycles over the classes representing the four states of operation was highly unbalanced (see Table 3), each class was resampled to a size of 15,000 cycles by random selection with replacement. That means that the classes 'Steady1', 'Steady2', and 'Pre-critical' were downsampled, and a random selection of the cycles over all four experiments was used for training. However, the size of the class representing the 'Critical' state was only 1265 cycles and had to be upsampled by a factor of nearly 12, drawing each cycle multiple times from the dataset. The number of 15,000 cycles was chosen, as it seemed to be a good compromise between retaining as much information as possible from the three larger classes and keeping the upsampling factor of the 'Critical' class reasonably small.

Table 3. Number of cycles for each class present in the training dataset before resampling.

Class	No. of Cycles	Resampling Factor
Steady1	51,217	0.29
Steady2	83,678	0.18
Pre-critical	21,177	0.71
Critical	1265	11.86

Before training the RF algorithm, a randomised hyperparameter tuning was performed using scikit-learn's *RandomizedSearchCV* function in order to optimise the following hyperparameters. Randomised hyperparameter tuning has the advantage of a fixed, predefined number of trials, independent of the total number of combinations, which can be very large. This strategy will find a near-best combination of hyperparameters at the advantage of not spending too much time on unpromising candidates [50]. For the present work, the number of iterations was set to 100. The following hyperparameters were optimised using randomised hyperparameter tuning: *n_estimators* indicates the number of individual decision trees in the RF, *min_samples_split* is the minimum number of samples to split an internal node, *min_samples_leaf* is the minimum number of samples required to be in a leaf node, *max_features* is the maximum number of features to consider at each split and was always set to the square root of the total number of features; i.e., 10, *max_depth* indicates the maximum number of levels within an individual decision tree and finally, *bootstrap = True* means that bootstrap samples are used to build each tree rather than the whole dataset. Table 4 shows the best obtained set of hyperparameters, which were subsequently used for training the algorithm.

Table 4. Best hyperparameter grid for the RF after hyperparameter tuning.

Hyperparameter	Value
n_estimators	101
min_samples_split	2
min_samples_leaf	1
max_features	'sqrt'
max_depth	30
bootstrap	True

The RF algorithm was trained using 75% of the input dataset as training data and 25% as test data used for determination of the quality estimators described in Section 2.3. Then, the prediction accuracy of the trained RF algorithm was assessed by a 5-fold cross-validation with random selection of cycles for the training and test dataset. Data were again split into 75% training and 25% test data for each run, which were randomly selected from the input dataset. Finally, the algorithm was validated on a labelled experiment (number 8 in Table 1), which was not used for training.

3. Results

3.1. Frictional Behaviour

The average duration of an experiment until reaching the stop criterion was 14.5 ± 2.6 h, corresponding to roughly 44,170 ± 7400 cycles.

Although the temporal evolution of the measured signals varied between the experiments, a few characteristic features were observed throughout the experiments. Figure 5 shows the time series of coefficient of friction, temperature, and contact pressure of experiment 4 as an example for characteristic features observed during the series of experiments. For the coefficient of friction, the arithmetic means of the absolute values of the 10% and the 90% quantiles are displayed. The 10% quantile gives a characteristic value for the coefficient of friction in negative stroke direction, whereas the 90% quantile was used for the positive direction.

After the run-in, the system was operating in a stable condition at a mean coefficient of friction of around 0.06, with the temperature steadily increasing close to 90 °C. After typically 15,000 to 30,000 cycles, a region of pre-critical and critical operation was observed. This manifests itself in a sudden increase in the coefficient of friction and the temperature exceeding 100 °C. This may be attributed to locally inferior lubrication and consequently short-time metal-to-metal contact and adhesion. After a few minutes, the system was able to loosen the adhesive contact spot or to tear off a machining chip from the edge of a lubricant macrodepot. After that, the system remained in the pre-critical state, self-healed, and thus slowly returning to steady operation, albeit at a slightly higher coefficient of friction in most cases, typically between 0.07 and 0.10. The eventual steady increase of the coefficient of friction can be attributed to the gradually deteriorating lubricant supply due to capillary forces, which reduce due to the increasing number and depth of abrasive grooves caused by wear particles. The short-time critical states with subsequent stabilisation of the system due to self-recovery could be observed repeatedly in all experiments.

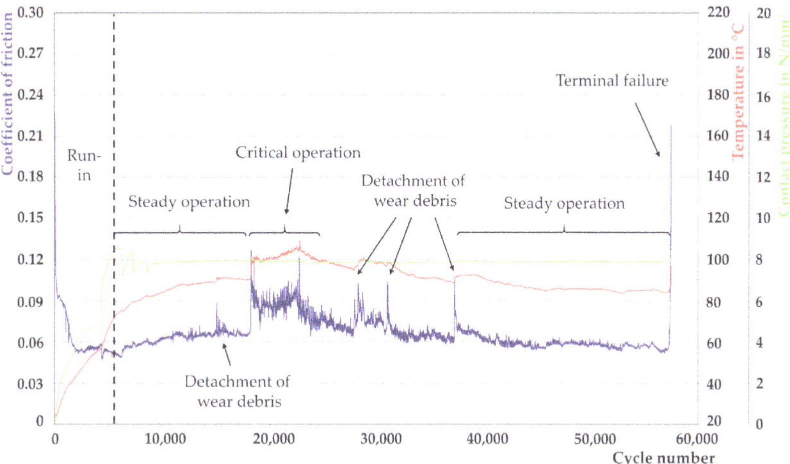

Figure 5. Temporal evolution of the coefficient of friction (blue), the sample temperature (red), and the applied contact pressure (green) from one selected experiment. The annotation of the observed features is based on a tribologist's expert opinion.

During the experiment, spikes in the friction curves were observed. In order to investigate the origin of these spikes, additional experiments were carried out and stopped manually when the first spike occurred. Investigation of the bearing revealed that these spikes were most likely caused by wear debris in the form of tiny machining chips detached from the edge of a lubricant macrodepot and subsequently transported further in the contact zone to be either embedded within another lubricant macrodepot or transported out of the contact at the edges of the sliding element; see Figure 6a. Figure 6b indicates the large extent of the clearly visible wear area on the self-lubricating journal bearing after the experiment.

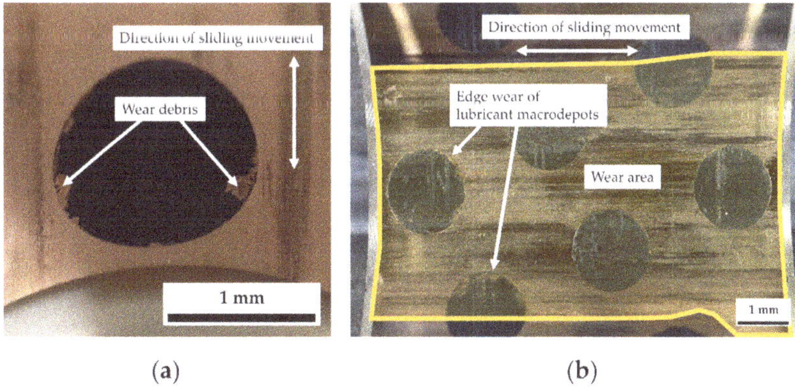

(a) (b)

Figure 6. (a) Macro image showing deposition of wear debris on the surface of a lubricant macrodepot, (b) Macro image of the wear area of a journal bearing after the experiment illustrating typically occurring grooves and shifting deposit material onto the bronze base structure and vice versa.

Prior to reaching the set threshold criteria of the system by reaching a given lateral force, four experiments exhibited an extended instable state, which lasted for up to several thousand cycles. However, in the other five experiments, the stop criterion was reached almost instantaneously, with instable operation of less than 10 min before termination of the experiment. Experiment 4, as shown in Figure 5, belongs to the latter category. In experiment 6, no intermediary critical operation was observed. The system remained steady for about 10 h, with a sudden increase of the lateral force in the end, exceeding the stop criterion. Experiment 3 was manually terminated after about 2 h of pre-critical operation, before reaching the stop criterion.

3.2. Classification of States of Operation

The prediction accuracy of the trained RF model was determined to be 0.991, corresponding to an OOB score of 0.996. Five-fold cross-validation yielded a mean prediction accuracy of the model of 0.993 ± 0.001. These values indicate a classification error rate between 0.5% and 1%.

Figure 7 shows the locations of the 20 most important features used for splitting nodes in the RF model. The *x*-axis label 'Relative position' refers to a relative position in time during the duration of one stroke rather than an actual physical position. One can see clearly that the most important areas are located around the two turning points of the stroke direction of the steady-state cycles, i.e., around 60 for the change between positive and negative stroke and around 100 or 0 for the change from negative to positive. Another region, where important features are located, can be found around 80, corresponding to the location of turning points of the critical cycles in the positive stroke direction. The feature around 50 may be associated with critical cycles in the negative stroke direction.

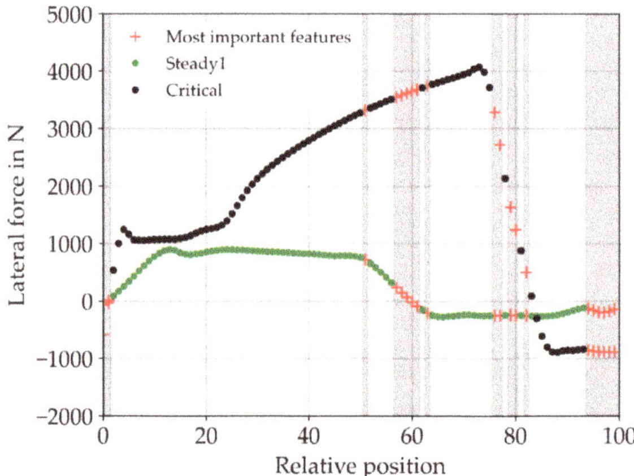

Figure 7. The 20 features with the highest importance are marked as red crosses.

In order to assess the prediction quality of the RF algorithm on other datasets, the dataset of experiment 8, which was not used for training the RF algorithm, was labelled according to the procedure described above, and classification metrics were calculated. A comparison between the labels assigned to each cycle and the labels predicted by the RF algorithm is shown in Figure 8.

The overall classification accuracy of experiment 8 was 0.939. Table 5 shows the precision and recall values for the four classes. Both steady states as well as the pre-critical state were recognised with high precision and recall. Of the cycles classified by the algorithm as 'Critical', only 78% were actually labelled as 'Critical'. The remaining 12%, or 188 cycles, had the true label 'Pre-critical'. However, 88% of the actual states labelled as 'Critical' were identified correctly. The corresponding absolute values are shown in the confusion matrix in Figure 9. The colour scale indicates the fraction between predicted labels and the total number of true labels assigned to the respective class, summing up to 1 for each row. For the diagonal elements, this corresponds to the recall. The last row and column, labelled as 'None', indicates cycles, for which the algorithm was not able to issue a prediction. This was predominantly the case for the two steady states, with about 4.5%

of the cycles labelled 'Steady2' not classified. This is also the reason for the relatively low recall of 0.9 for this class.

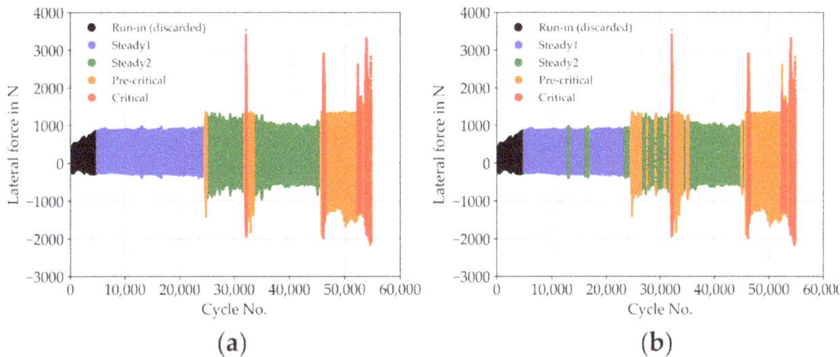

Figure 8. (**a**) Manually labelled dataset from Experiment 8, (**b**) the classification made by the trained RF algorithm.

Table 5. Precision and recall for experiment 8.

Class	Precision	Recall
Steady1	0.98	0.98
Steady2	0.97	0.90
Pre-critical	0.90	0.95
Critical	0.78	0.88

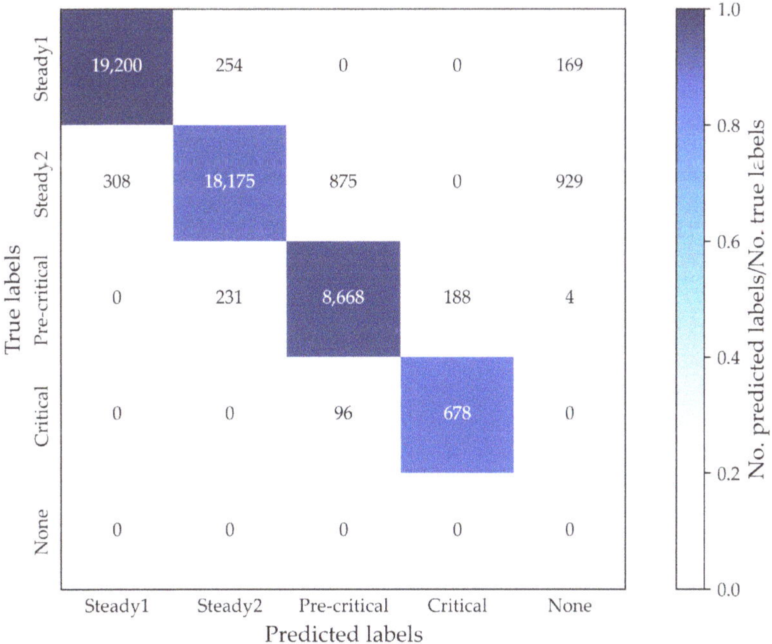

Figure 9. Confusion matrix of the labelled dataset of experiment 8.

Based on the results of the RF classification, Table 6 shows a summary of the lengths of the pre-critical phases preceding the end of the respective experiment. As already mentioned, the experiments could be divided in two distinctly different groups according to their behaviour towards the end of the experiment. In the first group, an extended pre-critical phase was observed before the termination of the experiment. This pre-critical phase was found to last between 60 and 211 min or between 7.4 and 19.2% of the total running time. Before reaching the stop criterion, individual critical cycles were observed during the pre-critical phase, with an increasing abundance of critical cycles towards the end, as shown in Figure 10.

Table 6. Start of pre-critical phase for each experiment.

Experiment No.	Total Running Time (Hours)	Start Pre-Critical Phase (Minutes before End)	Fraction of Total Running Time (%)
Experiment 1	15.4	7.5	0.8
Experiment 2	14.8	73	8.3
Experiment 3 [1]	12.3	113	15.3
Experiment 4	19.1	2.5	0.2
Experiment 5	14.7	4	0.5
Experiment 6	11.1	1.5	0.2
Experiment 7	13.5	60	7.4
Experiment 8	18.3	211	19.2
Experiment 9	11.7	5	0.7

[1] Experiment 3 was stopped manually, before the stop criterion was reached.

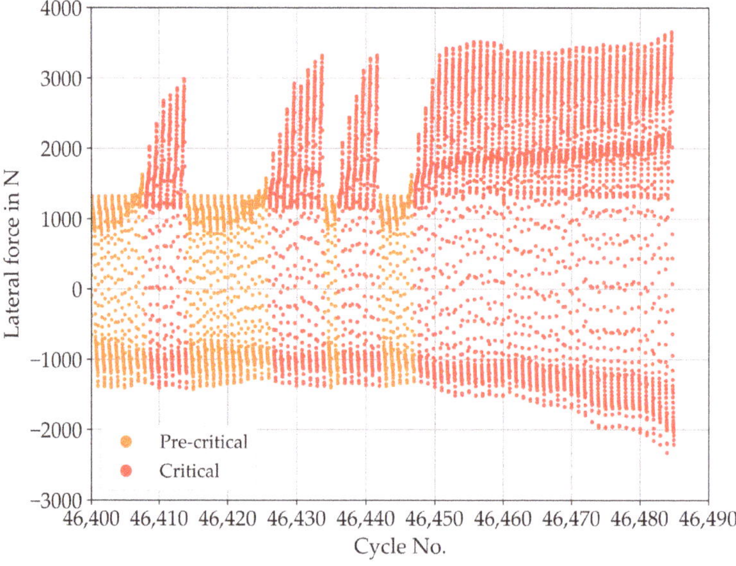

Figure 10. Change between pre-critical and critical operation at the end of experiment 1. Critical operation started with a pronounced increase in the friction force at one turning point.

The second group shows a pre-critical and critical operation rather suddenly. The stop criterion was exceeded within less than 10 min. Experiment 6 reached pre-critical operation as few as 1.5 min before termination of the experiment. This sudden critical behaviour may be due to a sudden loss of the lubricant supply, resulting in a pronounced increase in the lateral force, whereas in the first group, lubricant supply was sufficient to keep the system in an operable state over a longer period.

4. Discussion

This paper presents a semi-supervised method for the classification of states of operation during a tribological sliding experiment in oscillating, translatory motion using an RF classifier.

An RF classifier was selected due to its low complexity regarding implementation, its good prediction accuracies, and the low requirements for model tuning. The RF model can be easily trained, validated, and applied on a local machine, with the capability of real-time classification. RF classifiers are especially well suited for industrial applications, as no AI expert is required to set up and tune sophisticated ANN-based algorithms [24].

The algorithm was trained on the basis of individual cycles. This is only possible if the force data are recorded with high temporal resolution. The trained algorithm was able to classify the state of operation with an accuracy of 0.939 for data of a labelled test experiment, using samples from four different experimental runs (i.e., four different journal bearings with otherwise identical experimental setup) as the training dataset.

The proposed methodology can be extended to similar systems, with different dimensions and materials of the involved bodies. However, datasets from these systems are necessary for training the algorithm. The transfer of an already trained algorithm to other systems remains an aspect for further investigation.

As a future perspective, online classification of the current status of the system will help to identify critical operation conditions. This will allow taking real-time countermeasures to assist the self-recovering process of the system, such as reduction of oscillation frequency or normal load, up to stopping the experiment to prevent major damage. The experiment can be stopped during critical operation for ex post analysis, e.g., material or surface analysis of the sliding bodies. Detailed knowledge of the system and its history can be used to define more complex stopping criteria, additionally to simple threshold values.

The presented approach may be extended to applications in industrial machinery, provided that a continuous force measurement and a sufficient amount of training data from ex post analysis are available. Examples for potential industrial applications range from journal bearings mounted in industrial equipment or drive trains to hydraulic presses, pistons, and manufacturing tools, especially where the accessibility of the system is limited for optical inspection.

In a further step, the presented algorithm may form a basis for lifetime prediction. Experimentally determined durations until reaching the stop criteria and thus termination of the experiment may be used as additional input for training the algorithm. This is a challenging task, as terminal failure often occurs suddenly without showing progressive deterioration in advance [9]. In the present work, sudden terminal failure occurred in about half of the analysed experiments. In the other experiments, terminal failure was preceded by pre-critical operation of up to 3.5 h. Inclusion of further continuous sensor data, such as temperature, acceleration, airborne or structure-borne AE, may serve to improve labelling and provide additional information for lifetime prediction. With a combination of these sensors, training of a similar RF algorithm is possible, even if no continuous force data are available.

In contrast to most studies regarding ML in tribological applications, in the current study, a self-recovering system was analysed. Thus, the system may stabilise after a pre-critical or even critical phase and return to steady operation. For conventional tribological systems, pre-critical or critical operation indicates an impending failure of the system, and stopping the experiment is the only way to prevent major damages. For self-recovering systems, an online ML algorithm will have to distinguish between transient and terminal critical operation. To achieve that, additional datasets such as AE or acceleration data have to be included.

A high quality of the labels assigned to the training dataset has proven to be the key for a high prediction accuracy of the RF algorithm. The presented semi-supervised approach—labelling by unsupervised k-means clustering with manual refinement—offers the flexibility to choose within a range between fully automated, unsupervised labelling and entirely manual labelling based on expert knowledge. In order to provide high-quality labelled training datasets, tribological and engineering expertise have to be included in the classification process in any case.

There are several papers on friction and wear monitoring as well as failure classification using data from AE sensors, e.g., [9,14,51] or image data, including optical [22] and thermal imaging [23]. In contrast, the proposed method focuses on time-series data from a force sensor collected at sampling rate of 5 kHz, similar to e.g., [2], as a data source for training a ML algorithm. This has the great advantage that high classification accuracy can be reached by using only force data, recorded by default in any tribological experiment. However, time-series data from other sensors, such as AE or acceleration, and optical or thermal image data, can provide useful additional information, which can be used to increase the algorithm's classification accuracy.

The focus of the current work was set to the overall health condition of the bearing, which can be characterised by its state of operation, ultimately related to wear and lubrication in the contact area. As a system of self-lubricating journal bearings exhibits the ability for self-recovery during usage, it could be shown that the presented RF classifier allows detecting critical conditions prior to the onset of machine failure, solely based on the lateral force data. Future research will address the prediction of useful remaining lifetime and ultimate system failure.

5. Conclusions

In this paper, an oscillating, translatory sliding experiment of a self-lubricating bronze journal bearing, which provides the system with the ability to self-recover minor damages, was studied to elaborate a semi-supervised ML algorithm predicting critical operating conditions. An RF classifier was trained on the basis of single cycles of lateral force signals acquired with high resolution and including expertise knowledge of tribologists. Four different states of operation were identified based on the shape of the cycles. The main findings of the present paper are as follows:

- An RF algorithm, trained with high-resolution force signals of four experiments, showed a high degree of classification accuracy (0.939) after validation against a labelled dataset of another experiment.
- The labelling step is essential and preferably includes tribological expert knowledge. The proposed method offers the flexibility to choose within a range between fully automated and fully expert-related labelling.
- The application of a pre-trained algorithm to unlabelled data is very efficient and therefore can be used for immediate countermeasures to assist the self-recovering process of the system or to prevent major damage.

Author Contributions: Conceptualisation: U.C.-B. and F.P., data curation: J.P. and G.V., formal analysis: J.P., G.V. and I.A.N., methodology: U.C.-B., R.G., J.P. and G.V., project administration: J.P., software: J.P., supervision: G.V., validation: J.P., visualisation: J.P., writing—original draft preparation: J.P., R.G. and I.A.N., writing—review and editing: U.C.-B., F.P. and G.V. All authors have read and agreed to the published version of the manuscript.

Funding: This work was funded by the Austrian COMET Program (project InTribology1, No. 872176) via the Austrian Research Promotion Agency (FFG) and the Provinces of Niederösterreich and Vorarlberg and has been carried out within the Austrian Excellence Centre of Tribology (AC2T research GmbH).

Institutional Review Board Statement: Not applicable.

Informed Consent Statement: Not applicable.

Data Availability Statement: The data presented in this work are available on request from the corresponding author.

Conflicts of Interest: The authors declare no conflict of interest.

References

1. Pech, M.; Vrchota, J.; Bednář, J. Predictive Maintenance and Intelligent Sensors in Smart Factory: Review. *Sensors* **2021**, *21*, 1470. [CrossRef] [PubMed]
2. Gouarir, A.; Martínez-Arellano, G.; Terrazas, G.; Benardos, P.; Ratchev, S. In-process Tool Wear Prediction System Based on Machine Learning Techniques and Force Analysis. *Procedia CIRP* **2018**, *77*, 501–504. [CrossRef]
3. Pandiyan, V.; Caesarendra, W.; Tjahjowidodo, T.; Tan, H.H. In-process tool condition monitoring in compliant abrasive belt grinding process using support vector machine and genetic algorithm. *J. Manuf. Process.* **2018**, *31*, 199–213. [CrossRef]
4. Avendano, D.N.; Caljouw, D.; Deschrijver, D.; Van Hoecke, S. Anomaly detection and event mining in cold forming manufacturing processes. *Int. J. Adv. Manuf. Technol.* **2020**, 1–16. [CrossRef]
5. Hastie, T.; Tibshirani, R.; Friedman, J. *The Elements of Statistical Learning: Data Mining, Inference, and Prediction*, 2nd ed.; Springer: New York, NY, USA, 2009.
6. Saeidi, F.; Shevchik, S.; Wasmer, K. Automatic detection of scuffing using acoustic emission. *Tribol. Int.* **2016**, *94*, 112–117. [CrossRef]
7. König, F.; Sous, C.; Chaib, A.O.; Jacobs, G. Machine learning based anomaly detection and classification of acoustic emission events for wear monitoring in sliding bearing systems. *Tribol. Int.* **2021**, *155*, 106811. [CrossRef]
8. Pandiyan, V.; Prost, J.; Vorlaufer, G.; Varga, M.; Wasmer, K. Identification of Abnormal Tribological Regimes Using a Mi-crophone And Semi-Supervised Machine-Learning Algorithm. *Friction* **2021**. accepted for publication on 3 April 2021.
9. Shevchik, S.A.; Saeidi, F.; Meylan, B.; Wasmer, K. Prediction of Failure in Lubricated Surfaces Using Acoustic Time–Frequency Features and Random Forest Algorithm. *IEEE Trans. Ind. Inform.* **2016**, *13*, 1541–1553. [CrossRef]
10. Moder, J.; Bergmann, P.; Grün, F. Lubrication Regime Classification of Hydrodynamic Journal Bearings by Machine Learning Using Torque Data. *Lubricants* **2018**, *6*, 108. [CrossRef]
11. Thankachan, T.; Prakash, K.S.; Kamarthin, M. Optimizing the Tribological Behavior of Hybrid Copper Surface Composites Using Statistical and Machine Learning Techniques. *J. Tribol.* **2018**, *140*, 031610. [CrossRef]
12. Bhaumik, S.; Mathew, B.R.; Datta, S. Computational intelligence-based design of lubricant with vegetable oil blend and various nano friction modifiers. *Fuel* **2019**, *241*, 733–743. [CrossRef]
13. Deshpande, P.; Pandiyan, V.; Meylan, B.; Wasmer, K. Acoustic emission and machine learning based classification of wear generated using a pin-on-disc tribometer equipped with a digital holographic microscope. *Wear* **2021**, *203622*, 203622. [CrossRef]
14. Mokhtari, N.; Pelham, J.G.; Nowoisky, S.; Bote-Garcia, J.-L.; Gühmann, C. Friction and Wear Monitoring Methods for Journal Bearings of Geared Turbofans Based on Acoustic Emission Signals and Machine Learning. *Lubricants* **2020**, *8*, 29. [CrossRef]
15. Bustillo, A.; Pimenov, D.Y.; Mia, M.; Kapłonek, W. Machine-learning for automatic prediction of flatness deviation considering the wear of the face mill teeth. *J. Intell. Manuf.* **2021**, *32*, 895–912. [CrossRef]
16. Boidi, G.; Da Silva, M.R.; Profito, F.J.; Machado, I.F. Using Machine Learning Radial Basis Function (RBF) Method for Predicting Lubricated Friction on Textured and Porous Surfaces. *Surf. Topogr. Metrol. Prop.* **2020**, *8*, 044002. [CrossRef]
17. Sun, S.; Przystupa, K.; Wei, M.; Yu, H.; Ye, Z.; Kochan, O. Fast bearing fault diagnosis of rolling element using Lévy Moth-Flame optimization algorithm and Naive Bayes. *Eksploatacja Niezawodn. Maint. Reliab.* **2020**, *22*, 730–740. [CrossRef]
18. Argatov, I. Artificial Neural Networks (ANNs) as a Novel Modeling Technique in Tribology. *Front. Mech. Eng.* **2019**, *5*, 5. [CrossRef]
19. Souza, R.M.; Nascimento, E.G.; Miranda, U.A.; Silva, W.J.; Lepikson, H.A. Deep learning for diagnosis and classification of faults in industrial rotating machinery. *Comput. Ind. Eng.* **2021**, *153*, 107060. [CrossRef]
20. Rosenkranz, A.; Marian, M.; Profito, F.J.; Aragon, N.; Shah, R. The Use of Artificial Intelligence in Tribology—A Perspective. *Lubricants* **2020**, *9*, 2. [CrossRef]
21. Kateris, D.; Moshou, D.; Pantazi, X.-E.; Gravalos, I.; Sawalhi, N.; Loutridis, S. A machine learning approach for the condition monitoring of rotating machinery. *J. Mech. Sci. Technol.* **2014**, *28*, 61–71. [CrossRef]
22. Bergs, T.; Holst, C.; Gupta, P.; Augspurger, T. Digital image processing with deep learning for automated cutting tool wear detection. *Procedia Manuf.* **2020**, *48*, 947–958. [CrossRef]
23. Glowacz, A. Fault diagnosis of electric impact drills using thermal imaging. *Measurement* **2021**, *171*, 108815. [CrossRef]
24. Bustillo, A.; Pimenov, D.; Matuszewski, M.; Mikolajczyk, T. Using artificial intelligence models for the prediction of surface wear based on surface isotropy levels. *Robot. Comput. Manuf.* **2018**, *53*, 215–227. [CrossRef]
25. Erdemir, A. Solid Lubricants and Self-Lubricating Films. In *Modern Tribology Handbook, Volume One: Principles of Tribology*; Bhushan, B., Ed.; CRC Press: Boca Raton, FL, USA, 2001; pp. 787–825.
26. Neacșu, I.A.; Scheichl, B.; Vorlaufer, G.; Eder, S.J.; Franek, F.; Ramonat, L. Experimental Validation of the Simulated Steady-State Behavior of Porous Journal Bearings1. *J. Tribol.* **2016**, *138*, 031703. [CrossRef]
27. Eder, S.; Ielchici, C.; Krenn, S.; Brandtner, D. An experimental framework for determining wear in porous journal bearings operated in the mixed lubrication regime. *Tribol. Int.* **2018**, *123*, 1–9. [CrossRef]
28. Boidi, G.; Krenn, S.; Eder, S.J. Identification of a Material–Lubricant Pairing and Operating Conditions That Lead to the Failure of Porous Journal Bearing Systems. *Tribol. Lett.* **2020**, *68*, 108. [CrossRef]
29. Guo, J.; Du, H.; Zhang, G.; Cao, Y.; Shi, J.; Cao, W. Fabrication and tribological behavior of Fe-Cu-Ni-Sn-Graphite porous oil-bearing self-lubricating composite layer for maintenance-free sliding components. *Mater. Res. Express* **2020**, *8*, 015801. [CrossRef]

30. Rodiouchkina, M.; Berglund, K.; Mouzon, J.; Forsberg, F.; Shah, F.U.; Rodushkin, I.; Larsson, R. Material Characterization and Influence of Sliding Speed and Pressure on Friction and Wear Behavior of Self-Lubricating Bearing Materials for Hydropower Applications. *Lubricants* **2018**, *6*, 39. [CrossRef]

31. Zhang, C.; Pan, L.; Wang, S.; Wang, X.; Tomovic, M. An accelerated life test model for solid lubricated bearings used in space based on time-varying dependence analysis of different failure modes. *Acta Astronaut.* **2018**, *152*, 352–359. [CrossRef]

32. Jisa, R.; Monetti, C.; Kelman, P. Selbstschmierende Gleitsysteme aus schadensanalytischer Sicht. *Tribol. Schmier.* **2005**, *52*, 26–29.

33. Cihak-Bayr, U.; Steiner, H.; Glatzl, T.; Grundtner, R.; Pirker, F. Machine Learning Algorithms for Health Monitoring of Sliding Bearings. In Proceedings of the 60th German Tribology Conference, Göttingen, Germany, 23–25 September 2019; pp. 1–4.

34. Jisa, R. Tribologische Wechselwirkungen von Selbstschmierenden Gleitelementen Basierend auf Kupferlegierungen und Graphit-Öl-Schmierstoffen. Ph.D. Dissertation, Technische Universität Wien, Vienna, Austria, 2007.

35. Kluyver, T.; Ragan-Kelley, B.; Pérez, F.; Granger, B.; Bussonnier, M.; Frederic, J.; Kelley, K.; Hamrick, J.; Grout, J.; Corlay, S.; et al. Jupyter Notebooks—A Publishing Format for Reproducible Computational Workflows. In Proceedings of the 20th International Conference on Electronic Publishing, Göttingen, Germany, 7–9 June 2016; pp. 87–90. [CrossRef]

36. Harris, C.R.; Millman, K.J.; Van Der Walt, S.J.; Gommers, R.; Virtanen, P.; Cournapeau, D.; Wieser, E.; Taylor, J.; Berg, S.; Smith, N.J.; et al. Array programming with NumPy. *Nat. Cell Biol.* **2020**, *585*, 357–362. [CrossRef]

37. McKinney, W. Data Structures for Statistical Computing in Python. In Proceedings of the 9th Python in Science Conference, Austin, TX, USA, 28 June–3 July 2010; pp. 56–61. Available online: https://conference.scipy.org/scipy2010/slides/wes_mckinney_data_structure_statistical_computing.pdf (accessed on 26 April 2021).

38. The HDF Group. The HDF5® Library & File Format. Available online: https://www.hdfgroup.org/solutions/hdf5/ (accessed on 18 September 2020).

39. Savitzky, A.; Golay, M.J.E. Smoothing and Differentiation of Data by Simplified Least Squares Procedures. *Anal. Chem.* **1964**, *36*, 1627–1639. [CrossRef]

40. Breiman, L. Random Forests. *Mach. Learn.* **2001**, *45*, 5–32. [CrossRef]

41. Breiman, L. Bagging predictors. *Mach. Learn.* **1996**, *24*, 123–140. [CrossRef]

42. Louppe, G.; Wehenkel, L.; Sutera, A.; Geurts, P. Understanding Variable Importances in Forests of Randomized Trees. In Advances in Neural Information Processing Systems 26, Proceedings of the 27th Annual Conference on Neural Information Processing Systems 2013, Lake Tahoe, NV, USA, 5–10 December 2013. pp. 431–439. Available online: http://hdl.handle.net/2268/155642 (accessed on 3 May 2021).

43. Nembrini, S.; König, I.R.; Wright, M.N. The revival of the Gini importance? *Bioinformatics* **2018**, *34*, 3711–3718. [CrossRef]

44. Oshiro, T.M.; Perez, P.S.; Baranauskas, J.A. How Many Trees in a Random Forest? In Proceedings of the Transactions on Petri Nets and Other Models of Concurrency XV; Springer Science and Business Media LLC: Berlin/Heidelberg, Germany, 2012; pp. 154–168.

45. Probst, P.; Boulesteix, A.-L. To tune or not to tune the number of trees in random forest. *J. Mach. Learn. Res.* **2018**, *18*, 1–18.

46. Fawcett, T. An introduction to ROC analysis. *Pattern Recognit. Lett.* **2006**, *27*, 861–874. [CrossRef]

47. Powers, D.M.W. Evaluation: From precision, recall and F-measure to ROC, informedness, markedness and correlation. *arXiv* **2020**, arXiv:2010.16061. Available online: https://arxiv.org/abs/2010.16061 (accessed on 1 May 2021).

48. Pedregosa, F.; Varoquaux, G.; Gramfort, A.; Michel, V.; Thirion, B.; Grisel, O.; Blondel, M.; Prettenhofer, P.; Weiss, R.; Du-bourg, V.; et al. Scikit-learn: Machine Learning in Python. *J. Mach. Learn. Res.* **2011**, *12*, 2825–2830.

49. Wiley, M.; Wiley, J.F. *Advanced R Statistical Programming and Data Models: Analysis, Machine Learning, and Visualization*; Apress: New York, NY, USA, 2019.

50. Agrawal, T. *Hyperparameter Optimization in Machine Learning: Make Your Machine Learning and Deep Learning Models More Efficient*; Apress: New York, NY, USA, 2021.

51. Elforjani, M.; Shanbr, S. Prognosis of Bearing Acoustic Emission Signals Using Supervised Machine Learning. *IEEE Trans. Ind. Electron.* **2018**, *65*, 5864–5871. [CrossRef]

Article

A Digital Twin for Friction Prediction in Dynamic Rubber Applications with Surface Textures

Valentina Zambrano [1], Markus Brase [2], Belén Hernández-Gascón [1,*], Matthias Wangenheim [2], Leticia A. Gracia [1], Ismael Viejo [1], Salvador Izquierdo [1] and José Ramón Valdés [1]

[1] Instituto Tecnológico de Aragón—ITAINNOVA, C/María de Luna 7-8, 50018 Zaragoza, Spain; vzambrano@itainnova.es (V.Z.); lgracia@itainnova.es (L.A.G.); iviejo@itainnova.es (I.V.); sizquierdo@itainnova.es (S.I.); jrvaldes@itainnova.es (J.R.V.)

[2] Institut für Dynamik und Schwingungen—IDS, Leibniz Universität Hannover—LUH, An der Universität 1, 30823 Garbsen, Germany; brase@ids.uni-hannover.de (M.B.); wangenheim@ids.uni-hannover.de (M.W.)

* Correspondence: bhernandez@itainnova.es

Abstract: Surface texturing is an effective method to reduce friction without the need to change materials. In this study, surface textures were transferred to rubber samples in the form of dimples, using a novel laser surface texturing (LST)—based texturing during moulding (TDM) production process, developed within the European Project MouldTex. The rubber samples were used to experimentally determine texture-induced friction variations, although, due to the complexity of manufacturing, only a limited amount was available. The tribological friction measurements were hence combined with an artificial intelligence (AI) technique, i.e., Reduced Order Modelling (ROM). ROM allows obtaining a virtual representation of reality through a set of numerical strategies for problem simplification. The ROM model was created to predict the friction outcome under different operating conditions and to find optimised dimple parameters, i.e., depth, diameter and distance, for friction reduction. Moreover, the ROM model was used to evaluate the impact on friction when manufacturing deviations on dimple dimensions were observed. These results enable industrial producers to improve the quality of their products by finding optimised textures and controlling nominal surface texture tolerances prior to the rubber components production.

Keywords: reduced order modelling; dynamic friction; rubber seal applications; tensor decomposition; laser surface texturing; texturing during moulding; digital twin; machine learning; artificial intelligence

Citation: Zambrano, V.; Brase, M.; Hernández-Gascón, B.; Wangenheim, M.; Gracia, L.A.; Viejo, I.; Izquierdo, S.; Valdés, J.R. A Digital Twin for Friction Prediction in Dynamic Rubber Applications with Surface Textures. *Lubricants* **2021**, *9*, 57. https://doi.org/10.3390/lubricants9050057

Received: 30 March 2021
Accepted: 16 May 2021
Published: 20 May 2021

Publisher's Note: MDPI stays neutral with regard to jurisdictional claims in published maps and institutional affiliations.

Copyright: © 2021 by the authors. Licensee MDPI, Basel, Switzerland. This article is an open access article distributed under the terms and conditions of the Creative Commons Attribution (CC BY) license (https://creativecommons.org/licenses/by/4.0/).

1. Introduction

1.1. Surface Texturing in Tribological Applications

In most dynamic rubber applications, low friction is desired to increase the energy efficiency of the tribological system. Therefore, improved friction behaviour is an important design objective in the development process of components like dynamic seals. Within the industrial and scientific communities, significant improvements in dynamic seal performance have been achieved through a series of technological advances, such as the introduction of low-friction polymers [1]. However, this approach is limited by the operating conditions, such as temperatures or the chemical resistance of the new materials.

Surface texturing is an effective method to modify the friction level without changing materials or the lubricant in the dynamic seal contact. An early method of surface texturing, used to improve the tribological performance of mechanical components, is cylinder honing in internal-combustion engines [2]. Further experimental and theoretical studies revealed a significant reduction in friction due to grooves [3] and, in particular, micro dimples in the reciprocating contact of piston rings and cylinder bores of combustion engines [4,5].

The analysis of laser surface textured (LST) piston rings have been extended to model-based and experimental investigations of mechanical seals, identifying optimised dimple dimensions with lowest friction for specific operating conditions [6,7]. In this context, the dimples are advantageous as they serve as hydrodynamic bearings and exhibit the ability to store lubricant [8]. Moreover, the surface textures reduce the real area of contact and trap wear particles [9,10].

Other authors even highlighted the positive effect of surface textured rubber seals in soft elasto-hydrodynamic lubrication (SEHL), resulting in a significant friction reduction compared to the untextured references [11–13]. However, it has also been found that an inappropriate selection of the surface texture dimensions leads to a detrimental increase in friction due to excessive enlargement of cavitation zones in the lubricant, causing a reduced local film thickness and load carrying capacity [14,15]. Furthermore, the influence of the real dimple shape of textured mechanical seals has been identified as an important factor that needs to be considered for valid friction determination [16].

1.2. Reduced Order Modelling

The introduction of extended methods, based on artificial intelligence (AI), which go further than usual tribological measurements, is beneficial for friction predictions. The selected AI technique is based on the digital twin (DT) paradigm, i.e., a virtual representation of reality [17–19]. Several techniques belong within the DT paradigm, the most popular being Reduced Order Modelling (ROM) [20,21] and machine learning (ML) [22]. Both techniques are based on mathematical models for real-time simulations. In particular, ROM consists of a set of numerical strategies for multi-variable problem simplification to solve complex numerical systems and it aims to describe and, hence, predict a system's behaviour through a mathematical approximation, by preserving its main characteristics, as described in [23]. Tensor rank decomposition (TRD) approach is a non-intrusive, i.e., completely data-driven, method that allows to describe a complex system's behaviour, where its variables influence each other, as a simplified mathematical function that describes each variable's effect separately. As previously detailed in [23], TRD is based on the assumption that a problem of N, not necessarily independent, variables can be rewritten as the product of N one-dimensional functions, one for each of the variables of the system, as shown in Equation (1):

$$F(v_1, \ldots, v_N) = \sum_{m=1}^{M} \alpha_m \prod_{n=1}^{N} f_{m,n}(v_n) \tag{1}$$

where M is the order of approximation of the ROM model and $\alpha_m, m = 1, \ldots, M$ are weighting coefficients. The functions $f_{m,n}$, in their most simple form, are piecewise linear functions; hence the adjustment parameters are the positions and the values at which the functions change slope. The adjustment of all these parameters is carried out through a least square optimisation.

The first term of the sum in Equation (1) represents a first approximation of the system, being its corresponding coefficient α_1 the largest one, while the following terms would be corrections to it and will generally have lower coefficient values unless the correction only applies to a specific outliers population and does not affect the general trend of data.

1.3. Objectives

Although many experimental and model-based studies have been accomplished on the subject of friction reduction of surface textured components, all investigations require extensive series of experiments or complex contact- or fluid-mechanic simulations. Therefore, this paper aims to introduce a novel approach, in which a limited number of friction measurements of surface textured rubber samples are combined with ROM to identify optimal surface textures as a function of the prevailing operating conditions in real-time. Besides the nominal dimple texture parameters, the real dimple dimensions, defined by diameter, distance, and depth, are taken into account for the friction values computed by the ROM. Therefore, ROM technique is further used to determine nominal surface texture values uncertainties for valid friction prediction.

2. Materials and Methods

The following Sections are dedicated to the description of the experimental studies and the used software, data analysis (dataset publicly available under [24]), and statistical tools.

2.1. Rubber Specimen Geometry and Surface Texture Parameters

The objective of the experimental testing procedure is to measure the friction between surface textured rubber specimens and a rotating counter surface by utilising a pin-on-disc tribometer. The grinded steel counter surface exhibits a surface roughness of $R_a = 0.50$ μm, but no further texturing. Due to the variety of seal geometries available on the market, a simplified rubber sample geometry was chosen for the experiments, in order to test the surface texture-induced friction variation independent of a specific seal geometry. The corresponding geometry of the rubber samples is shown in Figure 1a.

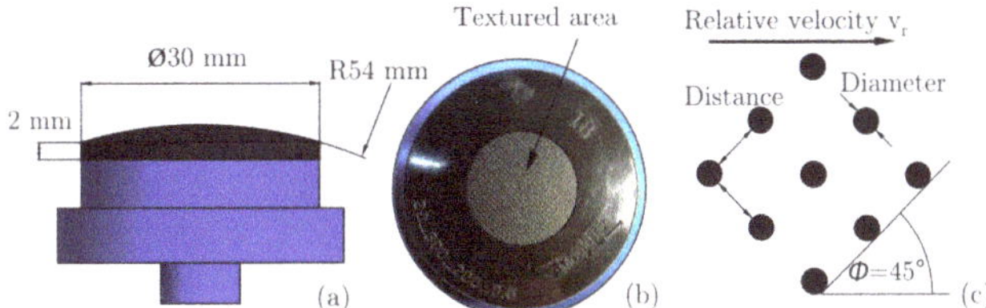

Figure 1. (**a**) Geometry of the rubber specimens including the most relevant dimensions, (**b**) picture of the rubber specimen with focus on the dimple texture and (**c**) positioning of the dimples in relation to the relative velocity vector.

A 2 mm thick layer of a fluorelastomer with a shore hardness of 80A (FKM 80A) is vulcanised onto a blue anodised aluminium specimen holder. Furthermore, the contact zone of the 30 mm diameter rubber sample has a spherical shape to avoid edge effects in the dynamic contact.

The surface textures are applied to the contact areas of the rubber samples in the form of deterministic positioned dimples, see Figure 1b. The geometry of the circular dimples is defined by the diameter, the distance and the depth. The corresponding parameters and the alignment of the dimples are shown in Figure 1c. During the test procedure, the positioning of the dimples, within a squared area, was rotated by an angle of $\phi = 45°$, since preliminary experiments showed about 20% greater friction reduction in this arrangement compared to $\phi = 0°$. The reason for this is the avoidance of continuous flow channels without dimples in the direction of relative motion for $\phi = 45°$. The angle ϕ between the square shaped texture arrangement and the relative velocity vector is visualised in Figure 1c.

In order to perform the experiments, test specimens with eight different surface textures were manufactured by texturing during moulding (TDM). The associated sample number i and the nominal texture parameters are listed in Table 1.

Table 1. Nominal dimple texture parameters defined by diameter, distance and depth.

Sample i	Dimple Diameter [μm]	Dimple Distance [μm]	Dimple Depth [μm]
1	100	300	10
2	200	100	10
3	200	200	10
4	300	100	30
5	300	200	20
6	300	200	30
7	300	300	20
8	-	-	-

The parameter range of the dimple dimensions was selected to ensure a transferability of the textures to dynamic seal applications. For example, if the diameter of a dimple is larger than the contact width of a seal, the system would leak. Since the contact width of many lip seals is only 0.80 mm or less [25], the maximum size of the dimple diameter and distance was set to 300 µm. In contrast, e.g., the lower limit of the dimple diameter of 100 µm was determined by the accuracy of the LST process of the TDM manufacturing method. Based on the texturing results of the rubber samples, the transfer of the textures to real seals has already been successfully realised, but will not be discussed within the scope of this work.

The design of experiment (DoE) on the variation of the surface texture parameters is chosen according to the requirements of the ROM, described in Section 2.4. While seven specimens exhibit a dimple texture, one specimen was produced without dimples to serve as a reference for the dimple-induced friction variation of the other textures. The surface roughness of all eight samples was adjusted to an identical value of R_a = 0.50 µm by laser surface processing of the mould. Therefore, the influence of dimple textures on friction is investigated independently of different surface roughnesses.

2.2. Test Rig Setup and Experimental Procedure for Determining the Coefficient of Friction

A pin-on-disc tribometer is utilised to measure the friction forces between the rubber specimens and the rotating steel counter surface. The design of the test rig is shown in Figure 2a.

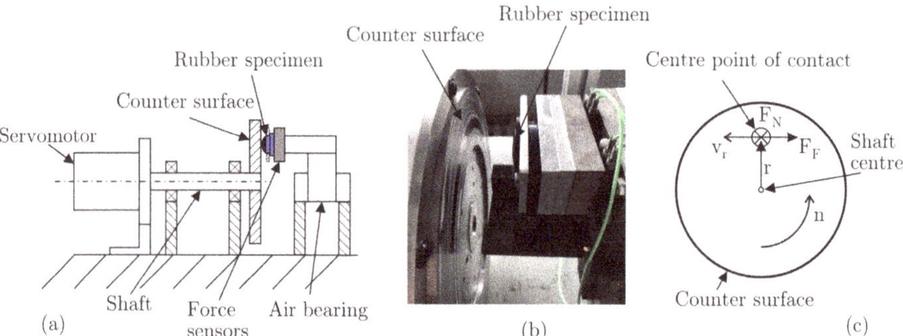

Figure 2. (**a**) Pin-on-disc tribometer design, (**b**) picture of the tribometer with focus on the rubber specimen and the rotating counter surface, and (**c**) contact conditions between the rubber sample and the counter surface.

The rotation of the counter surface is realised by a servomotor. The force sensors measure both the force F_N in the normal direction of the sample, as well as the friction force F_F in the circumferential direction of the rotating counter surface, compare Figure 2c. The rotational speed n of the counter surface and the normal force F_N are kept constant during each measurement, in order to measure quasi-stationary friction values under steady operating conditions. Each relative velocity v_r, specified in Table 2a, is tested at each of the 3 different contact pressure levels $p_{c,max}$, given in Table 2b.

The relative velocity v_r between the rubber specimen and the counter surface is equal to the circumferential velocity v_c at the point of contact: $v_r = v_c$. The velocity v_r is calculated from the rotational speed n of the counter surface and the distance r = 100 mm between the shaft centre and the contact point, see Figure 2c, as defined in Equation (2).

$$v_r = \omega r = 2\pi n r \tag{2}$$

The focus of this paper is on dynamic friction, so static friction is not investigated. Thus, the rotational speed n of the servomotor was varied between 0.6 and 24.0 min^{-1}, resulting in relative velocities v_r of 6 to 251 mm/s.

Table 2. Operating parameters that are examined during the test procedure. (**a**) Rotational speeds n of the servomotor and corresponding relative velocities v_r between the rubber sample and the counter surface (2), (**b**) together with the normal force F_N, the related maximum contact pressure $p_{c,max}$, the contact diameter d_c between the rubber sample and the counter surface, as well as the nominal contact area $A_{nominal}$ (2).

a	
Rotational Speed n [min^{-1}]	Relative Velocity v_r [mm/s]
0.6	6
1.2	12
1.8	19
3.0	31
6.0	63
12.0	126
18.0	188
24.0	251

b			
Normal Force F_N [N]	Max Contact Pressure $p_{c,max}$ [MPa]	Contact Diameter d_c [mm]	Nominal Contact Area $A_{nominal}$ [mm^2]
3.9	0.5	4.2	13.8
7.9	0.7	5.0	19.6
13.3	0.9	5.8	26.4

The normal forces F_N were selected to achieve maximum contact pressures $p_{c,max}$ of 0.5, 0.7, and 0.9 MPa, which are typical values in pneumatic seal applications [26]. Because of the spherical shape of the rubber specimen, the variation of the normal force F_N influences not only the magnitude of the parabolic contact pressure distribution p_c, but also the dimensions of the nominal contact area between the rubber sample and counter surface, see Figure 3b. Therefore, not only the maximum of the contact pressure distribution $p_{c,max}$ is specified in Table 2b, but also the corresponding contact diameter d_c and the respective circular nominal contact area $A_{nominal}$. The contact pressure distribution p_c and the contact diameter d_c are computed by finite element analyses, taking into account the rubber thickness of 2 mm. Within the static simulations, the rubber sample is pressed against the counter surface with the defined normal force F_N, see Figure 3a. At this, the coloured spherical contact area of the rubber specimen is brought into contact with the grey flat counter surface, resulting in the specified contact area and contact pressure p_c.

All components are modelled as 2D axisymmetric parts. A hyper elastic Mooney-Rivlin material behaviour is assigned to the 2 mm layer of the FKM80A rubber material, which is specified by the temperature-dependent material parameters $C_{10} = 1{,}442{,}425.12$, $C_{01} = 208{,}308.34$, $D = 6.059933161 \times 10^{-10}$, considering a Poisson's Ratio of 0.4995. For both measurements and simulations, the temperature is equal to 20 °C. The material of the anodised aluminium sample holder is modelled as pure elastic part with a Young's modulus of 70 GPa and a Poisson's Ratio of 0.34. The counter surface is defined as rigid part.

All experiments are performed with an adherent silicone grease OKS 1155 in the dynamic contact, which exhibits a base oil viscosity of 100 mm^2/s at 25 °C [27]. The same lubrication and conditioning procedure was applied to every rubber specimen before the actual friction measurements to ensure comparability between the results. Every single measurement lasts for 5 s and each measurement was repeated 5 times to ensure a statistical certainty. In order to generate the quasi-stationary friction values μ_{untext} for rubber specimen 8 and $\mu_{text,i}$ for rubber samples $i = 1$–7, the mean value of each measurement is calculated over the entire measuring period of 5 s. The friction coefficients μ_{untext} and $\mu_{text,i}$ are evaluated to identify dimple textures with the highest friction reduction potential. In addition, the results are further processed to generate the ROM model.

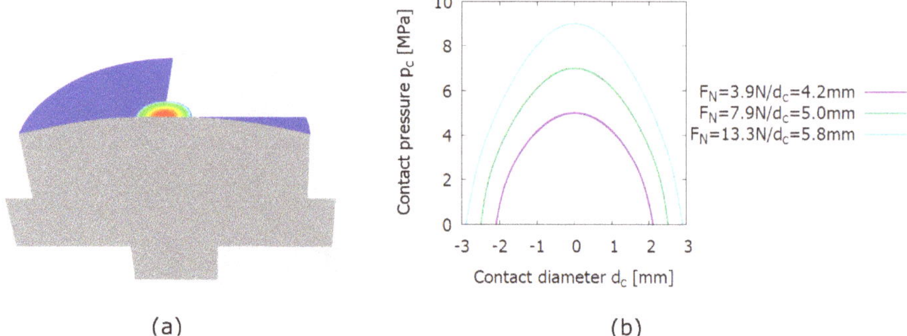

(a)   (b)

Figure 3. (**a**) Assembly of the finite element method (FEM). The coloured spherical contact area of the rubber specimen is brought into contact with the grey flat counter surface. (**b**) Parabolic contact pressure distribution p_c as a function of the contact diameter d_c.

2.3. Method of Measurement for Real Dimple Dimensions

An additional important aspect for the determination of the influence of the dimple textures on friction are the real dimensions of the dimples. While the nominal values of the investigated textures are given in Table 1, the real values of diameter, distance and depth vary due to imperfections in the innovative industrialised TDM production process of the rubber specimens. During this process, the negative of the desired texture is applied to a mould by LST ablation. A 5-axis 100 W pico-second laser is utilised for this purpose. During the injection moulding and vulcanisation procedure, the texture is directly transferred from the mould to the rubber surface. Protrusions in the mould become dimples in the elastomer. Due to imperfections in the laser machining of the mould, both a deviation from the nominal dimple parameters and ring-shaped cavities in the peripheral areas of the dimple valleys are identifiable, see Figure 4b. These deviations mainly result from re-solidifying metal vapours on the surface of the metallic vulcanisation mould.

In order to measure and visualise the actual dimple texture dimensions, a 3D optical microscope, based on focus-variation, is used. Figure 4a shows an exemplary 4×4 mm surface scan of rubber specimen 6 with a nominal dimple diameter of 300 µm, a distance of 200 µm and a depth of 30 µm.

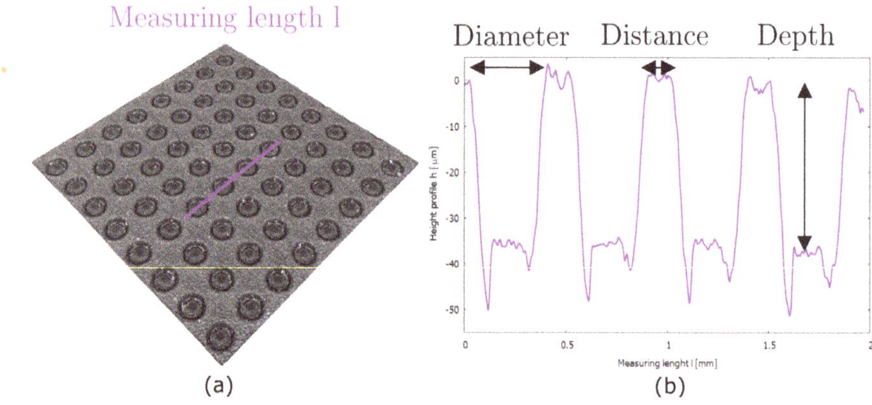

(a)   (b)

Figure 4. (**a**) Microscope image of rubber specimen 6, considering a measured area of 4×4 mm and (**b**) related height profile h as function of the measuring length l, considering the indicated line scan trough four adjacent dimples.

Figure 4b shows the corresponding line scan as height profile h over the measuring length l, which is measured through the centre of four adjacent dimples. The positions that are measured to determine the real dimple dimensions of diameter, distance, and depth are marked in the height profile. To ensure accurate measurement results of the dimple dimensions, the curvature of the contact surface of the rubber samples was subtracted, resulting in the flat surface scan depicted. Beyond this, no corrections were made to the surface scans. The dimple dimensions were measured at 16 different positions of the textured area of each rubber sample. The mean values of the 16 measurements and the corresponding dimensionless texture parameters are listed in Table 3.

Table 3. Real dimple dimensions, indicated by diameter, distance, and depth. The aspect ratio is the quotient of dimple depth and diameter. The textured area percentage is determined from the quotient of the textured area $A_{textured}$ and the nominal contact area $A_{nominal}$.

Sample i	Dimple Diameter [μm]	Dimple Distance [μm]	Dimple Depth [μm]	Aspect Ratio	Textured Area [%]
1	135	258	16	0.11	9
2	242	100	11	0.05	39
3	241	165	10	0.04	28
4	337	66	35	0.10	55
5	330	170	22	0.06	34
6	346	153	35	0.10	37
7	336	252	20	0.06	25
8	-	-	-	-	-

2.4. Software Development for Reduced Order Modelling

The software library (Twinkle), implemented and used to compute the ROM models, is available on GitHub platform [28] and described in [23], where its basic concept, structure and environmental dependencies are detailed.

Within this study, two separate ROM models were computed: one for untextured and one for textured tribometer samples data, consisting of 24 and 168 data points, respectively. The reliability of the two obtained ROM models was then validated using Python Scikit-learn [29] (version 0.22.1) k-fold cross validation technique, where a train-test procedure is performed k-times, randomly extracting a k-fraction sub-dataset for testing [30]. For the validation, the parameter k was set to 10, so that 10 different train-test validations were performed, randomly selecting the corresponding 90%–10% data fractions each time a new validation was performed.

Moreover, minimum values of the ROM function, i.e., Equation (1), were obtained using the ALGLIB Free Edition library [31], version 3.14.0, to find the surface texture dimensions that allow minimising friction, as described in Section 3.5.

Twinkle library can be described through the supervised learning algorithm concept in ML [32,33]. The input values used for ROM construction were the dimple dimensions, i.e., depth, diameter and distance, when available, together with pressure and velocity (please refer to Tables 1 and 2 or the available dataset under [24]), the output being the experimentally measured friction coefficient. Within the scope of this paper Equation (1) would represent the friction coefficient, expressed as the sum of 2 or 17 terms (in case of untextured or textured surface data, respectively). Each term being the product of one-dimensional functions, i.e., one function for each input separately, as shown for simplicity in Equation (3) for the untextured case.

$$\mu(pressure, velocity) = \alpha_1[f_1(pressure)f_1(velocity)] + \alpha_2[f_2(pressure)f_2(velocity)] \quad (3)$$

where α_1 and α_2 are the weighting coefficients of each computed term. The functions f_1 and f_2 are one-dimensional piecewise linear functions that describe the impact, on the friction outcome, of each input separately.

2.5. Reduced Order Modelling Data Pre-Processing for Friction Reduction

The friction variations (in%) obtained with the introduction of a dimple texture on the rubber surface are evaluated using Equation (4) in order to quantify the surface texture-induced friction variations.

$$\Delta\mu = 100\frac{\mu_{untext} - \mu_{text,i}}{\mu_{untext}}\%$$

(4)

In Equation (4) textured cases are compared to their corresponding untextured result, so that a negative value suggests a friction increase and a positive one reveals a friction reduction.

The final $\Delta\mu$ prediction was obtained using Equation (4), where the μ_{untext} and the $\mu_{text,i}$ values, predicted by the ROM models, were used, instead of using the experimentally measured results.

The two datasets, prepared and used in the statistical analysis, were created from two data samples of 48 data points each, using Python NumPy version 1.18.1 [34] and SciPy version 1.4.1 [35] libraries, as described in Sections 2.6 and 3.3.

2.6. Statistical Analysis of Real Dimple Dimensions

Due to manufacturing tolerances, the real dimple dimensions are not equal to the nominal ones. Experimental data were used to extract statistical information from 16 measurements of geometrical parameters of the textured surface, i.e., mean values and standard deviations for the three geometrical parameters were identified: dimple depth, diameter, and distance. The measuring procedure and mean values are described in Sections 2.3 and 3.1, respectively. By comparing the results to a foreseen normal distribution centred on the nominal value, having as standard deviation the value obtained from the different repetitions, it was possible to infer the type of probability density function (PDF) of each of the studied parameters.

Once the PDFs were known, the dimple depth was varied separately, while diameter and distance were varied together, as their variations are not independent and occur in the same plane, as defined by the textured area (Section 3.1). To do that, not all data were taken into account, but only those dimple parameters whose variations fell within the ROM's definition limits, minimum and maximum values in Tables 1 and 2, which resulted in 48 data points for both datasets, that were obtained when vertical, i.e., depth, and horizontal, i.e., diameter and distance, variations were applied.

The PDFs were hence created by means of a Python script and were used to generate the above mentioned datasets of 48 data points each, as it follows. Each original point was replicated 100 times where the nominal values were changed accordingly to their evaluated PDFs and a final amount of 4800 points were obtained. As mentioned before, the described approach was performed two times: one repeating depth values only and a second time replicating diameter and distance values together. Since dimple diameter and distance are dependent magnitudes, these were made to vary accordingly to one another as follows: at first a new diameter value was randomly generated through the known PDF, then the cumulative density function (CDF) was obtained for the specific diameter value and the corresponding distance value was calculated as the value holding the complement of the CDF in the distance PDF.

The new obtained dataset contains input parameters statistical fluctuations and was used to perform ROM predictions on the enlarged sets of points, where the dimple values experimental variations on dimple dimensions' nominal values are taken into account. The effects on friction due to the experimental texturing deviations were then analysed and a t-Student test was used for a statistical comparison of the obtained results to an ideal PDF around nominal surface texture values, as described in Section 3.4.

3. Results

In the following subsections the results obtained from the experimental tribotests and dimple dimensions nominal values assessment, the data pre-processing and the statistical analysis of textures dimensions are detailed.

3.1. Measurement Results of the Real Dimple Dimensions and Definition of Dimensionless Dimple Parameters

The aim of the real dimple dimensions measurement procedure, which has been previously described in Section 2.3, is the identification of the effective texture geometry. While the real dimple diameter and depth reveal larger values compared to the nominal dimensions, the dimple distances are smaller than the nominally specified values, compare Tables 1 and 3.

In addition to the diameter, distance, and depth dimple parameters, the dimensionless aspect ratio and textured area (equal to area density), which are often referred in literature [36–38], are further specified in Table 3. The aspect ratio is the quotient of the dimple depth and the dimple diameter and varies between 0.04 and 0.11 for the analysed textures. The area density is equal to textured area percentage. It is calculated by the quotient of the textured area $A_{textured}$ and the nominal circular contact area $A_{nominal}$, which are both visualised in Figure 5.

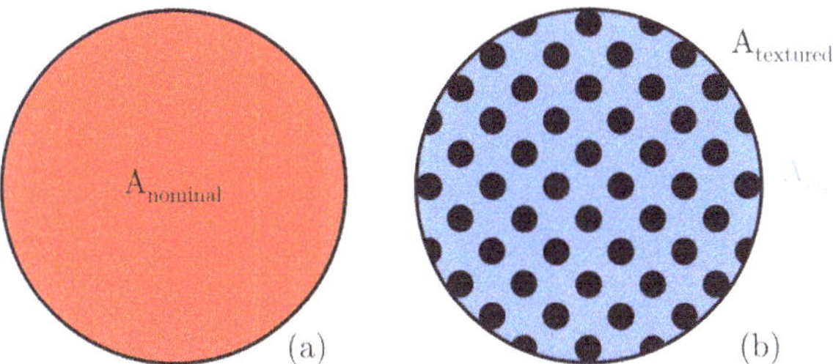

Figure 5. (**a**) Red marked nominal circular contact area $A_{nominal}$ between an untextured rubber specimen and the counter surface and (**b**) contact area of a textured sample. The textured area $A_{textured}$ is indicated by the black circular dimples, the untextured area that is in direct contact with the counter surface is coloured in blue.

The area density of the studied textures varies between 9% and 55%. Despite the fact that the three different normal forces F_N analysed result in three different nominal contact areas $A_{nominal}$, the percentage textured area is independent of the nominal contact area $A_{nominal}$, since more dimples come into contact with increasing normal force F_N, compare also Table 2b.

3.2. Reduced Order Modelling on Friction Coefficient Data

The two ROM functions (please refer to Equations (1) and (3)), obtained for untextured and textured friction coefficient datasets, converged to a stable solution, using only 2 and 17 terms, respectively. In order to assess the correctness and precision of the ROM results, the predicted value is compared to the corresponding experimental one, as ideally, both should be the same. The prediction lines are plotted, for both untextured and textured ROM models, in Figure 6a,b.

From Figure 6a,b it is possible to conclude that the ROM prediction is extremely accurate, being the obtained ROM models' standard deviations $\sigma_{untext} = 0.0014$ and $\sigma_{text} = 0.0012$ for untextured and textured data, respectively, and both regression lines (dashed black) match the goal prediction line (solid red).

Both ROM models were validated using the k-fold cross validation technique (see Section 2.4). The results obtained for the R^2 are shown in Table 4a,b for untextured and textured data, respectively.

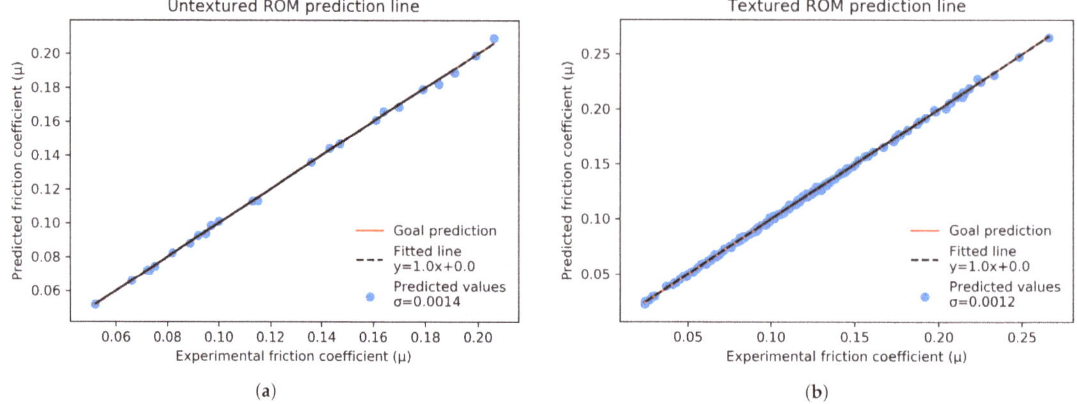

Figure 6. ROM prediction for (**a**) untextured and (**b**) textured friction coefficient data. The blue dots are obtained by evaluating the ROM with the experimental data, the dashed black line is the linear regression that fits the data, while the solid red line represents the ideal ROM result, where the ROM evaluation on each experimental friction value returns the same value.

Table 4. (**a**) Untextured and (**b**) Textured ROM R^2 for Train and Test datasets when the k-fold cross validation method is applied.

	a			b	
	Untextured ROM R^2			Textured ROM R^2	
	Train	Test		Train	Test
1	1.000	0.895	1	1.000	0.914
2	1.000	0.968	2	1.000	0.872
3	1.000	0.988	3	1.000	0.882
4	1.000	0.974	4	1.000	0.943
5	1.000	0.988	5	1.000	0.923
6	1.000	0.903	6	1.000	0.896
7	1.000	0.871	7	1.000	0.963
8	1.000	0.994	8	1.000	0.912
9	1.000	0.915	9	1.000	0.953
10	1.000	0.982	10	1.000	0.938
avg	1.000	0.948	avg	1.000	0.919

Table 4a,b show that for both ROM models the results of the validation are very precise and that there is always an excellent correlation between the input data, i.e., untextured or textured friction coefficient data (available under [24]), and the ROM prediction.

3.3. Reduced Order Modelling on Pre-Processed Data for Friction Variations

As described in Section 2.5, the friction variations were computed according to Equation (4), both for experimental and ROM predicted data. Similarly to Section 3.2, a prediction line was obtained to assess the accuracy of the results, obtained as the combination of the two (untextured and textured) ROM models. The results are shown in Figure 7.

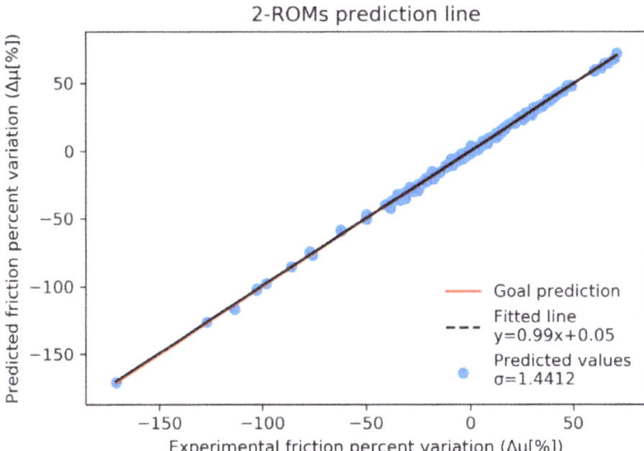

Figure 7. 2-ROM-models prediction for combined untextured and textured friction data, according to Equation (4). The blue dots are obtained by evaluating the combined 2-ROM predictions with the combined experimental data (Equation (4)), the dashed black line is the linear regression that fits the data, while the solid red line represents the ideal prediction result, where the predicted values perfectly match the experimental ones.

From Figure 7 it is observed that the ROM prediction is highly accurate, since the regression line (dashed black) is highly comparable to the goal prediction (solid red).

Moreover, an error propagation was performed on the friction percent variation variable (Equation (4)), in order to assess the maximum error assumed by the predictive model. For this purpose, experimental deviations on friction measurements were not included, since these are not considered intrinsic properties of the prediction accuracy, but depend on the a-priori goodness of the dataset solely. The results obtained when taking into account both the error propagation on $\Delta\mu$ and the standard deviation obtained for the 2-ROM prediction, i.e., $\sigma = 1.4412$ as shown in Figure 7, show values between 2% and 9%, with a mean value of 3% and 95% of data with a prediction deviation below 5%. Moreover, a study was performed to check the impact that each input parameter has on the measured friction, according to Equations (1) and (3), as shown in Figure 8 for textured ROM.

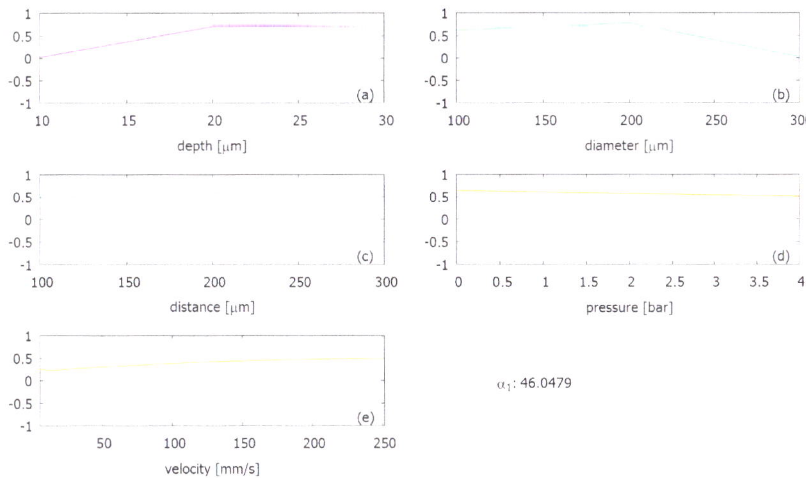

Figure 8. Textured data ROM's first term. The one-dimensional functions, for each variable are shown separately in the plots, according to Equation (1), labelled with letters from (**a–e**).

The one-dimensional functions (specified in Equation (1)) for the textured ROM, shown in Figure 8, represent the impact of each input on the friction variation. The textured ROM shows $\alpha_1 = 46.0479$ for the first term (see Equations (1) and (3)), being its weighting factor one or two orders of magnitude higher than the remaining terms. The first term of the ROM series expansion can be, hence, considered a fair approximation of the system, where other terms are corrections to it, as described in Section 1.2. From Figure 8 one can see that bigger friction variations occur at extreme surface texturing parameters, e.g., smaller dimple depth and bigger diameter values.

3.4. Statistical Analysis Results of Real Dimple Dimensions

As described in Section 2.6, PDFs were extrapolated for all texture parameters and were used to assess the measured differences from the desired nominal dimple values. The observed variations from the nominal dimple dimensions, described in Section 2.3, were introduced into the datasets in order to predict the corresponding friction variation using the previously computed ROM. The observed PDFs were normal distributions for both dimple depth and distance, while for diameter a skewed normal distribution was observed. dimple depth PDF showed a right shift of the mean equal to 3 μm and a standard deviation equal to 1.5 μm. Concerning the distance, the right shift of the mean was equal to 41 μm and 37 μm for corresponding nominal values of 200 μm and 300 μm, with a standard deviation of 13 μm and 9 μm, respectively. These findings led to significant differences in observed friction values, which means that deviations from nominal texture values, shown in Figures 9 and 10a,b, do actually affect friction as observed in Figure 11a,b.

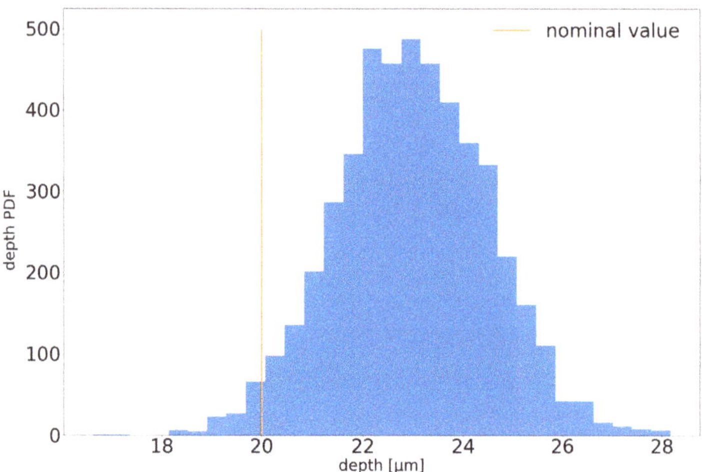

Figure 9. Dimple depth specific PDF and corresponding nominal value.

A ROM model allows verifying that the introduction of a statistical noise on nominal surface texture values affects friction, as shown in Figure 11a,b. From these Figures it is possible to observe significant variations in friction PDFs distributions when depth or diameter and distances, respectively, vary from nominal values, and to compare the obtained friction distribution to a normal PDF with zero mean and $\sigma = 0.01$.

A two-tailed t-Student test was performed to check the compatibility of the two PDFs, i.e., the noisy nominal values friction distribution (orange) and the ideal friction distribution (green), shown in Figure 11a,b. The test result showed a statistically significant difference between the noisy and the ideal PDF in case of dimple diameter-distance variations, according to what can be expected from Figures 9 and 10a,b.

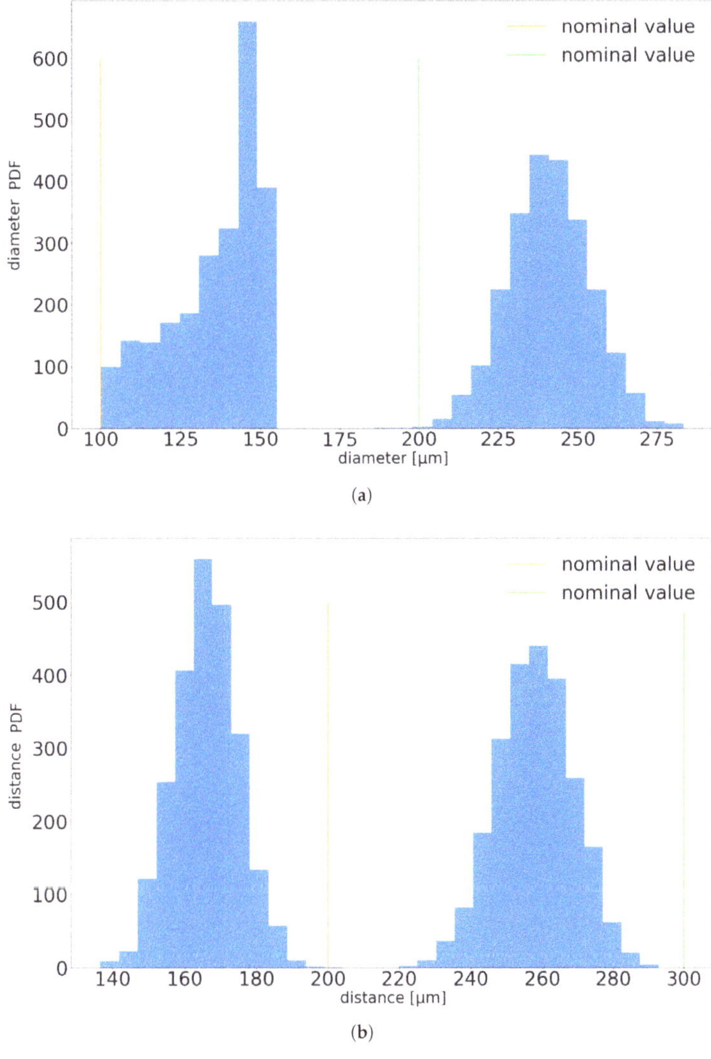

(a)

(b)

Figure 10. (**a**) Dimple diameter and (**b**) distance specific PDFs and respective nominal values.

As previously described in Section 2.5, two expanded datasets were generated introducing statistical noise to the textured data sample by varying the nominal dimple parameters (i.e., depth, diameter, and distance) according to the obtained PDFs. When depth was made to vary, only values equal to 20 μm could be used; in fact, nominal depth values of 10 μm and 30 μm would cause friction values to fall outside the ROM domain if a PDF was applied to generate data around these values. In this case the corresponding ranges for dimple diameter and distance were 200 μm and 200–300 μm, respectively. For the second dataset, given the ROM definition domain, feasible dimple diameter and distance ranges resulted in 100–200 μm and 200–300 μm, respectively, corresponding to depth values of 10 μm only. It is important to remark that, in this case, given the geometrical definition of the dimples, diameter and distance are inversely correlated and that the two PDFs are very different, being the first one a skewed distribution and the second a normal one (please refer to Section 2.5 for method details).

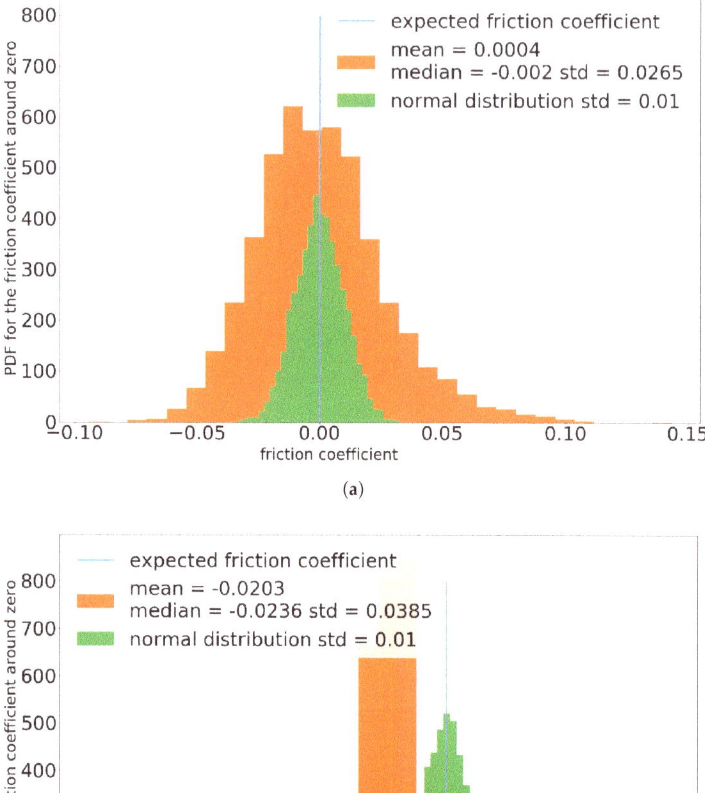

Figure 11. Friction coefficient distribution (orange) when (**a**) dimple depth or (**b**) diameter and distance are varried according to their specific PDFs (Figures 9 and 10a,b, respectively), centred and compared to a normal distribution (green) with zero mean and $\sigma = 0.01$.

Figure 12a,b show the findings for the friction prediction when the geometrical parameters of the seals deviate from nominal values.

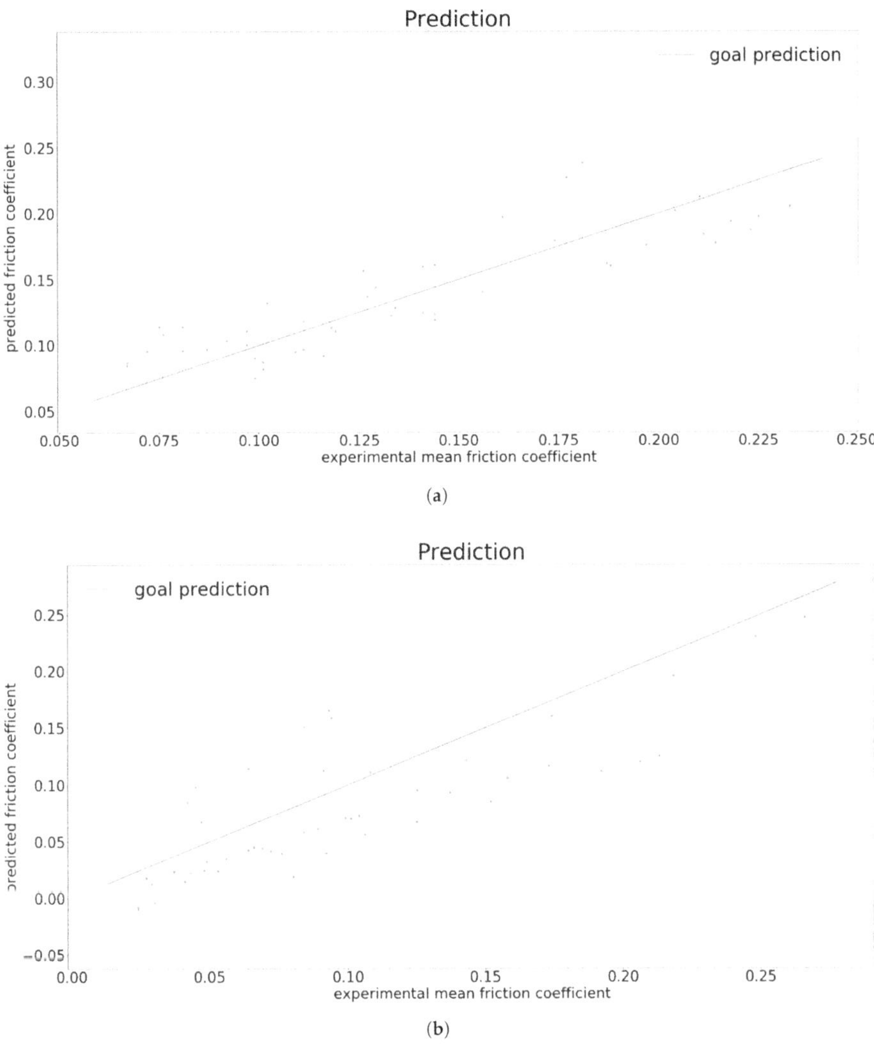

(a)

(b)

Figure 12. (a) ROM prediction with statistical noise introduction on depth and (b) on diameter and distance. The blue dots are obtained by evaluating the ROM on the statistically expanded dataset, the solid red line represents the ideal ROM result and the error bars show the friction variations linked to (a) depth or (b) diameter and distance deviations from nominal values.

According to the results shown in Figure 12a,b, when the textured ROM was evaluated on the expanded noisy dataset, friction prediction turned out to be completely affected by the statistical nominal texture values deviations. Once again one can observe that when diameter and distance were made to vary, these variations produced big uncertainties on the predicted friction values and statistically significant differences on mean values, which was corrected when an ideal PDF was used for texture parameters. In order to prove this, friction prediction was repeated using the green distribution in Figure 11a,b, where the texture parameters are centred in their nominal value, with a standard deviation of $\sigma = 0.01$. The obtained results are shown in Figure 13a,b.

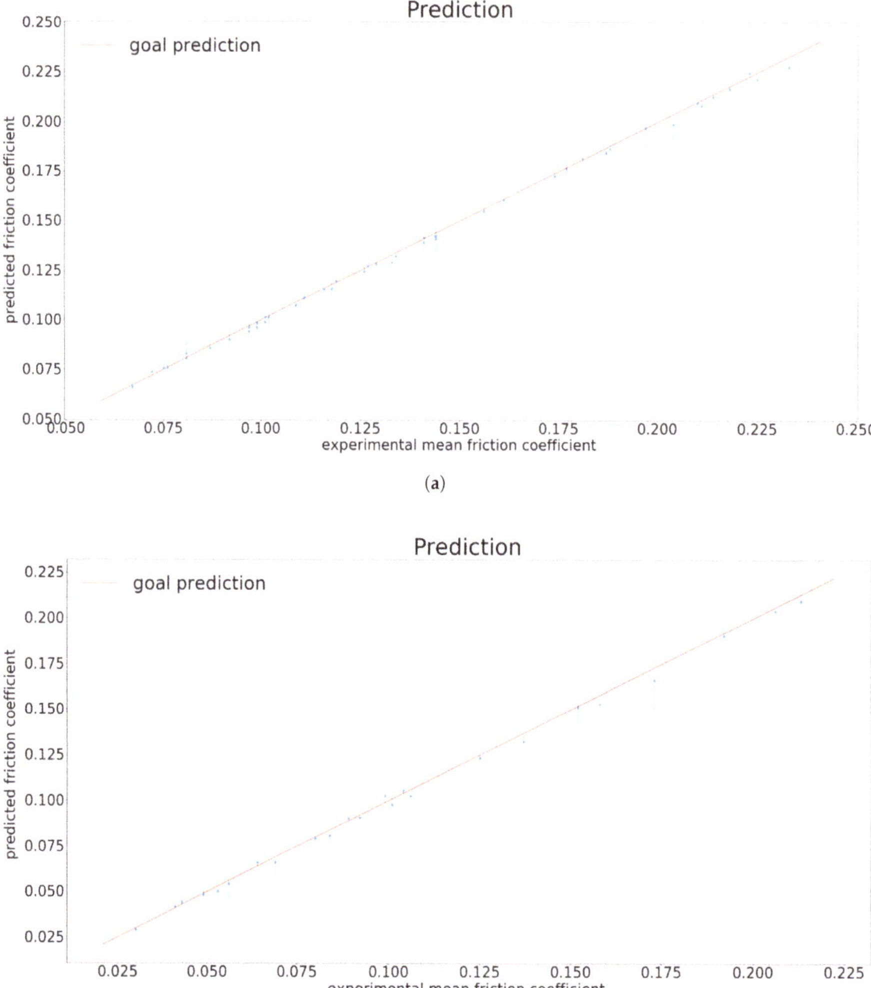

(a)

(b)

Figure 13. (**a**) ROM prediction with statistical noise introduction on depth and (**b**) on diameter and distance, with centred mean on nominal value and $\sigma = 0.01$. The blue dots are obtained by evaluating the ROM on the statistically expanded dataset, the solid red line represents the ideal ROM result and the error bars show the friction variations linked to (**a**) depth or (**b**) diameter and distance deviations from nominal values.

3.5. Experimental Friction Measurement Results and ROM Friction Prediction Outcome

The objective of the experimental testing procedure is to identify specific surface textures, which exhibit the lowest friction as a function of the relative velocity v_r and the contact pressure p_c. In Figures 14–16 the quasi-stationary friction coefficients μ of the 8 rubber specimens with different textures are shown as function of the relative velocity v_r. Data points depicted in Figures 14–16 are based on 5 measurements, whose mean values are shown. Error bars are not provided in the figures due to better visibility, although sigma ranges for the whole pool of data varied between $\sigma_{min} = 0.004$ and $\sigma_{max} = 0.196$ with a mean value of $\sigma_{mean} = 0.005$.

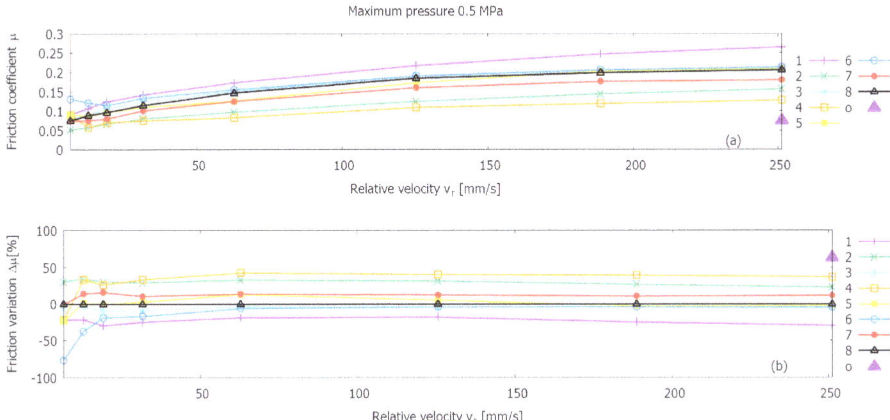

Figure 14. (**a**) Friction coefficient μ as function of the relative velocity v_r for the eight different rubber specimens and (**b**) the corresponding friction variation $\Delta\mu$ in relation to the untextured rubber sample 8, for contact pressures of $p_{c,max} = 0.5$ MPa.

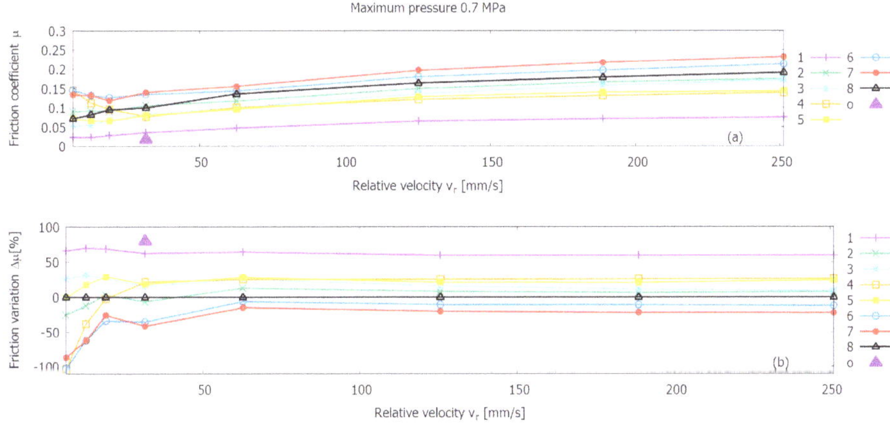

Figure 15. (**a**) Friction coefficient μ as function of the relative velocity v_r for the eight different rubber specimens and (**b**) the corresponding friction variation $\Delta\mu$ in relation to the untextured rubber sample 8, for contact pressures of $p_{c,max} = 0.7$ MPa.

Each figure visualises the measurement results for one of the three different contact pressures levels p_c considered. In addition, the friction variations of the different rubber samples are presented in relation to the untextured reference rubber specimen 8. Under all operating conditions, friction-reducing, as well as friction-increasing textures, can be identified. Positive values of friction variation indicate a texture-related reduction of friction, while negative friction variation values describe a friction increase compared to the untextured reference sample 8, see also Equation (4). Besides a texture related vertical shift of the friction characteristic, a horizontal shift of the minimum can be observed as well. Hence, surface texturing modifies friction in all regimes, which will be discussed in another publication.

In the following, the textures with the highest and lowest friction are discussed for each contact pressure level $p_{c,max}$. For $p_{c,max} = 0.5$ MPa rubber specimens 2 and 4 exhibit the lowest friction when considering the entire velocity range, while sample 1 reveals the highest friction level. Samples 2 and 4 have the highest area densities of 39% and 55%, respectively, while sample 1 has the lowest area density of 9% within all textures examined. Therefore, a clear trend is evident for $p_{c,max} = 0.5$ MPa, where the largest area densities analysed lead to the lowest friction levels, while the lowest area density leads to the highest friction level. The aspect ratio does not appear to have a major influence in this pressure condition as it fluctuates for samples 1, 2, and 4 between 0.11, 0.05, and 0.10, so that no clear trend is apparent.

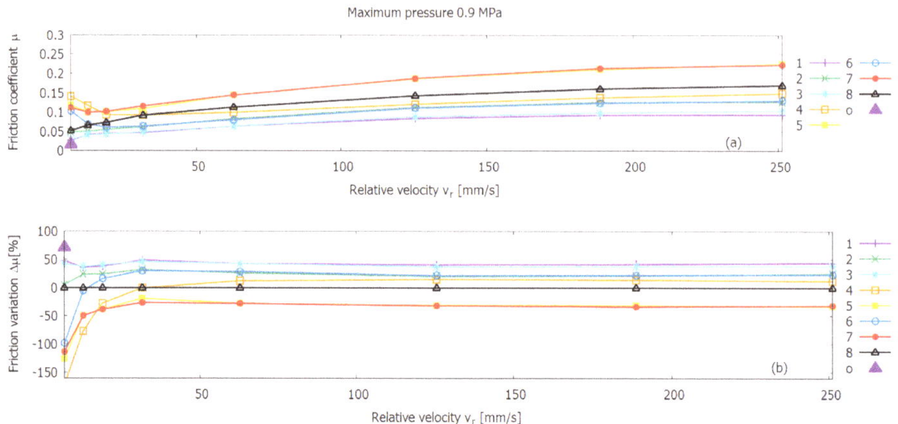

Figure 16. (**a**) Friction coefficient μ as function of the relative velocity v_r for the eight different rubber specimens and (**b**) the corresponding friction variation $\Delta\mu$ in relation to the untextured rubber sample 8, for contact pressures of $p_{c,max} = 0.9$ MPa.

For $p_{c,max} = 0.7$ MPa and $p_{c,max} = 0.9$ MPa rubber sample 1 and 3 exhibit the lowest friction level over the entire range of velocities. In contrast, specimen 7 shows the highest friction. Despite this agreement for both pressure levels, no clear trend can be derived in terms of dimple dimensions, area density, or aspect ratio. For example, sample 3 and 7 have similar area density values of 28% and 25%, respectively, as well as aspect ratios of 0.04 and 0.06. However, as mentioned above, the measured friction results of both textures differ greatly from each other.

Since there is no clear trend for dimple texture, it is concluded that the texture needs to be individually adapted to the existing operating conditions to minimize friction. For this purpose, ROM is an extremely efficient tool. In order to demonstrate the benefits of the ROM in vivid examples, four exemplary use cases were defined. The first theoretical use case is a friction contact intended to operate at the lowest contact pressure level $p_{c,max} = 0.5$ MPa and the highest velocity $v_r = 251$ mm/s within the operational conditions experimentally analysed. During the friction measurement procedure, a maximum friction reduction of 37% was already achieved with sample 4. However, the theoretical customer requires a further enhanced friction reduction of at least 60%. By using the ROM, a friction reduction of 63% can be predicted with the texture parameters given in row 1 of Table 5.

Table 5. Dimple dimensions predicted by the ROM, which further reduce friction based on the use cases.

Use Case Num.	Relative Velocity v_r [mm/s]	Max Contact Pressure $p_{c,max}$ [MPa]	Dimple Diameter [µm]	Dimple Distance [µm]	Dimple Depth [µm]	Aspect Ratio	Textured Area [%]	Predicted Friction Reduction [%]
1	251	0.5	270	100	10	0.04	42	63
2	31	0.7	300	186	11	0.04	30	81
3	6	0.9	274	111	11	0.04	39	72
4	100	0.6	300	140	11	0.04	36	79

The second use case is an application that operates at a contact pressure $p_{c,max} = 0.7$ MPa and a velocity of $v_r = 31$ mm/s. Friction measurements reveal a maximum friction reduction of 63% for sample 1. Again, the friction can be further reduced by a suitable texture, which is predicted by the ROM. The ROM indicates a friction reduction of 81% compared to reference seal 8 for the values specified in row 2 of Table 5.

The third use case is an imaginary technical requirement for a seal, which operates under the highest contact pressure level of $p_{c,max}$ = 0.9 MPa and the lowest velocity of v_r = 6 mm/s within the considered conditions. The experimental friction results show a maximum friction reduction of 48% for seal sample 1. With the use of the ROM, a reduced friction of 72% can be predicted in relation to the untextured case with the values given in row 3 of Table 5.

Based on the ROM results, one additional data point, indicated by a triangle, is added to each of the three plots of Figures 14–16.

In addition to friction predictions, considering the experimentally analysed operating conditions, one of the major advantage of the ROM is the ability to freely interpolate within the parameter space of the operating conditions. This is why the fourth use case shows the ROM prediction for experimentally untested working conditions, i.e., p_c = 0.6 MPa and velocity of v_r = 100 mm/s, where the friction reduction is about 79%, for the texture parameters shown in row 4 of Table 5.

The resulting aspect ratio for all four use case textures is equal to 0.04, while the textured area ranges from 30% to 42%. Compared to the experimentally analysed dimple textures given in Table 3, the dimple diameters are rather at the upper edge of the investigated texture dimensions, while the dimple distances and the depths are found in the lower range of the dimple dimensions. However, the dimple values are not similar. Thus, again it is obvious that a surface texture has always to be determined as a function of the operating conditions in order to achieve a maximum friction reduction. For each distinct application and individual technical requirement, textures have to be identified which result in an optimum friction with respect to a defined reference. These ROM results are not experimentally confirmed within the scope of this paper, as it would be necessary to re-produce rubber samples with appropriate textures, which was not part of MouldTex project [39].

4. Discussion

As can be concluded from Section 3.5, large texture-induced friction reductions, of up to 70%, could already be found experimentally inside the investigated parameter range of textures, velocities, and contact pressures. However, within the physical available samples, it is not always possible to identify clear trends for dimple dimensions that explain a high friction reduction compared to the untextured reference. Based on this finding, it is inferred that the surface texture dimensions need to be individually adapted to the given operating condition, in order to ensure a low friction level.

One possibility to identify optimal surface textures is the examination of a great number of dimple textures, which is significantly larger than the experimentally analysed set. However, this approach would be time-consuming and expensive due to TDM production requirements. The use of additional methods based on AI, such as ROM, are therefore advantageous. In this context, ROM is an effective method for finding the most suitable textures for specific operating conditions, as shown in Section 3.5, where the ROM is used to predict friction reduction values for both experimentally tested and untested working conditions, see Table 5 rows 1–3 and 4, respectively. Within this paper, ROM models have been used to predict and explore friction behaviour of surface textured rubber specimens by training the model on the experimental friction measurement results. Moreover, the model is able to quantify the measured friction variations that occur when deviations from nominal surface texture values are observed (Figures 9, 10a,b, 11a,b and 12a,b). ROM is extremely useful for simulating and predicting a system behaviour, especially when the physics behind its phenomenology are completely unknown or difficult to solve. Thanks to the ROM algorithms, users can predict the behaviour of their system in real-time, and specifically seal manufacturers can assess the parametric conditions that show the desired optimised results and select them before the rubber seal production. The efficiency of a data-driven ROM, as it is for Twinkle [23], is particularly dependent on the collected data, i.e., their associated error and their resulting manifold coverage.

With a thoroughly planned DoE it is possible to achieve an optimal space coverage, which leads to a reliable ROM's prediction result. The promising results, presented in Section 3.3 for the 2-ROM error propagation (Figure 7) strongly support ROM's accuracy on the given dataset. Nevertheless, if larger amounts of data were available, the computed ROM could be even more reliable in those space regions where less information was collected and, hence, confidently retrieve the best texture that allow minimising friction, according to friction variations computed in Equation (4). Moreover, if a ROM model was computed using simulation data, a higher sample space coverage and hence better reliability in the results could be achieved.

In addition, a ROM model allows verifying that the introduction of a statistical noise on nominal surface texture values affects friction, as shown in Figures 11a,b, where the two-tailed t-Student test showed a statistically significant difference between the experimental friction distribution (orange) and the ideal one (green) for dimple diameter-distance variations, see Figures 12a,b, were the difference on friction measurement are linked to manufacturing deviations in the nominal values of dimple textures. These variations were later compared to the effects obtained from an ideal PDF for dimple values experimental variations, i.e., mean value centred on nominal surface texture values and standard deviation $\sigma = 0.01$, as shown in Figures 13a,b, opening the way to potential surface texture manufacturing quality and tolerance investigations.

5. Conclusions

Within the scope of this paper, a pin-on-disc tribometer was used to measure dynamic friction in the contact of surface textured rubber specimens that were manufactured by texturing during moulding (TDM). The operating conditions, defined by relative velocity (v_r ranging from 6 to 251 mm/s) and contact pressure level ($p_{c,max}$ ranging from 0.5 to 0.9 MPa), were chosen to correspond to the operating conditions of pneumatic seals. Based on the different experimentally tested texture dimensions, a maximum friction reduction of 70% was determined compared to the untextured reference, with a dimple diameter of 100 μm, a dimple depth of 10 μm and an area density of 9%.

However, since no global surface texture optimum could be found experimentally that exhibits the lowest friction under all operating conditions, it is concluded that a surface texture needs always to be determined individually, based on the prevailing operating conditions, in order to achieve maximum friction reduction. For this purpose, a novel methodology was applied, combining friction measurements with Reduced Order Modelling (ROM). The objective of the ROM was the computation of optimal surface texture parameters that provide the highest friction reduction within the given parameter space of textures and operating conditions. By feeding the ROM with microscope-based texture measurements, it also takes into account deviations of the real dimple dimensions from the nominal dimple values for the output friction value. In summary, friction measurements are suitable as input parameters for the ROM so that the use of ROM for friction prediction has been proven successful. For demonstration, exemplary use cases were defined, from which it can be concluded that ROM enables the identification of optimal surface texture parameters that were not available for experiments, obtaining friction reductions from 63% to 81%, which are significantly higher than the experimentally tested surface textures for the same operating conditions. In addition, the value of ROM was further highlighted, as it is also able to freely interpolate between tested operating conditions to determine optimal textures for each operating condition within the range considered, resulting in a predicted friction reduction of 79%.

Moreover, ROM applicability has been extended further, showing that the method can be also used for statistical analysis, to evaluate the impact of manufacturing uncertainties, observed on surface texture nominal values, on friction measurements. Thus, ROM is not only an extremely powerful technique for scientific users to compute the friction reduction of surface textured components but also for industrial manufacturers of rubber components, to design rubber surfaces and evaluate the impact of manufacturing deviations on dynamic friction.

Author Contributions: Conceptualization, B.H.-G., M.B. and V.Z.; methodology, B.H.-G. and V.Z.; software, V.Z.; validation, B.H.-G., J.R.V., L.A.G., M.B., M.W. and V.Z.; formal analysis, M.B. and V.Z.; investigation, M.B. and V.Z.; resources, I.V., L.A.G., M.W. and S.I.; data curation, M.B. and V.Z.; writing—original draft preparation, M.B. and V.Z.; writing—review and editing, B.H.-G., J.R.V., L.A.G., M.B., M.W., S.I. and V.Z.; visualization, B.H.-G., J.R.V., L.A.G., M.B., M.W., S.I. and V.Z.; supervision, L.A.G., M.W. and S.I.; project administration, B.H.-G., J.R.V. and L.A.G.; funding acquisition, J.R.V. All authors have read and agreed to the published version of the manuscript.

Funding: This work has been funded by the European Union's research program Horizon2020, under grant agreement No. 768705.

Institutional Review Board Statement: Not applicable.

Informed Consent Statement: Not applicable.

Data Availability Statement: Publicly available datasets were analysed in this study. This data can be found here: http://devex.ita.es/mouldtex/optimizer/Data.zip (accessed on 22 March 2021).

Acknowledgments: We would like to thank ORP Stampi srl and SKM Aeronautics Ltd for providing the rubber specimens and ML Engraving srl for the surface texturing. We would also like to thank Carmen Alfaro, Rafael Rodríguez, and Susana Calvo for useful discussions.

Conflicts of Interest: The authors declare no conflicts of interest.

Abbreviations

The following abbreviations are used in this manuscript:

AI	Artificial Intelligence
CFD	Cumulative Density Function
DoE	Design of Experiment
DT	Digital Twin
FEM	Finite Element Method
LST	Laser Surface Texturing
ML	Machine Learning
PDF	Probability Density Function
PGD	Proper Generalized Decomposition
POD	Proper Orthogonal Decomposition
ROM	Reduced Order Modelling
SEHL	Soft Elasto-Hydrodynamic Lubrication
SVD	Singular Value Decomposition
TDM	Texturing During Moulding
TRD	Tensor Rank Decomposition

References

1. Korane, K. Low-Friction Seals Give High Fluid-Power Efficiency. Available online: https://www.machinedesign.com/news/article/21829567/lowfriction-seals-give-high-fluidpower-efficiency (accessed on 22 March 2021).
2. Jeng, Y.R. Impact of plateaued surfaces on tribological performance. *Tribol. Trans.* **1996**, *39*, 354–361. [CrossRef]
3. Wróblewski, P.; Iskra, A. Geometry of shape of profiles of the sliding surface of ring seals in the aspect of friction losses and oil film parameters. *Combust. Engines* **2016**, *55*. [CrossRef]
4. Ryk, G.; Kligerman, Y.; Etsion, I. Experimental Investigation of Laser Surface Texturing for Reciprocating Automotive Components. *Tribol. Trans.* **2002**, *45*, 444–449. [CrossRef]
5. Tomanik, E.; Profito, F.J.; Zachariadis, D.C. Modelling the hydrodynamic support of cylinder bore and piston rings with laser textured surfaces. *Tribol. Int.* **2013**, *59*, 90–96. [CrossRef]
6. Etsion, I.; Burstein, L. A Model for Mechanical Seals with Regular Microsurface Structure. *Tribol. Trans.* **1996**, *39*, 677–683. [CrossRef]
7. Etsion, I.; Kligerman, Y.; Halperin, G. Analytical and experimental investigation of laser-textured mechanical seal faces. *Tribol. Trans.* **1999**, *42*, 511–516. [CrossRef]

8. Etsion, I. Improving Tribological Performance of Mechanical Components by Laser Surface Texturing. *Tribol. Lett.* **2004**, *17*, 733–737. [CrossRef]

9. Pettersson, U.; Jacobson, S. Influence of surface texture on boundary lubricated sliding contacts. *Tribol. Int.* **2003**, *36*, 857–864. [CrossRef]

10. Grützmacher, P.G.; Profito, F.J.; Rosenkranz, A. Multi-Scale Surface Texturing in Tribology—Current Knowledge and Future Perspectives. *Lubricants* **2019**, *7*, 95. [CrossRef]

11. Shinkarenko, A.; Kligerman, Y.; Etsion, I. The effect of surface texturing in soft elasto-hydrodynamic lubrication. *Tribol. Int.* **2009**, *42*, 284–292. [CrossRef]

12. Etsion, I. Modeling of surface texturing in hydrodynamic lubrication. *Friction* **2013**, *1*. [CrossRef]

13. Zimmermann, M.; Wangenheim, M. Friction behaviour of surface textured elastomeric seals in pneumatic actuators. In Proceedings of the 23rd International Conference on Fluid Sealing, Manchester, UK, 2–3 March 2016; BHR Group: Cranfield, UK, 2016; p. 97.

14. Dobrica, M.B.; Fillon, M.; Pascovici, M.D.; Cicone, T. Optimizing surface texture for hydrodynamic lubricated contacts using a mass-conserving numerical approach. *Proc. Inst. Mech. Eng. Part J J. Eng. Tribol.* **2010**, *224*, 737–750. [CrossRef]

15. Vlădescu, S.; Olver, A.; Pegg, I.; Reddyhoff, T. Combined friction and wear reduction in a reciprocating contact through laser surface texturing. *Wear* **2016**. *358–359*, 51–61. [CrossRef]

16. Adjemout, M.; Aurore, A.; Bouyer, J.; Brunetière, N.; Marcos, G.; Czerwiec, T. Influence of the real dimple shape on the performance of a surface textured mechanical seal. *Tribol. Int.* **2014**, *115*, 409–416.

17. Tao, F.; Qi, Q.; Wang, L.; Nee, A. Digital Twins and Cyber-Physical Systems toward Smart Manufacturing and Industry 4.0: Correlation and Comparison. *Engineering* **2019**, *5*, 653–661. [CrossRef]

18. Tao, F.; Zhang, H.; Liu, A.; Nee, A.Y. Digital twin in industry: State-of-the-art. *IEEE Trans. Ind. Inform.* **2018**, *15*, 2405–2415. [CrossRef]

19. Chinesta, F.; Cueto, E.; Abisset-Chavanne, E.; Duval, J.; Khaldi, F. Virtual, Digital and Hybrid Twins: A New Paradigm in Data-Based Engineering and Engineered Data. *Arch. Comput. Methods Eng.* **2018**. [CrossRef]

20. Le Clainche, S.; Vega, J.M. A Review on Reduced Order Modeling using DMD-Based Methods. In Proceedings of the IUTAM Symposium on Model Order Reduction of Coupled Systems, Stuttgart, Germany, 22–25 May 2018; Springer: Berlin/Heidelberg, Germany, 2020; pp. 55–66.

21. Mignolet, M.P.; Przekop, A.; Rizzi, S.A.; Spottswood, S.M. A review of indirect/non-intrusive reduced order modeling of nonlinear geometric structures. *J. Sound Vib.* **2013**, *332*, 2437–2460. [CrossRef]

22. Wuest, T.; Weimer, D.; Irgens, C.; Thoben, K.D. Machine learning in manufacturing: Advantages, challenges, and applications. *Prod. Manuf. Res.* **2016**, *4*, 23–45. [CrossRef]

23. Zambrano, V.; Rodríguez-Barrachina, R.; Calvo, S.; Izquierdo, S. TWINKLE: A digital-twin-building kernel for real-time computer-aided engineering. *SoftX* **2020**, *11*, 100419. [CrossRef]

24. Dataset Generated and Used Within This Publication. Available online: http://devex.ita.es/mouldtex/optimizer/Data.zip (accessed on 22 March 2021).

25. Lee, C.Y.; Lin, C.S.; Jian, R.Q.; Wen, C.Y. Simulation and experimentation on the contact width and pressure distribution of lip seals. *Tribol. Int.* **2006**, *39*, 915–920. [CrossRef]

26. Parker Pneumatic Seals Catalogue. Available online: https://www.parker.com/Literature/Praedifa/Catalogs/Catalog_PneuSeals_PTD3351-EN.pdf (accessed on 22 March 2021).

27. OKS Adherent Silicone Grease Lubricant Specification. Available online: https://www.oks-germany.com/en/products/oks-1155-adherent-silicone-grease/ (accessed on 22 March 2021).

28. TWINKLE: A Digital-Twin-Building Kernel for Real-Time Computer-Aided Engineering. Available online: https://github.com/caelialTAINNOVA/Twinkle (accessed on 22 March 2021).

29. Pedregosa, F.; Varoquaux, G.; Gramfort, A.; Michel, V.; Thirion, B.; Grisel, O.; Blondel, M.; Prettenhofer, P.; Weiss, R.; Dubourg, V.; et al. Scikit-learn: Machine Learning in Python. *J. Mach. Learn. Res.* **2011**, *12*, 2825–2830.

30. Langford, J. The cross validation problem. In Proceedings of the International Conference on Computational Learning Theory, Bertinoro, Italy, 27–30 June 2005; Springer: Berlin/Heidelberg, Germany, 2005; pp. 687–688.

31. Bochkanov, S. ALGLIB. Available online: www.alglib.net (accessed on 22 March 2021).

32. Ayodele, T.O. Types of machine learning algorithms. *New Adv. Mach. Learn.* **2010**, *3*, 19–48.

33. Berry, M.W.; Mohamed, A.; Yap, B.W. *Supervised and Unsupervised Learning for Data Science*; Springer: Berlin/Heidelberg, Germany, 2019.

34. Harris, C.R.; Millman, K.J.; van der Walt, S.J.; Gommers, R.; Virtanen, P.; Cournapeau, D.; Wieser, E.; Taylor, J.; Berg, S.; Smith, N.J.; et al. Array programming with NumPy. *Nature* **2020**, *585*, 357–362. [CrossRef]

35. Virtanen, P.; Gommers, R.; Oliphant, T.E.; Haberland, M.; Reddy, T.; Cournapeau, D.; Burovski, E.; Peterson, P.; Weckesser, W.; Bright, J.; et al. SciPy 1.0: Fundamental Algorithms for Scientific Computing in Python. *Nat. Methods* **2020**, *17*, 261–272. [CrossRef]

36. Shinkarenko, A.; Kligerman, Y.; Etsion, I. The Validity of Linear Elasticity in Analyzing Surface Texturing Effect for Elastohydrodynamic Lubrication. *Trans. ASME J. Tribol.* **2009**, *131*. [CrossRef]

37. Wang, X.; Kato, K.; Adachi, K.; Aizawa, K. The effect of laser texturing of SiC surface on the critical load for the transition of water lubrication mode from hydrodynamic to mixed. *Tribol. Int.* **2001**, *34*, 703–711. [CrossRef]
38. Schneider, J.; Braun, D.; Greiner, C. Laser Textured Surfaces for Mixed Lubrication: Influence of Aspect Ratio, Textured Area and Dimple Arrangement. *Lubricants* **2017**, *5*, 32. [CrossRef]
39. Mouldtex Project. Available online: http://www.mouldtex-project.eu/ (accessed on 22 March 2021).

 *lubricants*

Article

Collaborative Optimization of CNN and GAN for Bearing Fault Diagnosis under Unbalanced Datasets

Diwang Ruan [1,*], Xinzhou Song [2], Clemens Gühmann [1] and Jianping Yan [3]

1. Chair of Electronic Measurement and Diagnostic Technology, Technische Universität Berlin, 10587 Berlin, Germany; clemens.guehmann@tu-berlin.de
2. School of Electrical Engineering and Computer Science, Technische Universität Berlin, 10587 Berlin, Germany; sxztthxq@gmail.com
3. School of Vehicle and Mobility, Tsinghua University, Beijing 100084, China; jianping_yan_de@yahoo.de
* Correspondence: diwang.ruan@campus.tu-berlin.de

Abstract: Convolutional Neural Network (CNN) has been widely used in bearing fault diagnosis in recent years, and many satisfying results have been reported. However, when the training dataset provided is unbalanced, such as the samples in some fault labels are very limited, the CNN's performance reduces inevitably. To solve the dataset imbalance problem, a Generative Adversarial Network (GAN) has been preferably adopted for the data generation. In published research studies, GAN only focuses on the overall similarity of generated data to the original measurement. The similarity in the fault characteristics is ignored, which carries more information for the fault diagnosis. To bridge this gap, this paper proposes two modifications for the general GAN. Firstly, a CNN, together with a GAN, and two networks are optimized collaboratively. The GAN provides a more balanced dataset for the CNN, and the CNN outputs the fault diagnosis result as a correction term in the GAN generator's loss function to improve the GAN's performance. Secondly, the similarity of the envelope spectrum between the generated data and the original measurement is considered. The envelope spectrum error from the 1st to 5th order of the Fault Characteristic Frequencies (FCF) is taken as another correction in the GAN generator's loss function. Experimental results show that the bearing fault samples generated by the optimized GAN contain more fault information than the samples produced by the general GAN. Furthermore, after the data augmentation for the unbalanced training sets, the CNN's accuracy in the fault classification has been significantly improved.

Keywords: fault data generation; Convolutional Neural Network (CNN); Generative Adversarial Network (GAN); bearing fault diagnosis; unbalanced datasets

Citation: Ruan, D.; Song, X.; Gühmann, C.; Yan, J. Collaborative Optimization of CNN and GAN for Bearing Fault Diagnosis under Unbalanced Datasets. *Lubricants* **2021**, 9, 105. https://doi.org/10.3390/lubricants9100105

Received: 23 August 2021
Accepted: 8 October 2021
Published: 15 October 2021

Publisher's Note: MDPI stays neutral with regard to jurisdictional claims in published maps and institutional affiliations.

Copyright: © 2021 by the authors. Licensee MDPI, Basel, Switzerland. This article is an open access article distributed under the terms and conditions of the Creative Commons Attribution (CC BY) license (https://creativecommons.org/licenses/by/4.0/).

1. Introduction

As an indispensable component in rotating machines, bearing health status directly affects or even determines the equipment service life. However, in practice, a bearing usually works under extreme and harsh conditions, which makes the bearing prone to faults [1]. Therefore, the timely and accurate fault diagnosis is crucial to reduce the maintenance costs and avoid serious accidents.

In recent years, the data-driven fault diagnosis has been attracting more and more attention from both academia and industry. Among the various data-driven methods, Convolution Neural Network (CNN) and Long Short Term Memory (LSTM) are the most widely used due to their powerful abilities in the complex feature extraction and nonlinear mapping. CNN was first employed in the bearing fault diagnosis by O. Janssens in 2016 [2], and, since then, many improvements have proposed to enhance the CNN's performance, such as 1D-CNN, 2D-CNN, multiscale CNN, and adaptive CNN [3–6]. Russell Sabir adopted LSTM for the bearing fault diagnosis based on the motor current signal and obtained a classification accuracy of 96% [7]. L. Yu and D. Qiu proposed the stacked LSTM and the bidirectional LSTM, respectively, and both LSTMs obtained an accuracy of more than 99% on the bearing fault diagnosis [8,9]. H. Pan combined 1D-CNN and 1D-LSTM into a unified structure by using the CNN's output into LSTM, achieving a satisfactory test accuracy up to 99.6% [10].

Although many sound results have been reported in the deep learning-based fault diagnosis, there are still many challenges to be solved. For example, all the studies mentioned above assume that there are plenty of high quality data for the deep network training. However, in many applications, the available history or experimental data is very limited or data provided is severely unbalanced. For example, the sample size under some fault classes is extremely smaller compared with the others. Either insufficient or unbalanced data will cause the serious performance reduction of deep networks. According to D. Xiao's work, when the training set samples were reduced from 1000 to 150, the CNN's accuracy declined correspondingly from 97.2% to 83.9% [11]. When the imbalance ratio increased from 2:1 to 40:1, the fault classification accuracy for the outer ring fault based on the GAN-SAE dropped sharply from 97.79% to 20.95% [12].

To address this problem, scholars have proposed diverse methods. Oversampling was first proposed to solve the data imbalance, where the direct replication was used to generate more samples for such labels that had very few ones [13,14]. Although this method is simple and efficient, it easily causes overfitting since no new information is incorporated. As another prospective method for data generation, GAN has been already used for new sample generation in the fault diagnosis. Both W. Zhang and S. Shao employed GAN to learn the mapping between the noise distribution and the actual machinery vibration data to expand available dataset. The results confirmed that the diagnosis accuracy could be improved once the imbalanced data was augmented by GAN [15,16]. However, when building and evaluating the GAN, published research studies only focus on the overall similarity between the generated data and the original one, which inevitably brings problems in the data quality. Small loss function in the general GAN only means that the generated data has a high similarity to the original signal, but it does not guarantee that the generated signal has captured the important characteristics of the original signal. When generating more samples for the unbalanced datasets in the fault diagnosis, it is important to ensure that the generated sample carries the same or nearly the same fault information as the original one, which includes both time and frequency domain characteristics. For this reason, an improved GAN is proposed in this paper and applied to generate samples for an unbalanced experimental dataset which is further used in the CNN-based fault diagnosis.

The main innovations of this paper include: (1) A GAN, together with a CNN, and two networks are optimized in cooperation. The GAN generates a more balanced dataset for the CNN, and the CNN evaluates the quality of the GAN's data generation. Both networks contribute to each other in performance improvement. (2) The fault characterization information is used to improve the general entropy-based loss function in the GAN. The amplitude and frequency errors in the envelope spectrum between the experimental and generated samples are taken as a correction term in the GAN's loss function to enable the GAN to produce samples with higher fidelity and identify more fault information.

The remaining part of this paper is organized as follows. Section 2 details the theory and methodology of the GAN, CNN, and loss function improvement. Section 3 describes the test bench and experimental dataset. Section 4 discusses and analyzes the results. Section 5 concludes the whole paper.

2. Methodology

2.1. Theory of the GAN

A GAN generates new data without any prior knowledge of the probability density function of original data. It mainly consists of a generator and a discriminator. The discriminator determines whether a sample comes from the original or generated dataset. On the contrary, the generator tries to produce data similar to the original one so that the discriminator can hardly make right decisions. In the general GAN, the loss functions of generator and discriminator are defined as Equations (1) and (2), respectively [15]:

$$L_G = -\frac{1}{K} \sum_{i=1}^{K} \log\left(D\left(x^i_{fake}\right)\right),\tag{1}$$

$$L_D = -\frac{1}{J} \sum_{m=1}^{J} \log(D(x^m_{real})) - \frac{1}{K} \sum_{i=1}^{K} \log\left(1 - D\left(x^i_{fake}\right)\right),\tag{2}$$

where J is the number of real samples, and K is the number of generated samples. x^m_{real} represents the data samples coming from the real training dataset, and x^i_{fake} denotes the data samples from GAN generator. $D(x^m_{real})$ designates the output of discriminator D with the input data sample x^m_{real}.

Based on the loss function L_G and L_D, the GAN can be trained as a minmax two-player game until the global optimum, $D(x_{real}) = D(x_{fake}) = 0.5$, is reached. This indicates that the generated data from the generator is so similar to the real one that the discriminator cannot tell the difference.

2.2. Fault Data Generation Based on GAN and CNN

The direct task of a GAN is to generate more samples for the labels with limited measurements. However, the ultimate goal is to improve the data-driven fault-diagnosis method performance when it deals with the imbalanced datasets. Therefore, it is reasonable to take the final fault-diagnosis results into consideration when constructing a GAN so that the data generated can indeed sharpen the algorithm's fault-diagnosis ability. In this paper, to facilitate research, a CNN is selected as a representative of the data-driven fault-diagnosis methods, and the diagnosis task is focused on the fault classification, so its performance is evaluated by the cross-entropy, as shown in Equation (3). The CNN's result is introduced as a correction term in the GAN's generator loss function as formulated in Equation (4):

$$L_{CNN} = -\sum_{i=1}^{N} x_i \log(p_i),\tag{3}$$

$$L_{G'} = L_G + \beta L_{CNN},\tag{4}$$

where N is the number of bearing fault types. $x_i = 1$, if the input sample belongs to the bearing fault type i; otherwise, $x_i = 0$. p_i is the output of softmax function, which represents the probability that the input data belongs to the bearing fault type i. The formulation for p_i is given in Equation (5), and it satisfies $\sum_{i=1}^{N} p_i = 1$ [17]. β is a scale factor to keep the loss functions of the GAN and CNN at the same range.

$$p_i = \frac{e^{a_i}}{\sum_{i=1}^{N} e^{a_i}}.\tag{5}$$

2.3. Improvement of Loss Function with Envelope Spectrum

The general GAN can produce data with high similarity to the original measurement, as stated in the last sub-section. In theory, the data fidelity can be even improved when a CNN is employed to collaboratively optimize a GAN. However, until now, all the data points in a sample are treated equally, and the GAN's target is to keep the generated data as similar to the original one as possible. However, in the fault diagnosis, some data points contain more information than others. For example, once a fault occurs on a certain component, such as the outer and inner ring or the balls, the corresponding fault characteristic frequencies (*FCF*) will appear in the acceleration spectrum. Compared with the overall similarity, the frequency and amplitude at the fault characteristic frequencies contain much more information about the bearing health condition. Therefore, the error of amplitudes and frequencies between the original signal and the generated one at the fault characteristic frequencies is defined as another correction term in the frequency domain as follows:

$$L_{frequency} = \sum_{i=1}^{N} \left(\left| M_{real}^i - M_{fake}^i \right| + \left| F_{real}^i - F_{fake}^i \right| \right),$$ (6)

where N denotes the maximum order of FCF, and $N = 5$ in this study. M_{real}^i and M_{fake}^i stand for the i-th order FCF amplitude from the real and generated sample. F_{real}^i and F_{fake}^i represent the i-th order FCF frequency from the real and generated sample. In addition, due to different value ranges of frequency and amplitude, in this study, the most widely used normalization method, MinMaxScaler [18], is applied to normalize the amplitudes and frequencies within the 5th-order FCF to the range of [0, 1].

Finally, $L_{frequency}$ is combined with L_{CNN} to construct the final loss function of the GAN's generator. As shown in Equation (7), the sum of L_{CNN} and $L_{frequency}$ is taken as a modification term in the general GAN's loss function L_G to ensure the generated data from GAN has a high similarity and captures the important information in detail at the same time. α is a weight factor.

$$L_{G''} = L_G + \alpha \left(L_{CNN} + L_{frequency} \right).$$ (7)

To obtain $L_{frequency}$, the first step is to calculate the theoretical FCF. The XJTU-SY dataset [19] introduced in the following section includes only three kinds of faults, namely the outer race fault, the inner race fault, and the cage fault. The theoretical $FCFs$ for the aforementioned 3 fault types are the $BPFO$ (Ball Passing Frequency on Outer race), $BPFI$ (Ball Passing Frequency on Inner race), and FTF (Fundamental Train Frequency), respectively. Their formulations are listed as follows [20]:

$$BPFO = \frac{n f_s}{2} \left(1 - \frac{d}{D} \cos \alpha \right),$$ (8)

$$BPFI = \frac{n f_s}{2} \left(1 + \frac{d}{D} \cos \alpha \right),$$ (9)

$$FTF = \frac{f_s}{2} \left(1 - \frac{d}{D} \cos \alpha \right),$$ (10)

where n is the number of rolling elements, and f_s means the shaft frequency. d represents the ball diameter, and D denotes the pitch diameter. α is the bearing contact angle.

After calculation of the theoretical FCF, the second step is to capture the actual FCF around corresponding theoretical values. The actual FCF can be affected by many factors, such as the shaft speed, external load, friction coefficient, raceway groove curvature, and the defect size [21,22]. Therefore, there exists bias between the theoretical FCF and the actual FCF in most cases. Besides, some harmonics of FCF influenced by modulation of other vibrations may not be detected in the test bench [22]. Thus, in this paper, the i-th order actual FCF is determined as the maximum peak in the interval of $[0.95, 1.05] \times FCF_{1st} \times i$, where FCF_{1st} is the first order theoretical FCF, and i is the current frequency order. The actual FCF of both the real measurement sample and generated sample are determined by above two steps. Once actual FCF is identified, the $L_{frequency}$ can be obtained by Equation (6).

2.4. Collaborative Training Mechanism of the GAN and CNN

Once the modification for the GAN loss function has been determined, the next step is to train a GAN in cooperation with a CNN. The collaborative training process is demonstrated in Figure 1. Generally, a GAN provides a more balanced dataset for CNN to improve its fault diagnosis accuracy. Whereas CNN evaluates the GAN's generated dataset and outputs its fault classification result as a correction term in the generator's loss function to improve the GAN's data-generation quality, under the collaborative training structure, both CNN and GAN performance can be enhanced. Specifically, as shown in Figure 1, the CNN is firstly built based on the unbalanced dataset, and its classification error is supposed to be high. Meanwhile, the discriminator, as well as the generator, of the GAN are established. Initially, the generator does not work so well, and the generated samples are not so similar to the original ones. The next step is to optimize the CNN and GAN collaboratively. During the optimization process, the GAN's generator learns to generate samples similar to the original signal. The newly generated samples are immediately added to the training dataset of the CNN so that the dataset imbalance can be reduced. When the Nash equilibrium is reached, which is defined as $D(x_{real}) = D(x_{fake}) = 0.5$, the optimization process stops. Lastly, the GAN's generator is used to extend the original dataset and fine-tune the CNN with the extended dataset. The architecture of the GAN proposed in this paper is detailed in Figure 2. Tables 1 and 2 summarize the hyperparameters of the GAN and CNN, respectively.

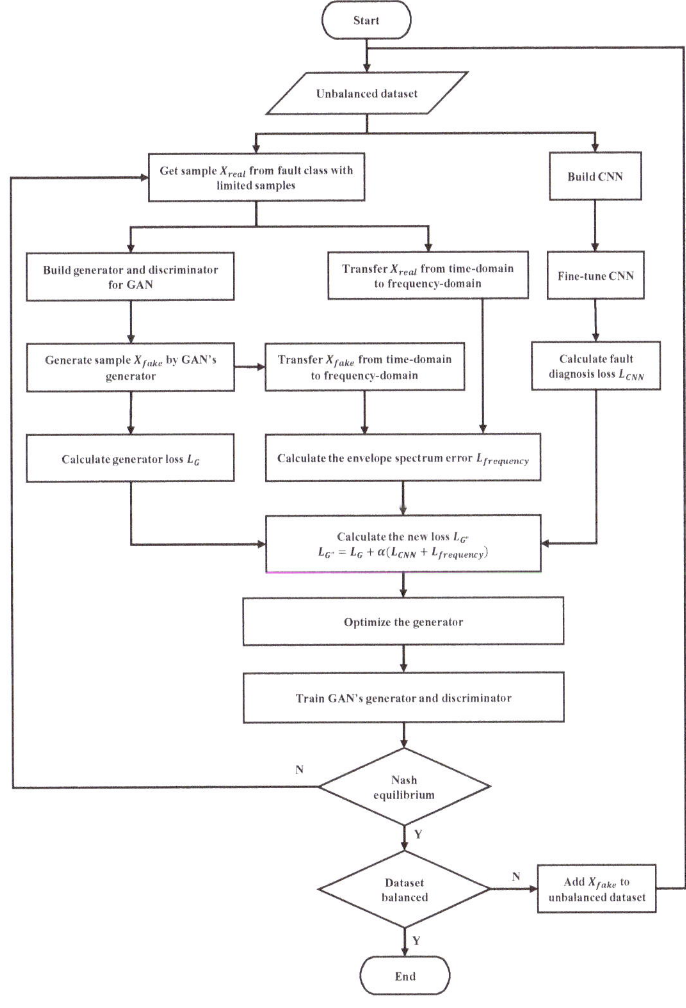

Figure 1. Collaborative training structure of the GAN and CNN.

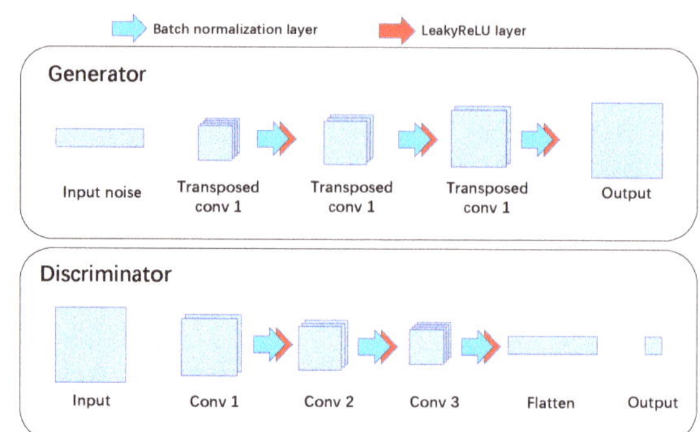

Figure 2. Architecture of generator and discriminator in the GAN.

Table 1. Hyperparameters of the GAN.

Hyperparameters	Values
Initial learning rate of generator	0.0001
Initial learning rate of discriminator	0.0001
Kernel size of discriminator's 1st layer	8×8
Kernel size of discriminator's other layers	4×4
Number of filters in discriminator's n-th layer	$16 \times 2^{n-1}$
Kernel size of generator's last layer	8×8
Kernel size of discriminator's other layers	4×4
Number of filters in generator's n-th layer	$512/2^{n-1}$
Max epochs	2000

Table 2. Hyperparameters of the CNN.

Hyperparameters	Values
Initial learning rate	0.0002
Max epochs	1000
Batch size	20
Kernel size of 1st layer	7×7
Kernel size of other layers	3×3
Number of filters in n-th layer	$16 \times 2^{n-1}$

3. Experimental dataset

3.1. Introduction of Bearing Test Bench and Dataset

Experimental data for validation comes from the Xi'an Jiaotong University (XJTU-SY) bearing test bench [19]. As shown in Figure 3, the bearing accelerated life test bench consists of an alternating current induction motor, motor speed controller, supporting shaft, supporting bearing, hydraulic loading system, and test bearing. The test bearing type is LDK UER204, and its basic parameters are summarized in Table 3. The bearing works under 3 different conditions, as specified in the first column of Table 4, where f_s stands for the shaft frequency, and F_r the radial loading force. Both the axial and radial accelerations are measured at a sampling frequency of 25.6 kHz, and the sampling interval between any two measurements is defined as 1 min, and each sampling lasts for 1.28 s. Under each condition, 5 bearings are tested, such as bearing 1_1–1_5 under condition 1. As each test bearing has a different lifetime, the measurement sample size of each test bearing varies from one to another.

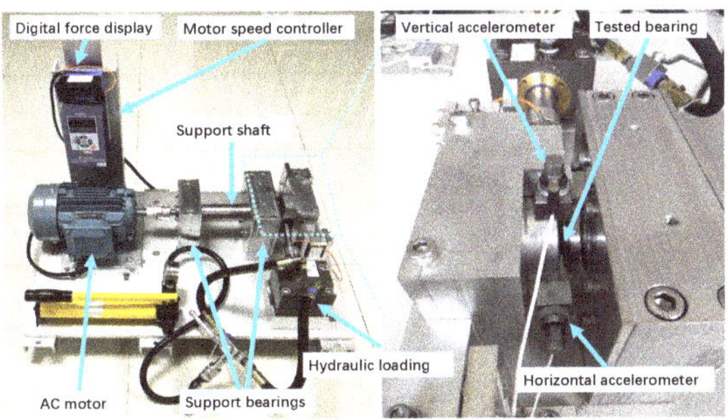

Figure 3. XJTU-SY experimental setup [19].

Table 3. Specifications of bearing parameters.

Parameters	Values	Parameters	Values
Inner raceway diameter	29.30 mm	Ball diameter	7.92 mm
Outer raceway diameter	39.80 mm	Number of balls	8
Pitch diameter	34.55 mm	Initial contact angle	0°

Due to the inherent micro-anisotropy and different working conditions, the lifetime and failure location of the test bearing differ from each other. For a single fault, there are 3 fault types in total, namely the outer race fault, the inner race fault, and the cage fault. Moreover, there are two datasets, bearing 1_5 and bearing 3_2, containing the measurements of compound fault. To simplify the labeling process, only a single fault is considered in this paper. As summarized in Table 4, the number of total samples is large enough for CNN training. However, the dataset is extremely unbalanced. For the most test bearings under all 3 conditions, the failure occurs on the outer ring, with very limited samples on the inner ring and the cage.

Table 4. Data specification of XJTU-SY bearing dataset.

Condition	Test Bearing	Measurement Sample Size	Fault Location
(1) $f_s = 35\,\text{Hz}$ $F_r = 12\,\text{kN}$	bearing 1_1	123	outer ring
	bearing 1_2	161	outer ring
	bearing 1_3	158	outer ring
	bearing 1_4	122	cage
	bearing 1_5	52	outer ring & inner ring
(2) $f_s = 37.5\,\text{Hz}$ $F_r = 11\,\text{kN}$	bearing 2_1	491	inner ring
	bearing 2_2	161	outer ring
	bearing 2_3	533	cage
	bearing 2_4	42	outer ring
	bearing 2_5	339	outer ring
(3) $f_s = 40\,\text{Hz}$ $F_r = 10\,\text{kN}$	bearing 3_1	2538	outer ring
	bearing 3_2	2496	inner ring & element & cage & outer ring
	bearing 3_3	371	inner ring
	bearing 3_4	1515	inner ring
	bearing 3_5	114	outer ring

3.2. Data Preprocessing

The XJTU-SY bearing dataset has recorded the bearing acceleration during the whole life cycle. The test bench runs continuously until the acceleration amplitude exceeds $10 \times A_{normal}$, which is defined as the failure point. Here, A_{normal} is the maximum amplitude of the horizontal or vertical vibration signals when the bearing runs in the normal operating stage. The fault location in Table 4 stands for position where the fault happens when bearing finally fails. In order to extract the sufficient measurement data for the fault classification while maintaining the correct labels, the signals with acceleration amplitude between $2 \times A_{normal}$ and $10 \times A_{normal}$ are regarded as the fault signals, as shown in Figure 4. All the measurement samples in the fault period are labeled with the corresponding final failure position, such as 1 for the cage fault, 2 for the inner race fault, and 3 for the outer race fault.

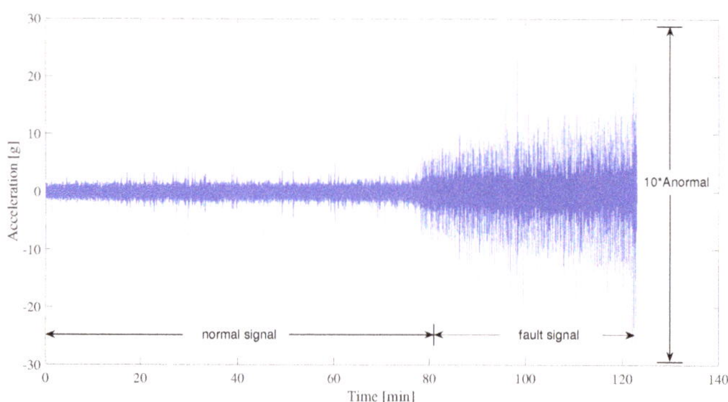

Figure 4. Complete life cycle of bearing 1_1.

After preparation for the valid source data and labels, the next step is the data preprocessing. At first, the original measurement is denoised by 3-level wavelet decomposition, with *Symlet*4 as the mother wavelet. After the noise cancellation for the high-frequency components, the data is normalized by z-score. Finally, the normalized data is transformed from 1D to 2D, which means that the acceleration series are sliced into fragments with the same length and then stacked row by row to build a matrix, as illustrated in Figure 5. In each sample, there are a total of 32,768 points of data in each sample. Therefore, the size of 2D matrix is determined as 181×181, and the reshaped 2D matrix is fed into GAN and CNN as images. All the work in this study is conducted in MATLAB Deep Network Designer.

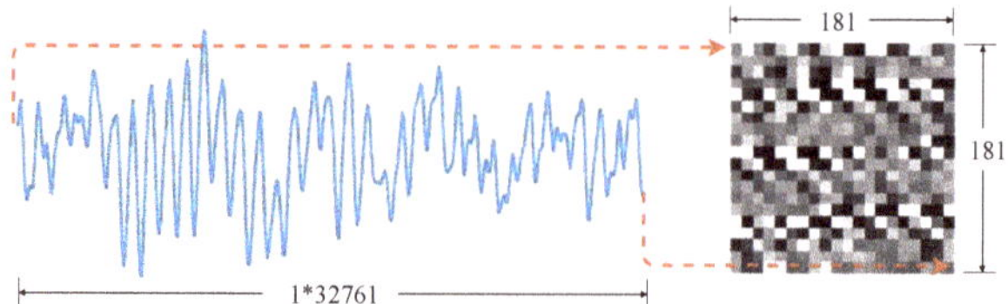

Figure 5. Illustration of data reshape.

4. Results and Analysis

4.1. Fault Data Generation Based on Optimized GAN

According to Table 5, there are significantly more samples for the outer race fault than for the inner race fault and the cage fault. Consequently, generating more samples for the inner race fault and the cage fault is paramount to reduce the dataset imbalance. It should be noticed that the inner race fault samples consist of data from bearing 2_1, bearing 3_3, and bearing 3_4, while the cage fault samples consist of data from bearing 1_4 and bearing 2_3. This means both the inner race and cage faults have measurement samples collected from different working conditions that define different data distributions. Furthermore, each test bearing has totally different aging dynamics, which can be deduced from their full life cycle trajectories [19]. As a result, the GANs for these datasets need to be trained individually. Bearing 1_4 has only one sample and is, hence, not feasible for the fault diagnosis. In total, 4 GANs need to be established for bearing 2_1, bearing 2_3, bearing 3_3, and bearing 3_4.

Table 5. Sample size of different fault types.

Fault Location	Test Bearing	Measurement Sample Size	Training Sets	Test Sets
Outer race	bearing 1_1	58	518	130
	bearing 1_2	108		
	bearing 1_3	69		
	bearing 2_2	77		
	bearing 2_4	12		
	bearing 2_5	173		
	bearing 3_1	55		
	bearing 3_5	106		
Inner race	bearing 2_1	26	110	28
	bearing 3_3	28		
	bearing 3_4	84		
Cage	bearing 1_4	1	167	42
	bearing 2_3	208		

The data samples generated by a general GAN and an optimized GAN are illustrated in Figure 6 and compared with the original ones after normalization. Specifically, Figure 6(a1) stands for the original signal of a measurement sample from bearing 2_1, Figure 6(a2) is the corresponding sample generated by the general GAN, and Figure 6(a3) shows the sample generated by the optimized GAN. Likewise, Figure 6(b1–b3) are the result for bearing sample 2_3, and Figure 6(c1–c3) for bearing sample 3_3. Take the inner race fault bearing 2_1 as an example; both GANs produce the samples with high similarity to the original ones measured in time domain, and even the peaks are accurately rebuilt. It can be further noticed that the optimized GAN generates a much more accurate peak amplitude than the general GAN. In order to evaluate the GAN's data-generation quality in time domain, every sample is regarded as a vector $\vec{x}$ ($\vec{x} \in \mathbb{R}_D$), and every sampling point x_i as an element in the vector.

The similarity between the generated sample and the original one can be measured by the angle between two corresponding vectors. Therefore, cosine similarity is adopted as a time domain similarity metric, which is defined as follows:

$$\cos\theta = \frac{\vec{m} \cdot \vec{n}}{|\vec{m}| \cdot |\vec{n}|},\tag{11}$$

where $\vec{m}$ and $\vec{n}$ stand for the acceleration series from the original measurement and the generated sample, respectively, with $\vec{m} = \{x_1, x_2, \cdots, x_L\}$ and $\vec{n} = \left\{x_1', x_2', \cdots, x_L'\right\}$. $|\vec{m}|$ and $|\vec{n}|$ identify the 2-norm of $\vec{m}$ and $\vec{n}$, respectively.

The cosine similarity results are summarized in Table 6. For all 3 cases, the sample generated by the optimized GAN has higher cosine similarity to the original one than that produced by the general GAN, which proves the superiority of the optimized GAN in the high-quality data generation. Additionally, the reason why the cosine similarity is relatively small can be explained as the acceleration values change within a big range of $[-5, 5]$, and the signal length is up to 32,761, which means any difference in acceleration amplitude or direction or time lag between counterpart points will bring big accumulative deviation. Besides, the assumption by taking the acceleration signal as 1D vector may not be so feasible when it contains too many elements, which needs further exploration in the future, such as using the Fréchet distance to replace the cosine similarity [23].

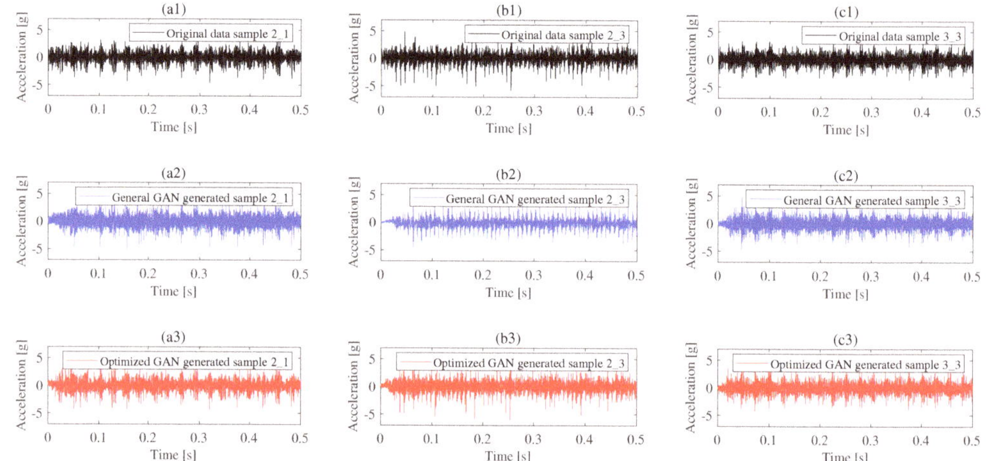

Figure 6. Comparison between original sample and generated sample in time domain; (**a**), (**b**) and (**c**) represent bearing 2_1, bearing 2_3 and bearing 3_3 respectively, while (**1**), (**2**) and (**3**) represent the original sample, general GAN and optimized GAN respectively.

Table 6. Cosine similarity of samples in time domain.

Generated Sample	Cosine Similarity	
	GAN	Optimized GAN
bearing 2_1	0.3214	0.3739
bearing 3_3	0.3374	0.3408
bearing 2_3	0.2009	0.2675

Apart from the overall similarity in time domain, the signal characteristics in the frequency domain are the same or even more important for the fault diagnosis. In this study, the envelope spectrum is processed on the original and generated samples. As only the 1st to 5th *FCFs* are considered in this study, the signal is first filtered by a low-pass filter of 1000 Hz, and then the envelope spectrum is extracted by Hilbert transform and Fast Fourier Transform. The results are displayed in Figures 7–9. Take Figure 7 as an example, which gives the envelope spectrum of bearing 2_1, where the black line is the result of the original measurement, the blue line stands for the sample generated by the general GAN, and the red line symbolizes the sample from the optimized GAN. The theoretical *BPFI* is also provided by the green dash line. We can find that the envelope spectrum of samples generated by the optimized GAN is similar to the original one, while it appears clearly different from that of the samples generated by the general GAN, especially the amplitudes at the real fault characteristic frequencies. Two locally enlarged views in Figure 7 show that the amplitude from the sample generated by the optimized GAN is much closer to that of the original sample, compared with the sample from the general GAN. The phenomenon is the same for the inner race fault (bearing 3_3), as well as the cage fault (bearing 2_3), which confirms that the optimized GAN can efficiently promote the generated signals to capture more accurate fault characteristics in the frequency domain. As for the other peaks besides fault characteristic ones, especially for the inner race fault, we can find that most of them are caused by the modulation from the shaft frequency and its harmonics, which is consistent with the previous research [24]. Additionally, the deviation between the actual FCF_s and the corresponding theoretical values can be explained by many factors, such as the frequency resolution of 0.7814 Hz, the occurrence of rolling element sliding, and the transient contact angles under high external load.

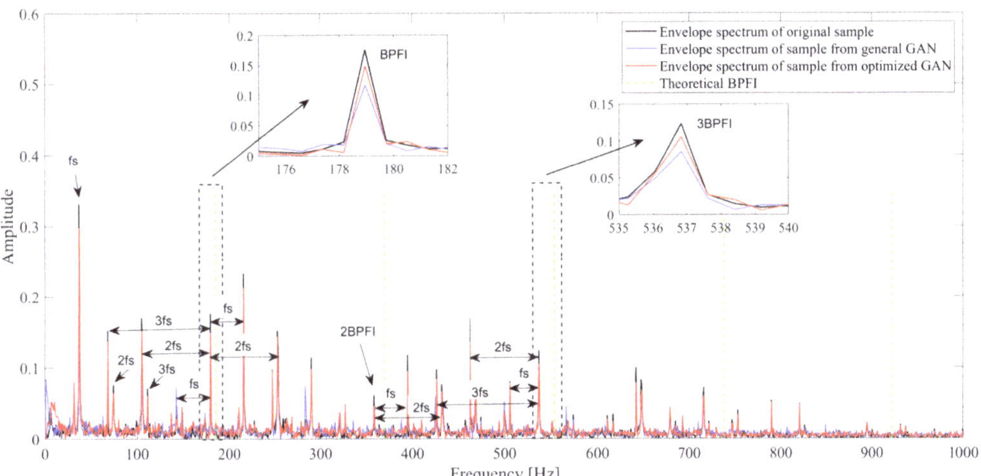

Figure 7. Envelope spectrum comparison: inner race fault of bearing 2_1.

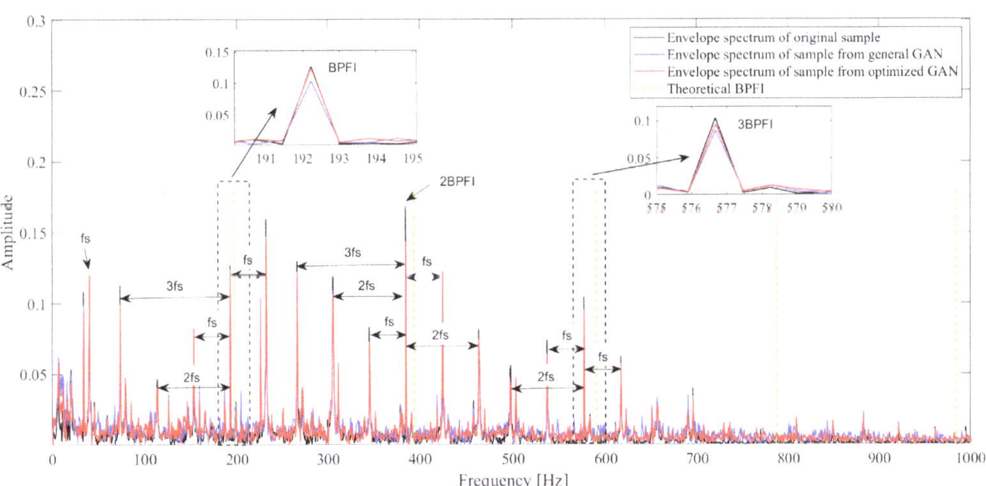

Figure 8. Envelope spectrum comparison: inner race fault of bearing 3_3.

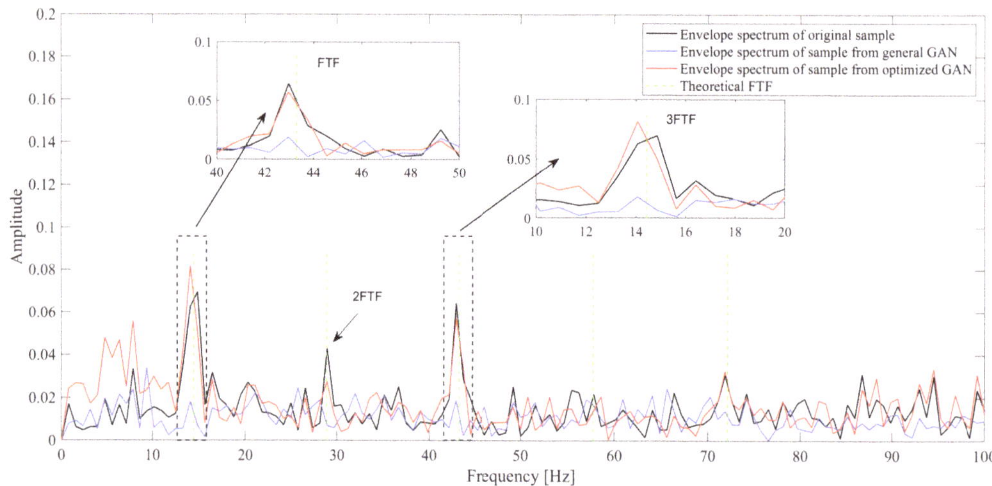

Figure 9. Envelope spectrum comparison: cage fault of bearing 2_3.

Tables 7–9 summarize the sample frequencies and amplitudes at the corresponding *FCF* and harmonics, as well as the relative error percentage of these two features between the generated and original samples. The comparison in Table 7 shows that, for all the 1st–5th order *BPFI*s, the frequencies and amplitudes of samples generated by the optimized GAN are much closer to the original ones than those of samples produced by the general GAN. For the sample generated by the optimized GAN, the frequency error percentage under all five orders of *BPFI* is zero, while the sample generated by the general GAN cannot fully capture the actual *BPFI* in the original ones, even though the deviation error is 0.34% and only exists in the 5th order *BPFI*. However, if we focus on the amplitudes under *BPFI*, the optimized GAN shows much more superiority over the general one. The amplitude errors under all 5 orders of *BPFI* from the samples generated by the optimized GAN are much smaller than those from the general GAN. Take the 2nd *BPFI* as an example; the actual amplitude from the original samples is 0.062, while the corresponding amplitudes of the samples from the general GAN and the optimized GAN are 0.023 and 0.047, respectively. The relative error percentage of amplitude drops from 62.0% to 23.8%. The above analysis confirms that the modification term $L_{frequency}$ in the GAN's generator loss function can enable the GAN to capture the fault information in the frequency domain. The same conclusion can be also drawn based on the results in Tables 8 and 9.

Table 7. Amplitudes and frequencies of bearing 2_1 at 1st–5th *BPFI*.

Sample Source	Parameter	1st—*BPFI*	2nd—*BPFI*	3rd—*BPFI*	4th—*BPFI*	5th—*BPFI*
Original sample	Frequency (Hz)	178.944	357.889	536.833	715.788	931.449
	Amplitude	0.176	0.062	0.124	0.072	0.020
Sample from general GAN	Frequency (Hz)	178.944	357.889	536.833	715.788	928.323
	Error (%)	0	0	0	0	0.34
	Amplitude	0.117	0.023	0.086	0.053	0.011
	Error (%)	33.3	62.0	30.1	26.1	44.1
Sample from optimized GAN	Frequency	178.944	357.889	536.833	715.788	931.449
	Error (%)	0	0	0	0	0
	Amplitude	0.149	0.047	0.106	0.060	0.018
	Error (%)	15.3	23.8	14.2	16.5	9.4

Table 8. Amplitudes and frequencies of bearing 3_3 at 1st–5th *BPFI*.

Sample Source	Parameter	1st—*BPFI*	2nd—*BPFI*	3rd—*BPFI*	4th—*BPFI*	5th—*BPFI*
Original sample	Frequency (Hz)	192.229	384.457	576.686	808.767	994.744
	Amplitude	0.127	0.168	0.104	0.015	0.013
Sample from general GAN	Frequency (Hz)	192.229	384.457	576.686	808.767	990.055
	Error (%)	0	0	0	0	0.5
	Amplitude	0.104	0.131	0.088	0.019	0.006
	Error (%)	18.0	22.1	15.9	26.0	49.5
Sample from optimized GAN	Frequency (Hz)	192.229	384.457	576.686	808.767	961.143
	Error (%)	0	0	0	0	0
	Amplitude	0.124	0.143	0.096	0.020	0.011
	Error (%)	2.2	14.7	7.9	26.7	14.4

Table 9. Amplitudes and frequencies of bearing 2_3 at 1st–5th *FTF*.

Sample Source	Parameter	1st—*FTF*	2nd—*FTF*	3rd—*FTF*	4th—*FTF*	5th—*FTF*
Original sample	Frequency (Hz)	14.847	28.912	42.978	57.825	71.890
	Amplitude	0.069	0.043	0.064	0.022	0.031
Sample from general GAN	Frequency	14.066	28.131	42.978	57.043	71.890
	Error (%)	5.3	2.7	0	1.4	0
	Amplitude	0.018	0.019	0.019	0.020	0.014
	Error	73.3	55.4	70.6	10.7	54.1
Sample from optimized GAN	Frequency (Hz)	14.066	28.912	42.978	58.606	71.890
	Error (%)	5.3	0	0	1.4	0
	Amplitude	0.082	0.028	0.057	0.021	0.033
	Error (%)	17.5	35.3	10.9	4.6	6.4

In summary, data generation results show that both the general GAN and the optimized GAN can generate similar samples compared to the original ones. However, the samples generated by the optimized GAN have higher similarity to the original one than that generated by the general GAN, especially at the FCF and harmonics in the frequency domain. More specifically, data generation for one fault type under different working conditions, such as bearing 2_1 and bearing 3_3, proves that the optimized GAN method can be applied to the bearings under the different working conditions. Furthermore, the results of bearing 2_1 (inner race fault) and bearing 2_3 (cage fault) demonstrate that the optimized GAN method adapts to the bearings with different defect types.

4.2. Fault Diagnosis Based on CNN_GAN

As introduced in Section 3, there are 648 outer race fault samples, 138 inner race fault samples, and 209 cage fault samples. In other words, the imbalance ratio of XJTU-SY bearing datasets is nearly 5:1:1.5 (outer race fault samples: inner race fault samples: cage fault samples). Besides, 80% of these samples are divided into the training set, with the remaining 20% as the test set. To fully evaluate the positive effect that the GAN has on CNN when dealing with the unbalanced datasets, two more training sets with the imbalance ratios of 10:1:2 and 20:1:2 are built by randomly selecting fewer inner race fault and cage fault samples from the XJTU-SY bearing datasets (the training dataset in Table 5), while the test set is fixed the same as the test set in Table 5. The sample composition of three training sets with different imbalance ratios and the test sets is illustrated in Figure 10.

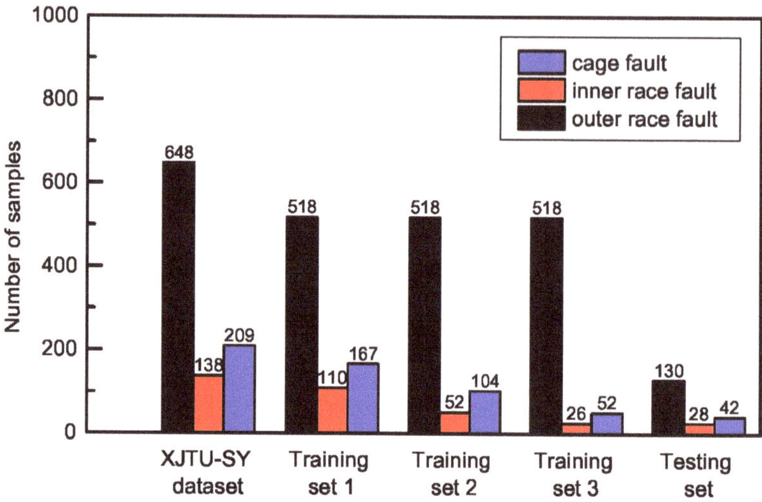

Figure 10. Composition of training sets and the test set.

Before validating the test set, on the one hand, CNN is trained on the training sets with the different imbalance ratios, in which the outer race fault has much more samples than the inner race fault and the cage fault. On the other hand, the unbalanced training sets are extended with the optimized GAN by generating more inner race fault and cage fault samples. After data generation, all 3 fault types in the extended training sets have the same sample size, with 518 samples individually. In other words, the ratios between the outer race fault samples, the inner race fault samples, and the cage fault samples become balanced. The general CNN and CNN_GAN mentioned above are validated with the same testing set. The difference between these two CNNs is that the former is trained with the imbalanced training set and then directly validated with the testing set, while the latter is trained with the extended dataset that has been balanced with the collaboration of the GAN and CNN and then validated with the testing set. The CNNs' performance comparison on the testing set is displayed in Table 10.

Table 10. Comparison of fault diagnosis performance between CNN and CNN_GAN.

Imbalance Ratio	CNN		CNN_GAN	
	Accuracy	Cross-Entropy Error	Accuracy	Cross-Entropy Error
Training set 1 (5:1:1.5)	98%	0.6071	100%	0.5645
Training set 2 (10:1:2)	88%	0.7013	90%	0.6642
Training set 3 (20:1:2)	68%	0.8478	88%	0.7012

For the general CNN, the fault diagnosis accuracy decreases from 98% to 88% when the imbalance ratio of training set increases from 5:1:1.5 to 10:1:2, and it sharply drops to 68% when the imbalance ratio further raises to 20:1:2. This confirms that the imbalance ratio of training datasets has a great influence on the CNN's performance. On the contrary, if a CNN is trained on the extended datasets that have been augmented with the generated samples from the optimized GAN, the CNN's performance can be significantly improved. For instance, when CNN_GAN is trained with the training sets 1 and 2 that have been extended and balanced, its fault classification accuracy on the testing sets achieves up to 100% and 90%, respectively. Even when the imbalance ratio raises up to 20:1:2, the CNN_GAN's fault classification accuracy still maintains 88%. Under all 3 training sets, the CNN_GAN has a smaller average cross-entropy error compared with the general CNN, which proves that the GAN can efficiently improve the CNN's fault diagnosis performance by generating new samples when dealing with the unbalanced datasets. Additionally, Table 10 shows that a training set with a higher imbalance ratio brings lower CNN classification accuracy, even after being balanced by data generation with a GAN. Though CNN_GAN performs better than CNN, the change tendencies of both two networks over increasing imbalance ratios are consistent, which indicates there exists an imbalance ratio limitation of the training set that CNN_GAN can handle with, especially for a predefined CNN's performance index. For example, in this case, if the target of the CNN's classification accuracy on the fixed imbalanced dataset is set as 90%, then, the CNN_GAN can deal with the training set with a maximum imbalance ratio of 10:1:2.

Besides the accuracy and cross-entropy, the confusion matrix gives more details of the classification for each label. As presented in Table 11, all these 3 cases are validated on the same dataset as the testing set in Figure 10 but trained with one of the three training sets with different imbalance ratios in Figure 10. Specifically, the general CNN is trained with the original unbalanced datasets, and the CNN_GAN is trained with the extended datasets that have been balanced by the optimized GAN. In these confusion matrices, the misclassified samples mainly come from the inner race fault and the cage fault because the outer race fault samples are dominant in each training set. Moreover, the higher the imbalance ratio is, the higher the prediction error is. With further comparison between the CNN and CNN_GAN, it can be found that the CNN_GAN achieves higher overall accuracy than the general CNN. In addition, the fault classification accuracy of both the inner race fault and the cage fault can be improved if the optimized GAN is employed to generate the inner race and cage fault samples. For example, under set 1 and set 2, the CNN's classification accuracy on the inner race fault increases from 85.7% to 100% and from 14.3% to 28.6%, respectively. With respect to the cage fault, the CNN's diagnosis accuracy increases remarkably from 4.8% to 90.5% under set 3. The result can be explained as: in the unbalanced dataset, the dominant fault type samples have much more influence on the loss function, which, therefore, push the CNN forward to extract more local features that are only shared by the dominant fault type, with CNN's ability lost to extract more general and robust features that can distinguish different fault types. This means that CNN has dropped into overfitting. While, for the CNN_GAN, the imbalanced data has been balanced, which means there are no dominant fault types in the training set. Therefore, the trained CNN_GAN can avoid overfitting and have the capability to capture fault features that can be used to recognize the fault types and be simultaneously robust enough. Based on the above analysis, it can be concluded that the balanced training dataset can effectively enhance the CNN's fault classification performance, and the optimized GAN can efficiently transform the unbalanced dataset into the balanced one by generating samples for the fault types that have limited data.

Table 11. Fault diagnosis confusion matrix under three training sets.

Diagnosis Network	Confusion Matrix on Testing Set											
	Training with Set 1 Unbalance Ratio (5:1:1.5)				Training with Set 2 Unbalance Ratio (10:1:2)				Training with Set 3 Unbalance Ratio (20:1:2)			
		CF	IRF	ORF			CF	IRF	ORF			CF	IRF	ORF
CNN	CF'	42 21.0%	4 2.0%	0 0.0%	91.3% 8.7%	CF'	42 21.0%	16 2.0%	0 0.0%	72.4% 27.6%	CF'	2 1.0%	0 0.0%	0 0.0%	100% 0.0%
	IRF'	0 0.0%	24 12.0%	0 0.0%	100% 0.0%	IRF'	0 0.0%	4 2.0%	0 0.0%	100% 0.0%	IRF'	0 0.0%	4 2.0%	0 0.0%	100% 0.0%
	ORF'	0 0.0%	0 0.0%	130 65.0%	100% 0.0%	ORF'	0 0.0%	8 4.0%	130 65.0%	94.2% 5.8%	ORF'	40 20.0%	24 12.0%	130 65.0%	67.0% 33.0%
		100% 0.0%	85.7% 14.3%	100% 0.0%	**98.0%** 2.0%		100% 0.0%	14.3% 85.7%	100% 0.0%	**88.0%** 12.0%		4.8% 95.2%	14.3% 85.7%	100% 0.0%	**68.0%** 32.0%

Table 11. *Cont.*

Diagnosis Network	Confusion Matrix on Testing Set		
	Training with Set 1 Unbalance Ratio (5:1:1.5)	Training with Set 2 Unbalance Ratio (10:1:2)	Training with Set 3 Unbalance Ratio (20:1:2)

CNN_GAN

Training with Set 1, Unbalance Ratio (5:1:1.5):

	CF	IRF	ORF	
CF'	42 / 21.0%	0 / 0.0%	0 / 0.0%	100% / 0.0%
IRF'	0 / 0.0%	28 / 14.0%	0 / 0.0%	100% / 0.0%
ORF'	0 / 0.0%	0 / 0.0%	130 / 65.0%	100% / 0.0%
	100% / 0.0%	100% / 0.0%	100% / 0.0%	**100%** / 0.0%

Training with Set 2, Unbalance Ratio (10:1:2):

	CF	IRF	ORF	
CF'	42 / 21.0%	4 / 2.0%	0 / 0.0%	91.3% / 8.7%
IRF'	0 / 0.0%	8 / 4.0%	0 / 0.0%	100% / 0.0%
ORF'	0 / 0.0%	16 / 8.0%	130 / 65.0%	89.0% / 11%
	100% / 0.0%	28.6% / 71.4%	100% / 0.0%	**90.0%** / 10.0%

Training with Set 3, Unbalance Ratio (20:1:2):

	CF	IRF	ORF	
CF'	38 / 19.0%	0 / 0.0%	0 / 0.0%	100% / 0.0%
IRF'	4 / 2.0%	8 / 4.0%	0 / 0.0%	66.7% / 33.3%
ORF'	0 / 0.0%	0 / 0.0%	130 / 65.0%	100% / 0.0%
	90.5% / 9.5%	28.6% / 71.4%	100% / 0.0%	**88.0%** / 12.0%

Target label: *CF*-cage fault, *IRF*-inner race fault, *ORF*-outer race fault; prediction label: *CF'*-cage fault, *IRF'*-inner race fault, *ORF'*-outer race fault.

5. Conclusions

To solve the CNN's performance reduction problem under the unbalanced datasets, an improved GAN is proposed to generate new data for the fault class with limited samples. The work can be summarized as follows:

- A collaborative network GAN_CNN is developed. The GAN generates an almost balanced dataset with data augmentation for the inner ring and the cage fault samples. Once the generated samples are added, the CNN evaluates the extended dataset quality and outputs the fault classification result to modify the loss function of the GAN's generator.
- Besides the overall similarity, the similarity on the envelope spectrum is considered when building the GAN. The envelope spectrum error from the 1st-5th order *FCF* between the experimental data and the generated data is taken as a correction term to the general cross-entropy based loss function of the GAN's generator.

Experimental validation is carried on the XJTU-SY bearing dataset. Results confirm the effectiveness of an optimized GAN and the collaborative structure of the CNN_GAN. The following are the main conclusions:

- When constructing the loss function for a GAN, the GAN performance can be improved by considering the envelope spectrum error. The generated samples have higher fidelity and contain more accurate fault information, which, in turn, contribute to the CNN's accuracy improvement.
- The collaborative network CNN_GAN performs better than the GAN or the CNN. The GAN generates more accurate data if the CNN's classification results are considered into the GAN's loss function. The CNN's fault classification accuracy can be significantly enhanced after the GAN generates more data for the unbalanced training dataset.

Though only the idea is validated with CNN_GAN in this paper, it can be extended with other methods. For example, the fault characteristic spectrum can be replaced by other metrics characterizing bearing fault status. With regard to the outlook, we will focus on the extension of this method and try to develop a physics-guided GAN. Validation with more experimental data and application cases will also be addressed in the future.

Author Contributions: Conceptualization, D.R. and C.G.; methodology, D.R.; software, X.S. and D.R.; validation, X.S.; formal analysis, D.R. and X.S.; investigation, D.R.; resources, D.R.; data curation, X.S.; writing—original draft preparation, X.S. and D.R.; writing—review and editing, C.G. and J.Y.; visualization, D.R. and J.Y.; supervision, C.G.; project administration, C.G.; funding acquisition, D.R. All authors have read and agreed to the published version of the manuscript.

Funding: This research was funded by CSC (China Scholarship Council) scholarship (201806250024) and Zhejiang Lab's International Talent Fund for Young Professionals.

Institutional Review Board Statement: Not applicable.

Informed Consent Statement: Not applicable.

Data Availability Statement: The original experimental data can be downloaded from: http:// biaowang.tech/xjtu-sy-bearing-datasets, and the data samples generated by optimized GAN for this study can be found in the following web-based repository: https://www.dropbox.com/sh/aqtzfb5 14x8hvmd/AAB-8cayG5dDsn0z_FFuiNosa?dl=0.

Acknowledgments: Acknowledgment is made for the XJTU-SY bearing dataset published by Xi'an Jiaotong University.

Conflicts of Interest: The authors declare no conflict of interest.

Abbreviations

The following abbreviations are used in this manuscript:

CNN	Convolutional Neural Networks
GAN	Generative Adversarial Networks
FCF	Fault characteristic frequency
LSTM	Long Short Term Memory
BPFO	Ball Passing Frequency on Outer race
BPFI	Ball Passing Frequency on Inner race
FTF	Fundamental Train Frequency

References

1. Li, N.; Lei, Y.; Lin, J.; Ding, S.X. An improved exponential model for predicting remaining useful life of rolling element bearings. *IEEE Trans. Ind. Electron.* **2015**, *62*, 7762–7773. [CrossRef]
2. Janssens, O.; Slavkovikj, V.; Vervisch, B.; Stockman, K.; Loccufier, M.; Verstockt, S.; Van de Walle, R.; Van Hoecke, S. Convolutional neural network based fault detection for rotating machinery. *J. Sound Vib.* **2016**, *377*, 331–345. [CrossRef]
3. Eren, L.; Ince, T.; Kiranyaz, S. A generic intelligent bearing fault diagnosis system using compact adaptive 1D CNN classifier. *J. Signal Process. Syst.* **2019**, *91*, 179–189. [CrossRef]
4. Zhang, W.; Peng, G.; Li, C. Bearings fault diagnosis based on convolutional neural networks with 2-D representation of vibration signals as input. *MATEC Web Conf.* **2017**, *95*, 13001. [CrossRef]
5. Guo, X.; Chen, L.; Shen, C. Hierarchical adaptive deep convolution neural network and its application to bearing fault diagnosis. *Measurement* **2016**, *93*, 490–502. [CrossRef]
6. Wang, D.; Guo, Q.; Song, Y.; Gao, S.; Li, Y. Application of multiscale learning neural network based on CNN in bearing fault diagnosis. *J. Signal Process. Syst.* **2019**, *91*, 1205–1217. [CrossRef]
7. Sabir, R.; Rosato, D.; Hartmann, S.; Guehmann, C. Lstm based bearing fault diagnosis of electrical machines using motor current signal. In Proceedings of the 18th IEEE International Conference On Machine Learning And Applications (ICMLA), Boca Raton, FL, USA, 16–19 December 2019; pp. 613–618.
8. Yu, L.; Qu, J.; Gao, F.; Tian, Y. A novel hierarchical algorithm for bearing fault diagnosis based on stacked LSTM. *Shock Vib.* **2019**. [CrossRef] [PubMed]
9. Qiu, D.; Liu, Z.; Zhou, Y.; Shi, J. Modified Bi-Directional LSTM Neural Networks for Rolling Bearing Fault Diagnosis. In Proceedings of the ICC 2019-IEEE International Conference on Communications (ICC), Shanghai, China, 20–24 May 2019; pp. 1–6.
10. Pan, H.; He, X.; Tang, S.; Meng, F. An improved bearing fault diagnosis method using one-dimensional CNN and LSTM. *J. Mech. Eng* **2018**, *64*, 443–452.
11. Xiao, D.; Huang, Y.; Qin, C.; Liu, Z.; Li, Y.; Liu, C. Transfer learning with convolutional neural networks for small sample size problem in machinery fault diagnosis. *Proc. Inst. Mech. Eng. Part C J. Mech. Eng. Sci.* **2019**, *233*, 5131–5143. [CrossRef]
12. Zhou, F.; Yang, S.; Fujita, H.; Chen, D.; Wen, C. Deep learning fault diagnosis method based on global optimization GAN for unbalanced data. *Knowl.-Based Syst.* **2020**, *187*, 104837. [CrossRef]
13. Cordón, I.; García, S.; Fernández, A.; Herrera, F. Imbalance: Oversampling algorithms for imbalanced classification in R. *Knowl.-Based Syst.* **2018**, *161*, 329–341. [CrossRef]
14. Ren, S.; Zhu, W.; Liao, B.; Li, Z.; Wang, P.; Li, K.; Chen, M.; Li, Z. Selection-based resampling ensemble algorithm for nonstationary imbalanced stream data learning. *Knowl.-Based Syst.* **2019**, *163*, 705–722. [CrossRef]
15. Shao, S.; Wang, P.; Yan, R. Generative adversarial networks for data augmentation in machine fault diagnosis. *Comput. Ind.* **2019**, *106*, 85–93. [CrossRef]

16. Zhang, W.; Li, X.; Jia, X.D.; Ma, H.; Luo, Z.; Li, X. Machinery fault diagnosis with imbalanced data using deep generative adversarial networks. *Measurement* **2020**, *152*, 107377. [CrossRef]
17. Yuan, B. Efficient hardware architecture of softmax layer in deep neural network. In Proceedings of the 29th IEEE International System-on-Chip Conference (SOCC), Seattle, WA, USA, 6–9 September 2016; pp. 323–326.
18. Kramer, O. Scikit-learn. In *Machine Learning for Evolution Strategies*; Springer: Berlin/Heidelberg, Germany, 2016; pp. 45–53.
19. Wang, B.; Lei, Y.; Li, N.; Li, N. A hybrid prognostics approach for estimating remaining useful life of rolling element bearings. *IEEE Trans. Reliab.* **2018**, *69*, 401–412. [CrossRef]
20. Randall, R.B.; Antoni, J. Rolling element bearing diagnostics—A tutorial. *Mech. Syst. Signal Process.* **2011**, *25*, 485–520. [CrossRef]
21. Niu, L.; Cao, H.; He, Z.; Li, Y. A systematic study of ball passing frequencies based on dynamic modeling of rolling ball bearings with localized surface defects. *J. Sound Vib.* **2015**, *357*, 207–232. [CrossRef]
22. Saruhan, H.; Saridemir, S.; Qicek, A.; Uygur, I. Vibration analysis of rolling element bearings defects. *J. Appl. Res. Technol.* **2014**, *12*, 384–395. [CrossRef]
23. Devogele, T.; Etienne, L.; Esnault, M.; Lardy, F. Optimized discrete fréchet distance between trajectories. In Proceedings of the 6th ACM SIGSPATIAL Workshop on Analytics for Big Geospatial Data, Redondo Beach, CA, USA, 7–10 November 2017; pp. 11–19.
24. Mishra, C.; Samantaray, A.; Chakraborty, G. Ball bearing defect models: A study of simulated and experimental fault signatures. *J. Sound Vib.* **2017**, *400*, 86–112. [CrossRef]

Article

A Semantic Annotation Pipeline towards the Generation of Knowledge Graphs in Tribology

Patricia Kügler [1,*], Max Marian [2], Rene Dorsch [1], Benjamin Schleich [1] and Sandro Wartzack [1]

[1] Engineering Design, Friedrich-Alexander-Universität Erlangen-Nürnberg (FAU), Martensstr. 9, 91058 Erlangen, Germany; rene.dorsch@fau.de (R.D.); schleich@mfk.fau.de (B.S.); wartzack@mfk.fau.de (S.W.)

[2] Department of Mechanical and Metallurgical Engineering, School of Engineering, Pontificia Universidad Católica de Chile, Macul, Santiago 6904411, Chile; max.marian@ing.puc.cl

[*] Correspondence: kuegler@mfk.fau.de

Abstract: Within the domain of tribology, enterprises and research institutions are constantly working on new concepts, materials, lubricants, or surface technologies for a wide range of applications. This is also reflected in the continuously growing number of publications, which in turn serve as guidance and benchmark for researchers and developers. Due to the lack of suited data and knowledge bases, knowledge acquisition and aggregation is still a manual process involving the time-consuming review of literature. Therefore, semantic annotation and natural language processing (NLP) techniques can decrease this manual effort by providing a semi-automatic support in knowledge acquisition. The generation of knowledge graphs as a structured information format from textual sources promises improved reuse and retrieval of information acquired from scientific literature. Motivated by this, the contribution introduces a novel semantic annotation pipeline for generating knowledge in the domain of tribology. The pipeline is built on Bidirectional Encoder Representations from Transformers (BERT)—a state-of-the-art language model—and involves classic NLP tasks like information extraction, named entity recognition and question answering. Within this contribution, the three modules of the pipeline for document extraction, annotation, and analysis are introduced. Based on a comparison with a manual annotation of publications on tribological model testing, satisfactory performance is verified.

Keywords: tribo-testing; tribo-informatics; machine learning; artificial intelligence; natural language processing; tribAIn; BERT

Citation: Kügler, P.; Marian, M.; Dorsch, R.; Schleich, B.; Wartzack, S. A Semantic Annotation Pipeline towards the Generation of Knowledge Graphs in Tribology. *Lubricants* 2022, 10, 18. https://doi.org/10.3390/lubricants10020018

Received: 14 December 2021
Accepted: 23 January 2022
Published: 25 January 2022

Publisher's Note: MDPI stays neutral with regard to jurisdictional claims in published maps and institutional affiliations.

Copyright: © 2022 by the authors. Licensee MDPI, Basel, Switzerland. This article is an open access article distributed under the terms and conditions of the Creative Commons Attribution (CC BY) license (https://creativecommons.org/licenses/by/4.0/).

1. Introduction

The emergence of efficient and sustainable technologies represents a major challenge for the 21st century. While renewable energy sources are increasingly replacing fossil fuels in order to reduce CO_2 emissions, the influence of friction and wear on the energy efficiency of a wide range of technical processes has hardly reached public awareness. However, these offer considerable potential for saving CO_2 and resources. Holmberg and Erdemir [1] estimated that roughly 23% of the global primary energy is consumed to overcome friction and to repair/replace worn components in tribo-technical systems. The authors predicted that these energy losses could be reduced by up to 40% through tribological advances. Accordingly, companies and research institutions are focusing on new concepts, materials, lubricants, or surface technologies in a wide range of applications. This is also reflected in the continuously growing number of publications related to the domain of tribology, which in turn serve as inspiration, guidance, and benchmark for researchers and developers, but which are almost impossible to keep up with due to their vast quantity and the associated complexity and diversity. Thereby, profound data bases in combination with machine learning (ML) and artificial intelligence (AI) approaches can support sorting through the complexity of patterns and identifying trends [2]. Therefore, they are more and more employed in the analysis, design, optimization, or monitoring of tribological systems in various fields [3], ranging from composite materials [4], drive technology [5,6], manufacturing [7], surface engineering [8,9], or lubricant formulation [10,11]. As pointed out by Marian and Tremmel [12], novel findings and additional value in the domain of tribology can especially be created by extracting knowledge from available literature and drawing higher-level conclusions. For example, Kurt and Oduncuoglu [13] trained artificial neural networks (ANNs) with data from literature to study the influence of normal load, sliding speed as well as the type and weight fraction of various reinforcement phases within a polyethylene matrix on the resulting friction and wear behavior. Similarly, Vinoth and Datta [14] utilized 153 data sets from literature to predict mechanical properties of carbon nanotube or graphene reinforced polyethylene in dependency of composition, particle size, and bulk properties by means of an ANN. Subsequently, multi-objective optimization by genetic algorithms and corresponding experimental validation actually demonstrated improved tribological properties compared to the references. Using 80 data sets from four-ball-tests and 120 data sets from pin-on-disk experiments with varying base oils and friction modifiers as reported in literature as well as an ANN and a genetic algorithm, Bhaumik et al. [15] optimized the lubricant formulation and experimentally validated their results. The aforementioned studies indicate the potential through leveraging knowledge from the available literature. However, the data acquisition and processing still are very manual in the field of tribology, involving the review of publications and the extraction of relevant (most frequently textually/descriptive) information, which limits the generation of sophisticated and broad databases and thus the further use of ML/AI [12].

High manual efforts to acquire and curate information and knowledge for further processing are not limited to the domain of tribology and is known as "knowledge acquisition bottleneck" [16]. Although the latter has been discussed since the rise of expert systems in the 1980s [17], for instance with the purpose of tribological design decisions [18] or failure diagnosis [19] to mention two examples from the tribological domain, knowledge acquisition and thus knowledge engineering are still quite manual and time-consuming tasks. Studer et al. [20] argue that knowledge engineering is a modeling activity, which goes beyond the simple transfer of directly accessible knowledge into an appropriate computer representation towards a model construction process [21]. In consequence, knowledge structuring and modeling plays an important role in the knowledge acquisition process. Hoekstra [22] therefore refers to a "knowledge reengineering bottleneck", which highlights the general difficulty of continuously reusing existing generic and assertional knowledge. The latter refers to data-level or object knowledge, while generic knowledge concerns schema-level describing conceptual knowledge and is represented as a domain theory to structure the respective domain. This includes the decision on used vocabulary to describe the domain and a representation form to formalize the model. Chandrasegaran et al. [23], as well as Verhagen et al. [24], emphasized the importance of semantic interoperability for knowledge reuse and sharing, which is frequently dealt with ontological models represented in formal logics. According to Gruber [25], an ontology is an "explicit specification of a conceptualization". This means that an ontology can be used to explicitly define a domain model for sharing and reusing structured knowledge by humans and machines. In other domains, e.g., bioinformatics, ontologies are widely used for knowledge structuring, data integration and decision support systems [26]. One successful example is the Gene Ontology (GO) [27], which provides broadly accepted vocabulary for annotating gene product data from different databases and sources. Exploiting ontologies for accessing and reusing experimental knowledge has also been pursued in the domain of tribology. One example is the "OntoCommons" project (https://ontocommons.eu/industrial-domain-ontologies, accessed on 14 December 2021), where a tribological use case aimed at reducing efforts in tribological experiments by reusing existing knowledge. Thereby, Esnaola–Gonzalez and Fernandez [28] argue, that semantic technologies, and more specifically ontologies propose a suited representation for the vaguely documented results of experiments. Within the domain of materials science, the "European Materials Modelling Ontology" (EMMO, https://emmc.info/emmo-info/, accessed on 14 December 2021) provides a representational ontology based on materials modelling and characterization knowledge. Furthermore, we recently introduced the tribAIn ontology [29] for reusing knowledge from tribological experiments. The domain ontology was built for the purpose of providing a common and machine-readable schema for structuring tribological experiments intending to improve reuse and shareability of testing results from different sources. Since this contribution relies on the tribAIn ontology, more detailed information is provided in Section 2.2. In addition to schema-level generic knowledge, assertional knowledge refers to specific knowledge objects, e.g., results from individual experiments. As mentioned before, assertional knowledge from experiments in the domain of tribology is usually published in natural language, thus publications are a well-suited knowledge source for acquiring the current state of tribological findings. Dealing with natural language sources is usually problematic since it is ambiguous and unstructured. Moreover, textual descriptions may be incomplete in the sense of formal models. Due to the time-consuming process of acquiring and structuring knowledge from textual sources in systematic literature studies or manual database construction, those knowledge bases are not suited for long-term reuse and continuous extension. A successful example for generating structured information from textual sources is the DBpedia project [30], which extracts structured data from Wikipedia content using templates and pattern matching techniques. The structured format then allows querying the vast content in a sophisticated way instead of searching articles by keywords and processing the information manually. In terms of the results from tribological experiments, publications—similar to Wikipedia—contain structured (e.g., operational parameters, wear rate, coefficient of friction, etc.) and unstructured knowledge (for example interpretive description and discussion of results). By extracting the information from text in a structured way, the knowledge can be queried, processed, and compared. Thus, one could query for tribological experiments on desired materials and testing conditions, for example dry-running pin-on-disk model tests with various reinforcement phases within composites or deposited coatings on the specimen surfaces.

A large-scale employment of aforementioned knowledge extraction approaches, however, strongly demands for strategies for (semi-)automatically streamlining data acquisition. Therefore, this contribution aims at the introduction of a semantic annotation pipeline based upon natural language processing (NLP) methods in order to overcome the "knowledge reengineering bottleneck" in the domain of tribology. The motivation behind this contribution is mainly inspired by the current practice in biomedical research, where a massive growth in published research articles led to increasing attention for automated information extraction methods to support human researchers [31,32]. Regarding similar challenges, like sharing research outcomes via natural language publications, semantic ambiguity and interdisciplinary nature of the domain, this contribution is a first attempt to apply (semi-)automatic knowledge acquisition techniques within the domain of tribology. Therefore, while the methods used within this contribution have already shown potential in similar knowledge acquisition and structuring issues within the biomedical domain, this paper aims at the effective use of these methods in tribology. The contribution is structured as follows: First, the applied methods for the acquisition pipeline are introduced, containing a description of the underlying domain theory of tribological test methods as a generic schema as well as the relevant semantic web and NLP techniques, especially named entity recognition and question answering under the use of the BERT language model [33]. The semantic annotation pipeline and packages used for implementation are summarized in Section 3. Subsequently, the access-level and performance of the pipeline are demonstrated in Section 4, including a description of the Web-User-Interface and a technical evaluation of the single modules of the pipeline. Finally, we discuss the potentials and limitations of the pipeline, as well as connections and outlooks to further approaches in Section 5.

2. Theory and Methods

2.1. Domain Theory from Tribology

As mentioned before, generic knowledge builds a domain theory, which can be represented as a formal ontology. In terms of semantic annotation, the domain theory is used as structured metadata, the unstructured resource is enriched with. Therefore, relevant concepts and relations from established methodologies of tribological testing are used to build the schema for the semantic annotation pipeline. Generally, a tribological system can be described by its system structure, input and output variables and their functional conversion within the open or closed system boundary [34] (Figure 1). The system structure consists of the relatively moving body and counter-body, which are rubbing against each other and may be completely or partially separated by an intermediate medium (liquid or gaseous). Operational input variables, such as loads, kinematics, duration, and temperatures, can be summarized in the stress collective. Depending on the latter, as well as any disturbance variables and the system structure, the body and counter-body physically and chemically interact at temporally and spatially varying locations. On the one hand, this results in loss variables such as friction and wear, which cause changes in the surface, loss of material and energy dissipation. On the other hand, this results in the actual functional variables of the tribological system. The mechanisms and applications of tribology extend over several size scales. This ranges from processes on the nano- or micro-level in the field of physics, chemistry, and material sciences, such as the formation of boundary layers or the shearing of nanoparticle layers and ends with machine elements and assemblies as well as multiple tribological contacts in the engineering sciences in the micrometer to meter range, for example in rolling bearings or gears. Accordingly, tribometry, i.e., tribological measuring and testing technology, covers all dimensional ranges of tribology determining friction and wear parameters of tribological systems. The significance of various quantifiable measured variables, e.g., a friction coefficient averaged over time or a wear coefficient, usually depends on the underlying mechanisms, the measurement method, and the objective of the study. Given the function and structure of tribological systems, tribological testing can be divided into six categories according to the simplification of the system structure, the stress collective or the environmental conditions. While original and complete systems are tested under real operating and environmental conditions in field tests (category I), this is carried out under laboratory conditions with merely practical operating conditions in test bench tests (II). In aggregate (III) and component tests (IV), this is further reduced to the investigation of original aggregates or components. Specimen tests (V) are conducted with specimens that are similar to the components and subjected to similar stresses as in the target application. Finally, model tests (VI) involve fundamental analyses of friction and wear processes with simplified specimens under defined loads. Typical representatives of the latter are disk-on-disk, cylinder-on-cylinder, ball-on-disk or pin-on-disk tribometer tests. The advantages of the individual test categories can be combined by a suitable test chain [34].

2.2. The TribAIn Ontology

Kügler et al. [29] introduced the tribAIn ontology as a schema for structuring, reusing and sharing experimental knowledge within the tribological domain. The ontology was modelled highly relying on existing tribological test methods (see Section 2.1 and [34]). The presence of a common and shared methodology as well as terminology are vital assumptions for specifying a formal ontology of a domain, since those build a strong and accepted conceptualization, the formal specification relies on. Furthermore, the ontology is based on the EXPO ontology (ontology of scientific experiments) introduced by Soldatova and King [37], which is a generic formal description of experiments. Since tribAIn shares the same purpose of efficient analysis, annotation and sharing of results from scientific experiments, EXPO concepts were reused and further specified for the domain of tribology. The tribAIn ontology is formalized in OWL (Web Ontology Language) [38], which is a common ontology language based on description logics (DLs) [39]. Knowledge formalized with an ontology language is expressed in form of triples: <subject> <predicate> <object>, which means the ontology can be visualized as a directed graph with named relations between two classes (concepts). In the following, we will use Turtle Syntax [40] for streamlining triples of tribAIn. Since every object within an OWL ontology has a unique identifier, the prefix *tAI* is used for the tribAIn IRI (Internationalized Resource Identifier), thus concepts and relations of the tribAIn namespace can be identified by this prefix. The ontology provides concepts to describe the three main working areas "Experimental Design", "Procedure" and "Experimental Results" (Figure 2). The concepts from these areas structure the information about a specific experiment, with the tribological system (*tAI:TriboSystem*) investigated, pre-processing procedures (*tAI:IndustrialProcess* and subclasses) as well as the test procedure (*tAI:TribologicalTesting*) itself and links that information with the outcome of the investigation (*tAI:OutputParameter* and subclasses). Due to the close relation to the underlying methodology (cf. Section 2.1), the concepts refer to common terms within the domain of tribology. Parameters or variables, for instance loads or temperatures, are described using a pattern containing the two triples: *Parameter hasValue xsd:float* and *Parameter hasUnit Unit*. The first triple links a value of the datatype float to an instance of the class (or some subclass of) *Parameter*, while the other triple links a unit to the same instance. In this manner, measurement series are generated in a consistent fashion, which can be compared and analyzed.

Due to the design of the tribAIn ontology, a knowledge base (KB) which uses the ontology as schema, can be queried in terms of the following example questions (cf. [29]):

- Which tribological systems were investigated under dry-running conditions using a solid lubricant coating?
- Which variables were tested regarding their influence on the behavior of a material pairing?
- Which wear rate was calculated of sample XY?

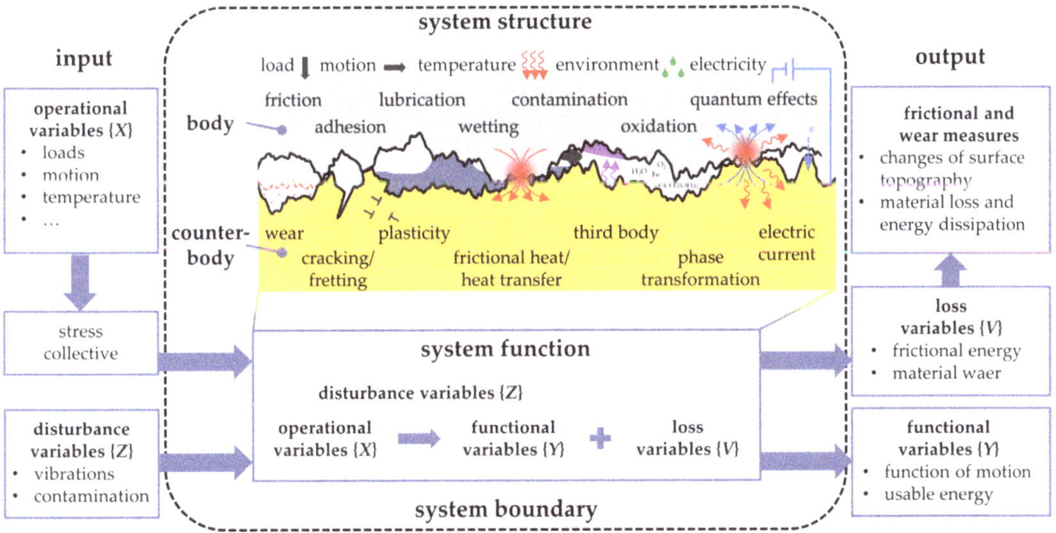

Figure 1. Overall representation of a tribological system, its target function, and interactions in tribological contacts. Redrawn and adapted from [29,34–36].

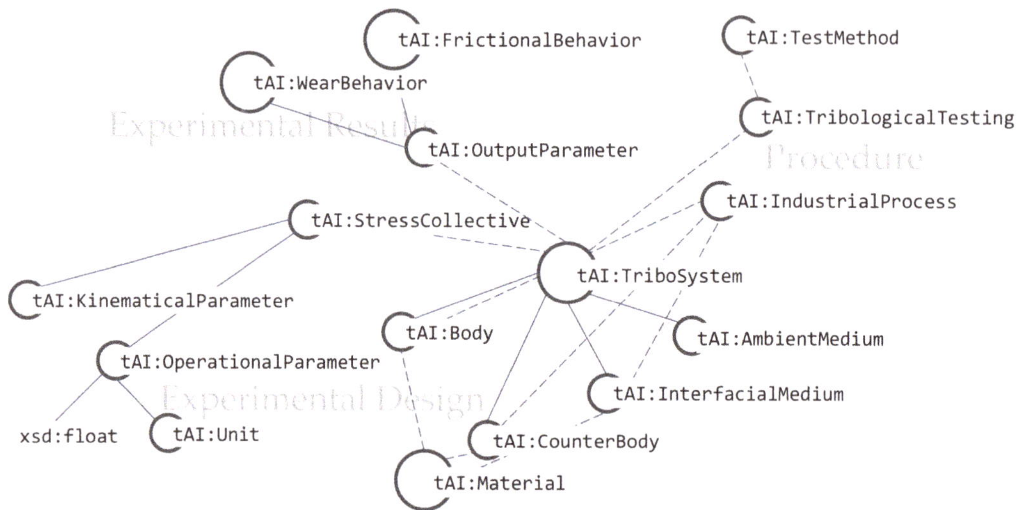

Figure 2. Excerpt of relevant tribAIn concepts. Redrawn and adapted from [29].

2.3. Ontologies, Knowledge Graphs and Semantic Annotation

As Gruber [41] states, within ontologies, definitions associate names of the entities within a universe of discourse. Therefore, the schema provided by an ontology can be shared among different knowledge graphs, which hold the actual data. This is referred to as ontological commitment and is a guarantee for consistency, even for incomplete knowledge, since there are agreements to use a shared vocabulary [41]. Those commitments to a specific vocabulary (or terminology) are also implicitly made within natural language communication. Within the domain of tribology, they exist for instance for the description of a tribological system (cf. Figure 1). Since tribological testing should enable reproduceable and comparable results, experiments must be built upon a common methodology, which defines the system structure as well as input- and output parameters. Describing experiments as well as results within a scientific publication under the use of a common terminology is a first step of knowledge formalization (Figure 3).

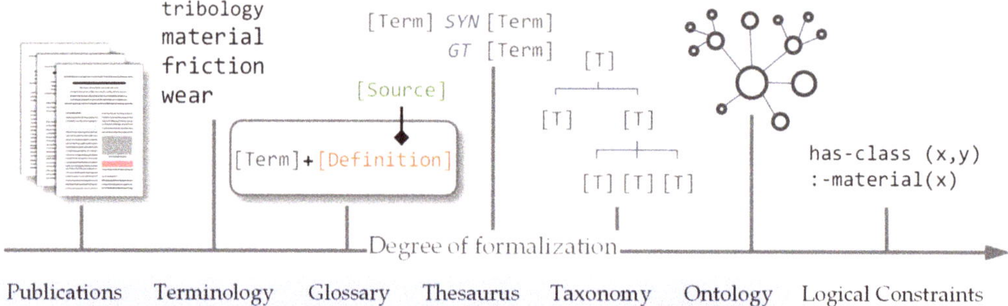

Figure 3. Different degrees of formalization from natural language text to logical constraints. Redrawn and adapted from [42].

Nevertheless, the challenge with sharing knowledge among natural language publications is the vague or even insufficient description, since often knowledge about the domain theory is assumed to be present to the human reader. An example is the following description from the materials section of an experimental study on $Ti_3C_2T_x$ nanosheets (MXenes) [43–45] investigated as solid lubricant for machine elements [46]:

> *"Commercially available thrust ball bearings 51201 according to ISO 104 […] consisting of shaft washer, housing washer and ball cage assembly were used as substrates (Figure 1a)."*

With background knowledge of the tribological domain, it becomes clear, that the studied tribological system here is a certain thrust ball bearing and with the information given by the referenced figure, the coated parts can be identified by a human reader. However, within the textual description, the link to the underlying methodology is not stated explicitly. Thus, the documentation of the experiment is incomplete and ambiguous from a formalization perspective. In other words, a semantic gap between textual descriptions from publications and the general knowledge models with a higher degree of formalization (Figure 3) prevents machine-supported processing of existing tribological knowledge from publications. In order to bridge such a gap, semantic annotation is the process of joining natural language and formal semantic models (e.g., an ontology) [47]. A semantic annotation of the example cited above associated with ontological concepts from the tribAIn ontology [29] is shown in Figure 4. In this example, the string "thrust ball bearings 51201" is recognized as referring to the instance *tbb_51201*, which is a tribological system ("*tbb_51201 a tAI:TriboSystem*" in the triple notation of Figure 4). Furthermore, the components of the thrust ball bearing are referred to the instances *sw_51201* (shaft washer), *hw_51201* (housing washer) and *as_bc_51201* (ball cage assembly) and are annotated as parts of the tribological system within the triple notation. Annotating text snippets semantically to instances of an ontology enriches the natural language text with machine-readable context. For example, the instance "*tbb_51201*" may not only be referred to the experimental testing described in the publication, but also be linked within the knowledge graph to information from the ISO 104 mentioned within the text snippet. Therefore, the semantic annotation process links mentions of entities from different sources to knowledge objects within a knowledge graph, which are further semantically defined by an ontological schema.

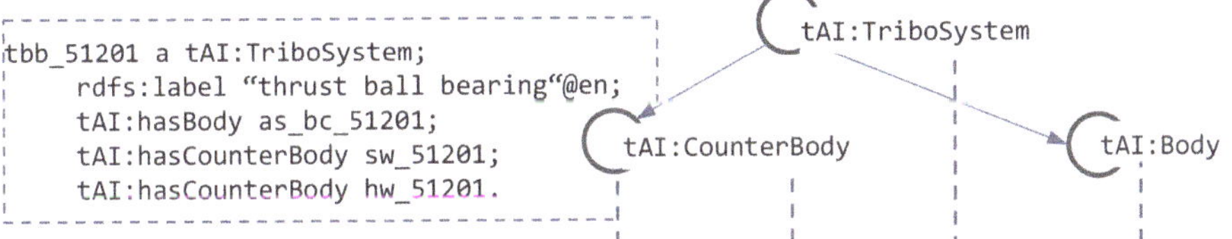

"Commercially available thrust ball bearings 51,201 according to ISO 104 [...] consisting of shaft washer, housing washer and ball cage assembly were used as substrates (Fig. 1a)."

Figure 4. Example of a semantic annotation of a text excerpt from [46] with concepts from the tribAIn ontology [29] graphically visualized and in triple notation (Turtle format).

Furthermore, some semantic annotation systems perform ontology population, which means not only annotating documents with respect to an existing ontology resulting in semantic documents but creating new instances from the textual source [47]. For example, the ball bearing from the example above is instantiated as a new knowledge object within a knowledge graph. One advantage of building knowledge graphs from textual sources is the direct link between mentions of knowledge objects within a source and the capability of generating structured data from those mentions, even if the facts about a knowledge object origin from different sources. A schematic architecture of a semantic knowledge base, which consists of a domain ontology on schema-level, as well as a knowledge graph that holds the data about knowledge objects, is shown in Figure 5. An example of structured information is given for the knowledge object "thrust ball bearings", once as its use in a tribological test setup and once as a rolling bearing with its specification.

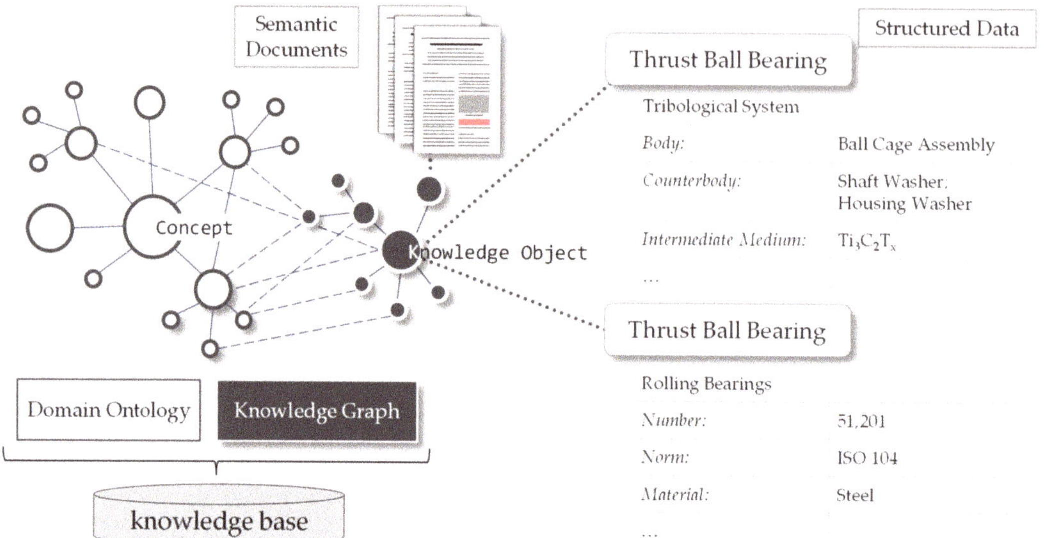

Figure 5. Schematic architecture of a semantic knowledge base, consisting of schema-level ontologies and a knowledge graph containing knowledge objects as structured data and mentions of knowledge objects from semantic documents.

Thus, different information from various sources is linked for an object of the knowledge graph. Moreover, the original textual sources are also linked nodes within the graph. Semantic annotation can be performed manually, semi-automatically, or automatically. Thereby, the semi-automatic approach is preferred since manual annotation is time-consuming and automatic approaches can lead to unreliable information within the resulting knowledge graph [47].

2.4. Natural Language Processing

A semi-automatically semantic annotation process is often conducted by methods from NLP. The main challenge of NLP is the representation of contextual nuances of human language, since the same matter can be described utilizing different wording and the same word can be used for different meanings depending on the context. Therefore, enabling machines to understand and process natural language demands the provision of a machine-readable model of language. However, Goldberg [48] describes a challenging paradox in this context: Humans are excellent in producing and understanding language and are capable to express and interpret strongly elaborated and nuanced meaning of language. In contrast, humans struggle at formally understanding and describing the rules, which govern our language [48]. Rules in this context are not only referred to syntax and grammar, but also to contextual concerns. For example, considering a classic NLP task of document classification into one of the four categories metals, fluids, ceramics, or polymers. Human readers categorize documents relatively easy into those topics guided by the words used within a publication but writing up those implicitly applied rules for categorization is rather challenging [48]. Therefore, machine learning models are trained to learn vectorized text representations from examples, which are suited input formats for NLP downstream tasks (e.g., document classification). The classic preprocessing steps for generating those text representations from a document corpus are summarized in Figure 6.

Almost any analysis of natural language starts with splitting the documents (e.g., plain text, charts, figures), removing noise (e.g., references, punctuation) and normalization of word forms [49]. Subsequently, the plain text is further split into minimal entities of textual representation, the tokens, on word- or character-level. Since ML models assume some kind of numerical representation as input, the tokens are replaced by their corresponding IDs [50]. If a text is split into tokens on word level, the question arises, what counts as a word. To answer this question, morphology deals with word structures and the minimal units a word is built from, such as stems, prefixes and suffixes. Those minimal units are important, if a tokenizer has to deal with unknown words (meaning words, which were not within the training corpus) [48]. Tokenizers like WordPiece [51] represent words as subword vectors [49], e.g., "nanosheets" can be separated in the subwords "nano" and "sheet" and the plural-ending "- s". However, the tokens are further transferred in so-called embeddings, which are an input representation a ML or deep learning architecture can handle for NLP tasks. An embedding is a representation of the meaning of a word; thus, they are learned under the premise, that a word with the same meaning has a similar vector representation [49]. A distinction is made between static embeddings and contextualized embeddings. One quite popular static word embedding package is Word2Vec [52,53]. A shortcoming of those static embeddings is that polysemantic is not properly handled since one fixed representation is learned for each word in the vocabulary even if a word has a different meaning in different contexts [49,54].

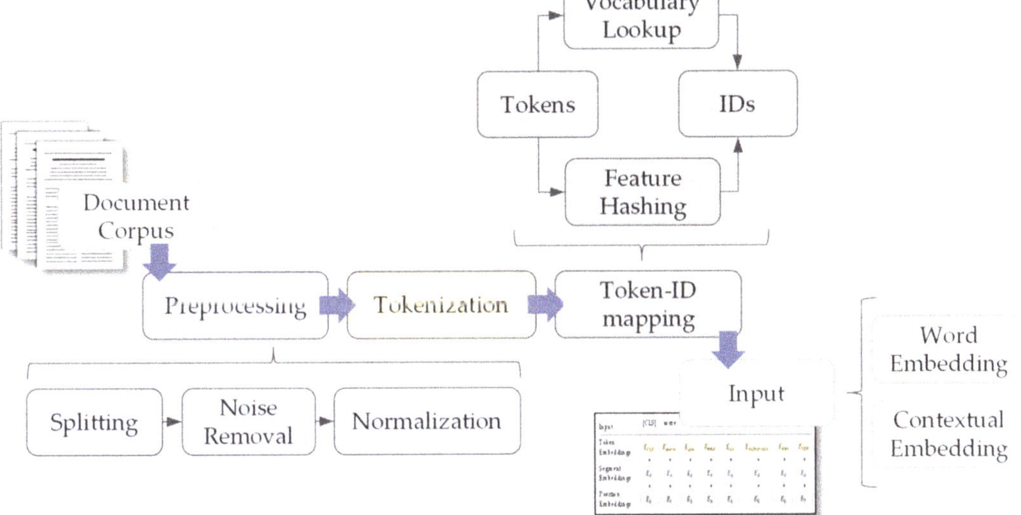

Figure 6. Preprocessing steps to generate embeddings from text as input to NLP downstream task. Redrawn and adapted from [50].

Therefore, contextualized (dynamic) embeddings provide different representations of each word based on other words within the sentence. State of the art representatives are ELMO (Embeddings from Language Models) [55], GPT & GPT2 (Generative Pre-Training) [56] and BERT (Bidirectional Encoder Representations from Transformers) [33], which are also referred as pre-trained language models. BERT is a multi-layer bidirectional transformer encoder [33,57], which is provided in a base version with 12 layers and large version with 24 layers. Most of the recent models for NLP tasks are pre-trained on language modeling (unsupervised) and fine-tuned (supervised) with task-dependent labeled data [58]. Thus, those models are trained to predict the probability of a word occurring in a given context [48]. BERT is pretrained on large amount of general-purpose texts from

BooksCorpus and English Wikipedia, which resulted in a training corpus of about 3300 M words [33]. Devlin et al. [33] differentiate BERTs pre-training from the other mentioned models, consisting of two unsupervised tasks: masked language modeling (LM) and next sentence prediction (NSP) (see also [59] for further information on BERTs pre-training). Fine-tuning BERT for downstream tasks, like Question Answering (QA) or Named Entity Recognition (NER), the same architecture is used apart from the output layer (see Figure 7). The input layer consists of the tokens (Tok 1...Tok n). The special token [CLS] signs the starting point of every input and [SEP] is a special separator token. For instance, question answering pairs can thus be separated within the input [33]. The contextual embeddings ($E_1 \ldots E_n$) further result in the final output ($T_1 \ldots T_n$), after being computed through every layer resulting in different intermediate representations (T_{rm}). For more information on Transformer architectures, the interested reader is referred to [60]. There are different extensions of the original model of BERT, which are specialized for certain downstream tasks or domain terminologies. The SciBERT model [61] is pre-trained on scientific papers improving the performance of downstream tasks with scientific vocabulary. BioBERT [32] is pre-trained on large-scale biomedical corpora and improves the performance of BERT especially in biomedical NER, relation extraction and QA. Furthermore, SpanBERT [62] is a pre-training approach, which is focused on a representation of text spans instead of single tokens. Both pre-training tasks from the original BERT are adapted for predicting text spans instead of tokens, which is especially useful in relation extraction or QA.

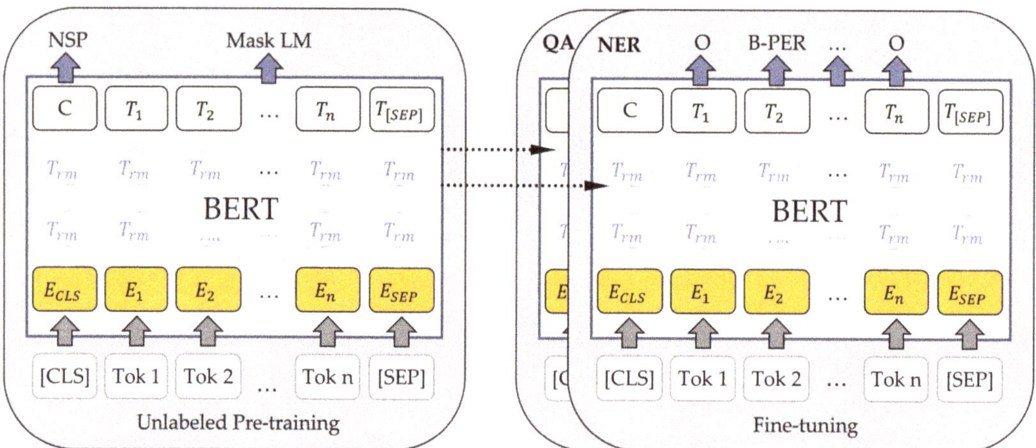

Figure 7. BERT pre-training and fine-tuning procedures using the same architecture for both. Only the output layer differs depending on the downstream task e.g., NER, QA. Redrawn and adapted from [33].

2.5. NLP Downstream Tasks

Information Extraction (IE) is a task of obtaining structured data from unstructured information, e.g., embedded in textual sources, by recognizing and extracting occurrences of concepts and relationships among them [49,63]. IE is often used to build knowledge graphs from textual representations (e.g., DBpedia), since those can be queried and are a common way of presenting information to users [49]. IE and semantic annotation (see Section 2.3) are often combined, since both share the subtask of NER. NER is a sequence-labeling task to recognize and tag words or phrases usually like "Person" (B-PER), "Location" (B-GEO) or "Organization" (B-ORG) within textual data. A named entity can be anything, which has a proper name, thus can be distinguished from other objects [49]. Therefore, NER is often based on a specific domain vocabulary, e.g., in biomedicine [32,64]. Moreover, relation extraction is also a subtask of IE in the context of building knowledge graphs and mainly deals with the extraction of binary relations like child-of or part-whole relationships used within taxonomies, ontologies and knowledge graphs [49]. IE can be used for template filling, meaning recognizing and filling a pre-defined template of structured data from the unstructured sources (cf. Figure 5) [49]. Question Answering (QA) is a task of information retrieval, but with a query, which is a question in natural language and a response as an actual answer [63]. QA is often used within Chatbots of customer services or within virtual speech assistants (e.g., Amazon Alexa or Apple Siri). The main difference from classic retrieval operations is the form of asking questions in natural language instead of formal database queries and the retrieval of a precise answer to the question instead of document retrieval. Therefore, QA can be exploited for generating structured data and template filling.

3. Semantic Annotation Pipeline

The developed semantic annotation pipeline consists of three separate modules and a graphical user interface (GUI). The modules communicate via REST API. Due to the modular architecture, the single module can be exchanged and thus the pipeline can be adapted to particular applications. The pipeline is trained for annotating and extracting information from tribological publications with the scope of model tests (cf. Section 2.1).

3.1. Document Extraction

The first module (Figure 8) performs the preprocessing step introduced in Section 2.4 (Figure 6). Since the source format (PDF) is not suited for further processing, plain text is extracted. By parsing through the documents, elements like figures, charts or tables are also detected.

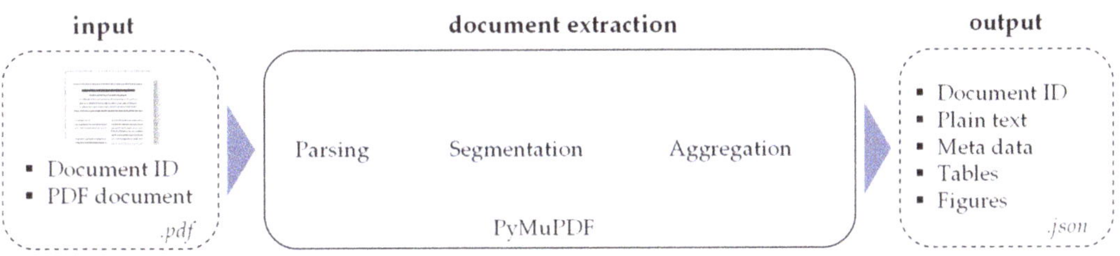

Figure 8. Document extraction module.

Besides detecting non-textual elements, the document is segmented into its paragraphs (e.g., abstract, introduction etc.), since–depending on the IE purpose–the relevant information may be provided mainly in a certain paragraph. For example, the introduction section usually contains information about the aim of the investigation, while the results section further provides a description of the outcome. Parsing is performed based on syntactical rules and pattern matching, e.g., indentations, blank spaces or different fonts, can be used as indicators for the detection process. Besides the content of the publication, meta data about the document (e.g., author, DOI, date, publishing information) is extracted. The last step is the aggregation of the previously segmented elements into JSON (Java Script Object Notation), which is a common and platform independent data sharing format. The document extraction module is implemented using the PyMuPDF Library [65].

3.2. Document Annotation

The document annotation module (Figure 9) performs the actual annotation process. The module expects the files in JSON format from the document extraction module as input and is capable of annotating plain text and table data. The annotation module uses the Flair library [66] and the embeddings of the NER model are trained on the tribological annotation categories displayed in Figure 10. Within a parameter study, embeddings from BERT (Base), SciBERT and SpanBERT were tested against each other. Thereby, SpanBERT were chosen, since those have shown the best results with an F1 (micro) score of 0.8065 and an F1 (macro) score of 0.8012.

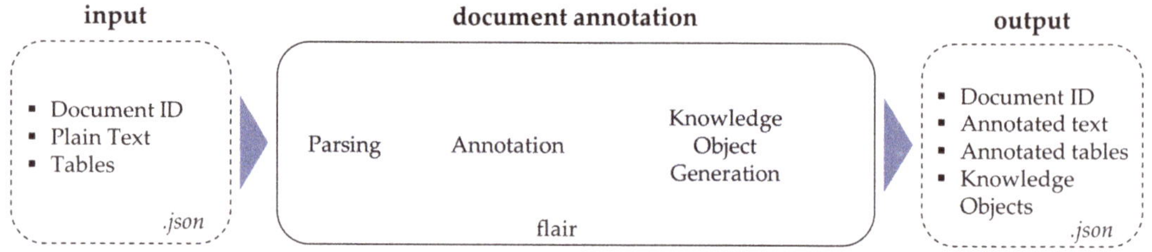

Figure 9. Document annotation module.

Sentence:	Annotation	Category	Tribological Annotation Model (Categories)
#1 Newly emerging Ti3C2Tx-nanosheets (MXenes) have attracted considerable attention in energy storage, catalysis and more recently tribology. ...	MXene	Composite Element	Bodystructure
	Ti3C2Tx	Composite Element	Composite Element
	1.47 GPa	Operational Parameter	Environmental Medium
	0.80 GPa	Operational Parameter	Geometry
#2 Thus, corresponding initial HERTZian pressures p Hertz at the contact center were about 1.47 GPa and 0.80 GPa, respectively. ...	Scanning electron microscopy	Test Method	Intermediate Medium
	SEM	Test Method	Kinematic Parameter
#3 Scanning electron microscopy (SEM; ZEISS, CrossBeam 1540 EsB) of selected samples was conducted to investigate the underlying friction and wear mechanisms as well as the formation of tribo-layers.			Manufacturing Process Opertional Parameter Test Method

Knowledge Object Generation

Knowledge Object	Category
MXene; Ti3C2Tx	Composite Element
1.47 GPa	Operational Parameter
0.80 GPa	Operational Parameter
Scanning electron microscopy; SEM	Test Method

Figure 10. Example of semantic annotation and knowledge object generation within the document annotation module and annotation categories for example sentences from [67].

The annotation step within the module recognizes entities of the tribological categories. An example is shown in Figure 10. The inputs are three different sentences (plain text), which are parsed. Then, entities of the different categories are annotated. In a second step, the annotations are aggregated to knowledge objects, thus for instance the two recognized entities MXene and $Ti_3C_2T_x$ refer to the same knowledge object (Figure 10). Due to the knowledge object generation, different terms used to describe the same entity within a text are aggregated to a single object. The generation of knowledge objects is mainly based on identifying acronyms and synonyms. The identified character strains are then compared. For character strains, which go beyond four, a fuzzy comparison using the Fuzzy-Wuzzy Library (https://pypi.org/project/fuzzywuzzy/, accessed on 14 December 2021) is conducted, which calculates the Levensthein distance to compare two-character strains. Output data from the document annotation module is again streamlined in JSON format and contains the annotated text and table data as well as the aggregated knowledge objects.

3.3. Document Analysis

The document analysis module (Figure 11) is a QA system that extracts answers from the text to questions about tribological model tests to create triples from the document. The QA system is built on the PyTorch framework (https://pytorch.org/, accessed on 14 December 2021) using a SciBERT-Model from the Hugging Face library (https://huggingface.co/, accessed on 14 December 2021).

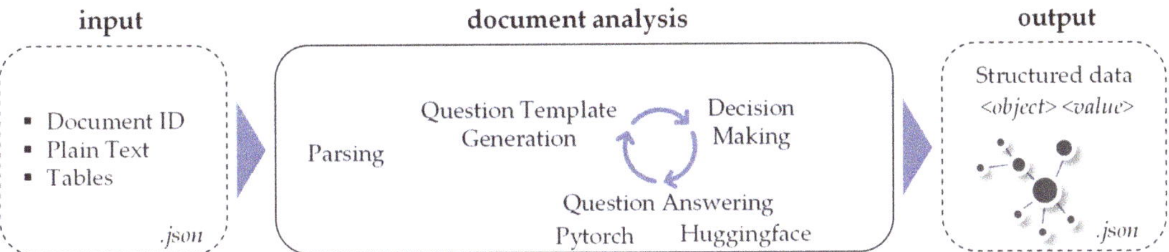

Figure 11. Document analysis module.

The BERT model is fine-tuned by question-answer pairs. Question templates (Figure 12) are generated, which contain the questions for extracting knowledge objects from the text. Those templates determine the structure of the information, which should be extracted from the text. This means, the question templates can be customized depending on the extraction task. The decision maker is an intermediate aggregation step containing multiple redundancies, which ensures higher reliability of an extracted answer. Therefore, the question template contains the same question rephrased several times. Furthermore, the answer space is restricted by using regular expressions (Regex) to define an expected answer pattern and by specifying an entity type (tribological category) of the extracted answer. The final result is an ID of a knowledge object and its textual annotation, for the case a knowledge object can be assigned. Otherwise, the textual passage is extracted as answer to the question.

```
                                testing_duration":{
                                    "question":[
Question(s)                             "What was the test duration for the test?",
                                        "How many minutes went the test?",
                                        "How long did the test go on?"
                                    ],
Answer space                        "expectedAnswerSpace":[
(Regex)                                 "([0-9]{0,5}\\.|)[0-9]+ *(min|h|s)"
                                    ],
Search space                        "presearchSpace":[
                                        " minutes ", "min ", " hour", " second",
                                        "sec ", " h ", "time"
                                    ],
Entity subcategory                  "specific_questionType":"TESTING_DURATION",
Entity category                     "broader_questionType":"OPERATIONAL_PARAMETER",
                                        ...    ]
                                }
```

Figure 12. Question template example for the extraction of the testing duration of an experiment.

4. Implementation and Evaluation

4.1. Web-Application

For testing the capability, the pipeline can be assessed through a web interface (see Data Availability Statement). The three modules communicate via REST API with the GUI. Publications have to be provided in PDF-format and can be delivered by drag and drop to the first module (1). As indicated in Figure 13, a manual control and adaptation step is integrated between the three separate modules. Thus, the acquisition of structure works in a semi-automatic manner with human supervision.

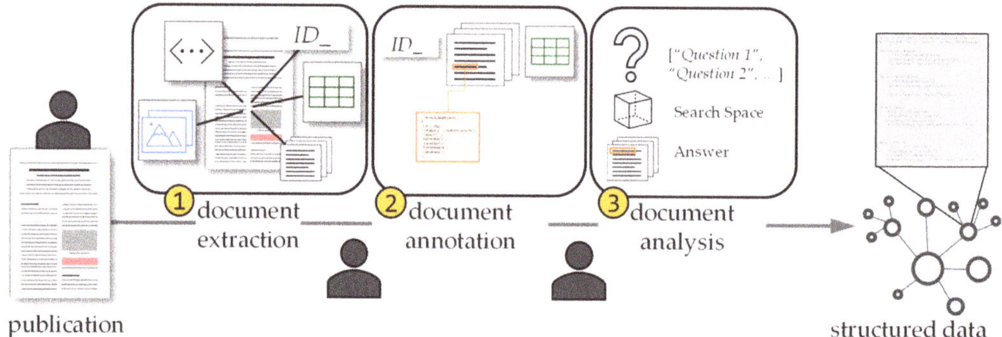

Figure 13. Process of document annotation via the web-interface.

The output from the first module is the extracted text and other entities (e.g., figures, tables, metadata) from the PDF. The text itself is split into chapters, paragraphs and sentences, which serve as input to the next modules. The user check enables adaptation and correction of the automatically generated output from the module (e.g., adaption of the paragraph separation). Subsequently, the pipeline can be continued via the GUI and the JSON-data produced by the first module forwarded to the document annotation module (2), which performs NER and knowledge object generation. The output from the second module can also be checked by the user after the automatic annotations are generated. The NER process thereby performs the annotation part while the knowledge objects are the semantic output of the annotation pipeline (see Figure 5 in Section 2.3 for the role of knowledge objects within a semantic knowledge base). The check especially contains proof of annotations and the aggregation to knowledge objects. Next, the pipeline is continued to the final context analyses via the document analysis module, which performs the QA and generates the structured data as final output.

4.2. Resulting Knowledge Graph

The output from the document annotation module is a linked data structure, containing the aggregated knowledge objects related to the mentions within paragraphs and tables of the publication. Therefore, the output can be visualized as a knowledge graph containing the structured data annotated within the respective publication. This is exemplified in Figure 14 for a representative publication [67]. The generated knowledge graph is a complex network of nodes and relationships. Thereby, the size of the nodes corresponds to the number of mentions of a knowledge object within the publication. This means that a knowledge object with only one mention has the minimum size while the size increases with the number of mentions. In this way, the important knowledge objects within the publication can be easily identified within the graph.

While the output graph from the document annotation module is particularly suited to identify the main topics of the analyzed publication in form of the most often mentioned knowledge objects, the output from the document analysis module is a graph generated from the question templates (Figure 15). The knowledge graph contains the identified and correctly classified answers (triangles) given by the decision maker. Thereby the triples are generated from the schema provided by the tribAIn ontology [29]. Thereby, the excerpts of the knowledge graph in Figure 15 refer to the example questions introduced in Section 2.2 regarding the tested variables (1) and the calculated wear rate (2). The generated linked data combined with an ontology provides a formally and semantically unambiguous representation which can be queried, filtered and further processed.

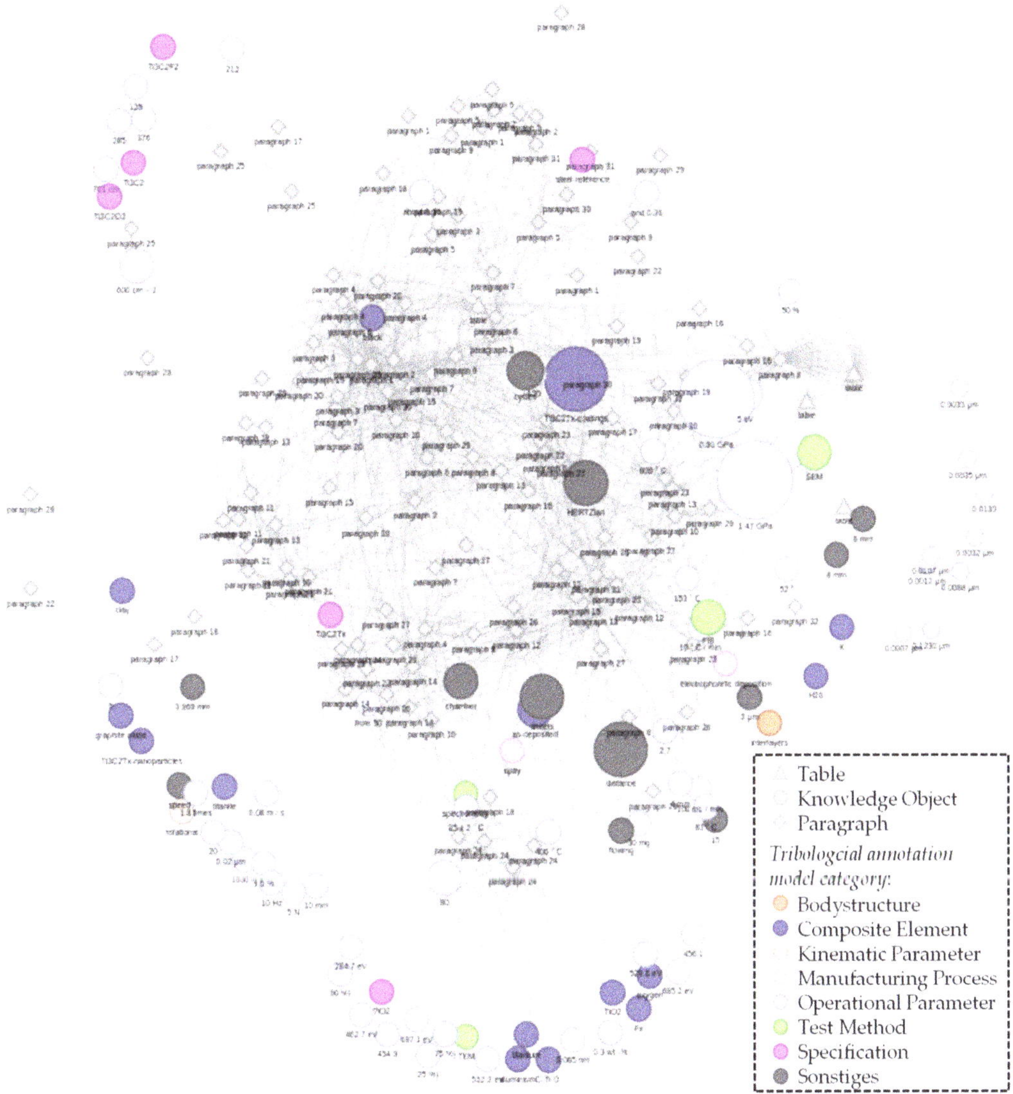

Figure 14. Schematic visualization of the resulting knowledge graph from the document annotation module for the processed representative publication [67].

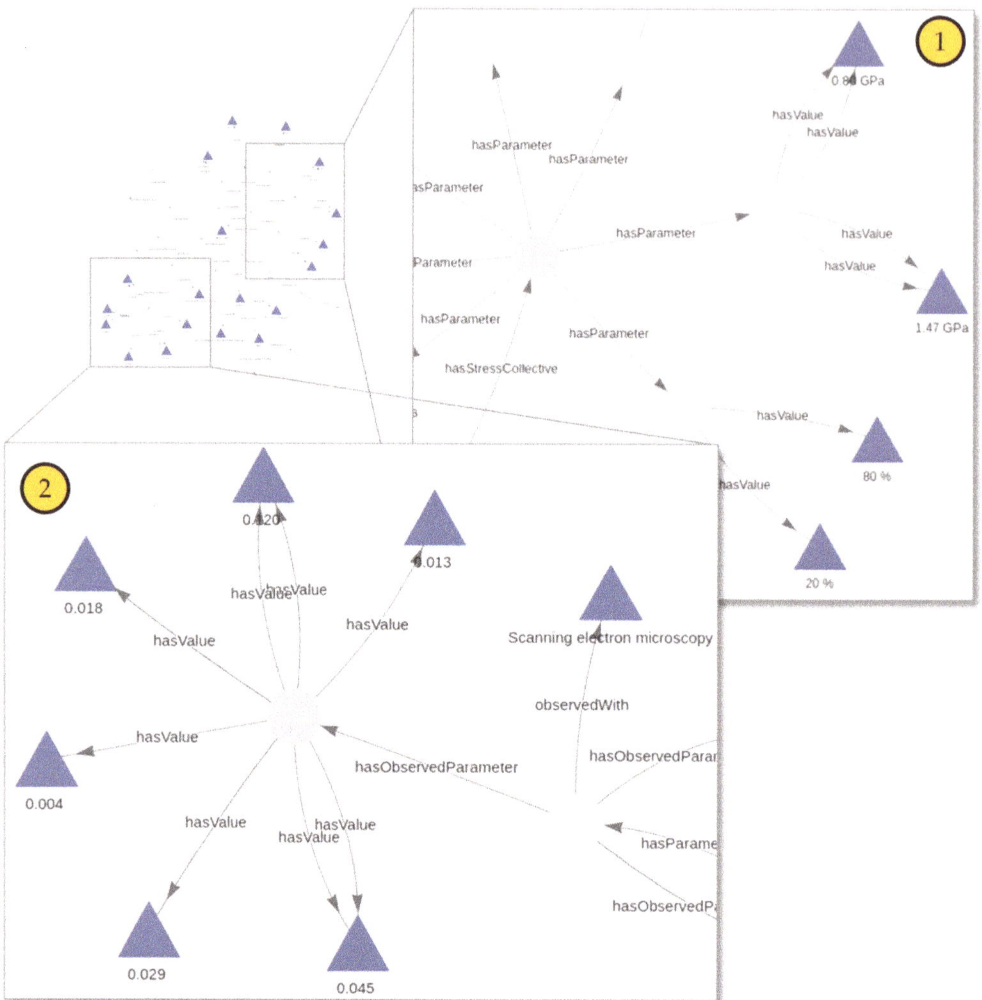

Figure 15. Schematic of the resulting knowledge graph from the document analysis module for the processed representative publication [67] with a detailed view on the varied operational parameters (1) and the outcome measurements of the wear rate (2).

4.3. Evaluation

For evaluation, five documents from a pool of existing publications on model-tests were used for an initial performance test of the pipeline:

- **Doc#1**: Marian et al., 2020 [67]
- **Doc#2**: Mekgwe et al., 2019 [68]
- **Doc#3**: Li & Xu 2018 [69]
- **Doc#4**: Wang et al., 2018 [70]
- **Doc#5**: Byeong-Choon & In-Sik 2017 [71]

The documents differentiate in length and format, for instance number of columns. Furthermore, Doc#5 is kind of defect since the PDF contains invisible text overlaps. Thereby, the three modules were evaluated separately depending on the evaluation aim. Since semantic annotation is not a common task within the domain of tribology, standard test documents, which are widely accepted for performance measurement of NLP tasks, do not exist. Therefore, the test documents were manually annotated for the special purpose of evaluating the pipeline introduced within this contribution. The document extraction module was analyzed with respect to its quality in extracting and separating text and other elements like figures and tables from the PDF documents. This resulted in a comparison of the data from the extraction module against the ground truth (GT) for text, figures and tables. This shows if the system works as it is intended. The smaller the deviations from GT, the more reliable is the PDF extraction. Within Table 1, the reliability of text extraction is assessed against the criteria, if chapters, paragraphs, sentences, words and chars are correctly detected and separated. The deviations from GT are relatively small for Doc#1 (chapter −12.5%; paragraph −8.6%, sentence −7.3%, word −7.3%, char −9.8%), Doc#2 (chapter 37.5%; paragraph −26.7%, sentence −3.0%, word −10.0%, char −18.2%, Doc#3 (chapter 7.1%; paragraph 11.5%, sentence −2.4%, word −1.3%, char −5.4%) and Doc#4 (chapter 27.3%; paragraph −22.2%, sentence 23.6%, word −5.0%, char −10.9%), while the deviation is substantially higher for Doc#5 (chapter 240%; paragraph 400%, sentence 149%, word 132%, char 118%). Those high deviations can be attributed to the defect PDF, which contains embedded textual and other elements, which overlap the intended content of the document. The results of the figure extraction analysis are shown in Table 2. Almost all figures within the test set were correctly detected. Only one figure was partly incorrect extracted in Doc#2 and two figure areas were incorrectly recognized in Doc#4. However, all figures were correctly extracted within the defect PDF Doc#5. Thereby, 14 additional figures were identified, which is due to the overlayed elements within the PDF. The extraction of tables seems also reliable since the majority of tables are correctly recognized (see Table 3). An exception is within Doc#1, which can be attributed to the table being rotated within the publication. This shows that the first module depends on the quality and regularity of the input files. Since the module provides a manual check, small deviations from the expected output can easily be fixed via the GUI.

Table 1. Quality of text extraction regarding the extracted chapters, paragraphs, sentences, words, chars and if an abstract was detected (true/false). The GT is given in the brackets.

Document	Doc#1	Doc#2	Doc#3	Doc#4	Doc#5
Chapter	14 (16)	11 (8)	15 (14)	14 (11)	31 (9)
Paragraph	32 (35)	11 (15)	29 (26)	14 (18)	40 (18)
Sentence	179 (193)	65 (67)	124 (127)	68 (55)	237 (95)
Word	4213 (4547)	1304 (1449)	2898 (2937)	1205 (1276)	5192 (2241)
Char	26,127 (28,881)	7497 (9160)	17,119 (18,089)	6998 (7759)	30,928 (14,198)
Abstract	true	true	true	true	true

Table 2. Quality of figure extraction regarding detected figures, incorrectly detected figure area and additional extractions. The GT is given in the brackets.

Document	Doc#1	Doc#2	Doc#3	Doc#4	Doc#5
Figure	12 (12)	5 (5)	12 (12)	5 (5)	12 (12)
Incorrect area	0	1	0	2	0
Additional Figure	0	0	0	0	14

Table 3. Quality of tables extraction regarding detected tables, additional extractions and correct number of cells. The GT is given in the brackets.

Document	Doc#1	Doc#2	Doc#3	Doc#4	Doc#5
Table	0 (1)	1 (1)	0 (0)	0 (0)	1 (1)
Additional Table	0	0	0	0	0
Incorrect cells	-	0	-	-	0

The document annotation module was evaluated with respect to the ability of NER and knowledge object generation. Three language models (BERT, SciBERT and SpanBERT) were trained with the hyperparameters shown in Table 4, which were an outcome of a previously conducted parameter study. Thereby, an RNN (recurrent neuronal network) architecture was used with one layer and a hidden size of 128. Dropout [72,73] is a method to reduce overfitting by deactivating a number of neurons randomly from the neural network. The learning rate defines the step size of the optimization and thus controls how quickly the model learns the given problem. The batch size specifies the number simultaneously evaluated examples. Since the used language models have already been pre-trained on large-scale general language data (cf. Section 2.4), the training includes only fine-tuning, which is computationally less expensive. The training of the models took about 20 to 30 min each on a NVDIA RTX 2070 and 8 GB RAM.

Table 4. Hyperparameters for training language models.

Language Model	RNN Layers	Hidden Size	Dropout Rate	Learning Rate	Mini Batch Size
BERT	1	128	0.0479	0.1	32
SciBERT	1	128	0.0020	0.1	32
SpanBERT	1	128	0.1454	0.15	32

Micro and Macro F1 scores were calculated to select the best of the three models for the recognition task. Therefore, the five documents were manually annotated due to the tribological annotation model categories. For every category the precision, recall and F1 score were calculated three times for each of the trained language models with regard to the manual annotations (see Table 5). The test set contained 986 annotated sentences for the tribological annotation model categories already introduced in Figure 10.

Table 5. Precision (P), Recall (R) and F1 score for each tribological annotation model category.

Category	Score	BERT			SciBERT			SpanBERT		
		1	2	3	1	2	3	1	2	3
Body structure	P	0.8000	0.7037	0.7692	0.7037	0.6667	0.7097	0.7600	0.7600	0.7241
	R	0.8333	0.7917	0.8333	0.7917	0.8333	0.9167	0.7917	0.7917	0.8750
	F1	0.8163	0.7451	0.8000	0.7451	0.7407	0.8000	0.7755	0.7755	0.7924
Composite element	P	0.7833	0.7241	0.7272	0.7846	0.7286	0.7429	0.7576	0.8030	0.7812
	R	0.7833	0.7000	0.8000	0.8500	0.8500	0.8667	0.8333	0.8833	0.8333
	F1	0.7833	0.7118	0.7619	0.8160	0.7846	0.8000	0.7936	0.8412	0.8064
Environmental medium	P	1.0000	0.8000	1.0000	1.0000	1.0000	1.0000	0.8333	0.8333	1.0000
	R	0.8000	0.8000	0.8000	0.8000	0.6000	0.6000	1.0000	1.0000	0.8000
	F1	0.9000	0.8000	0.9000	0.9000	0.8000	0.8000	0.9167	0.9167	0.9000
Geometry	P	0.9375	0.8824	0.8750	0.7500	0.8235	0.8235	0.8824	0.8889	0.8325
	R	0.7895	0.8950	0.7368	0.6316	0.7368	0.7368	0.7895	0.8421	0.7368
	F1	0.8572	0.8887	0.8000	0.6857	0.7777	0.7777	0.8334	0.8649	0.7817
Intermediate medium	P	0.8462	0.7143	1.0000	0.9286	1.0000	0.8571	1.0000	1.0000	1.0000
	R	0.6111	0.5556	0.6111	0.7222	0.7222	0.6667	0.6667	0.6667	0.6667
	F1	0.7097	0.6250	0.7586	0.8125	0.8387	0.7500	0.8000	0.8000	0.8000
Kinematic parameter	P	0.6667	0.6667	0.6364	0.7273	0.6923	0.6923	0.8000	0.8000	0.7273
	R	0.8000	0.8000	0.7000	0.8000	0.9000	0.9000	0.8000	0.8000	0.8000
	F1	0.7273	0.7273	0.6667	0.7619	0.7826	0.7826	0.8000	0.8000	0.7619

Table 5. *Cont.*

Category	Score	BERT			SciBERT			SpanBERT		
		1	2	3	1	2	3	1	2	3
Manufacturing process	P	0.6250	0.5000	0.6250	0.8182	0.6923	0.7143	0.7143	0.7500	0.6667
	R	0.4545	0.5455	0.4545	0.8182	0.8182	0.9091	0.4545	0.5455	0.5455
	F1	0.5263	0.5218	0.5263	0.8182	0.7500	0.8000	0.5555	0.6316	0.6000
Operational parameter	P	0.7222	0.7571	0.7647	0.7656	0.7937	0.8197	0.7059	0.7500	0.7424
	R	0.8525	0.8689	0.8525	0.8033	0.8197	0.8197	0.7869	0.8361	0.8033
	F1	0.7820	0.8092	0.8062	0.7840	0.8065	0.8197	0.7442	0.7907	0.7717
Specification	P	0.7143	0.7500	0.7368	0.6842	0.8235	0.7647	0.6190	0.7778	0.7778
	R	0.7895	0.7895	0.7368	0.6842	0.7368	0.6842	0.6842	0.7368	0.7368
	F1	0.7500	0.7692	0.7368	0.6842	0.7777	0.7222	0.6500	0.7567	0.7567
Test method	P	0.8947	0.8571	0.8571	0.8500	0.8095	0.8182	0.8889	0.8421	0.8497
	R	0.8947	0.9474	0.9474	0.8947	0.8947	0.9474	0.8421	0.8421	0.8497
	F1	0.8947	0.9000	0.9000	0.8718	0.8500	0.8781	0.8649	0.8421	0.8497

The resulting F1 scores are summarized in Table 6 for BERT, SciBERT and SpanBERT, which were each calculated in triplicate. As mentioned before, SpanBERT featured the best scores within the second run, which may be due to the annotated entities referring to the tribological categories, that are often spans of words instead of single tokens (e.g., "Scanning electron microscopy").

Table 6. Evaluation and selection of the NER model. F1 scores for BERT, SciBERT and SpanBERT.

	BERT			SciBERT			SpanBERT		
	1	2	3	1	2	3	1	2	3
F1 (micro)	0.7823	0.7570	0.7782	0.7847	0.7905	0.7992	0.7702	0.8065	0.7879
F1 (macro)	0.7736	0.7443	0.7645	0.7868	0.7859	0.7880	0.7726	0.8012	0.7851

The annotations generated through NER were further aggregated to knowledge objects within the document analysis module. The resulting number of aggregations is shown in Table 7. Annotations are considered incorrectly aggregated if at least two annotations are assigned to the same knowledge objects, although they do not belong together (false positive). Furthermore, if at least two annotations which belong to a knowledge object are not aggregated, they count as false negative. This criterion captures the reductivity of the knowledge object generation while the counts of correctly and incorrectly aggregated annotations provide an insight into the precision of the generation. A precision of 89.5% is reached for the test pool while the recall is about 84.4%. This can be considered as sufficient for the quality of knowledge object generation.

Table 7. Evaluation of knowledge object generation containing the number of annotations, of all knowledge objects as well as correctly aggregative (true positive), incorrectly aggregated (false positive) and not aggregated (false negative) knowledge objects.

Document	Annotations	Knowledge Objects	Correctly Aggregated	Incorrectly Aggregated	Not Aggregated
Doc#1	944	236	59	12	11
Doc#2	296	68	21	1	11
Doc#3	609	98	57	4	2
Doc#4	323	75	32	5	0
Doc#5	400	70	36	2	14
Σ	2572	547	205	24	38

Finally, the document analysis module was evaluated due to its quality of answering the questions from the templates. The criteria for assessing the quality were grouped to the quality of question answering itself and if the decision maker prefers the right answer. The final results over all questions are shown in Table 8. The GT is counted, if at least one answer within the text can be given to the question. The criteria for question answering itself are split into the cases if the expected answer is found in text and/or if at least one additional answer was found independent of the expected answer. The need for the decision maker can be seen from the fact that additional answers besides the expected one were found for all documents. The criteria for the decision maker were thereby split into the cases if the correct answer was preferred by the decision maker, if an incorrect answer was preferred, or if no answer was found or preferred. When the text contains at least one correct answer (GT), the question answering itself found the correct answer with a probability of 60.4%. The decision maker found the correct answer with a probability of 62.3%. At this point it should be noted that the quality of answers is highly influenced by the question templates. This means what questions are asked of the publication to get a desired answer.

Table 8. Evaluation results of all answers to question templates regarding the input parameters (e.g., kinematical parameters), structural information of the tribological system (e.g., geometry) and output parameters (e.g., friction and wear).

Document	GT	Question Answering		Decision Maker		
		Answer Found	Additional Answers	Correct Answer	Incorrect Answer	No Answer
Doc#1	26	16	15	13	1	10
Doc#2	20	13	18	7	11	9
Doc#3	20	9	13	7	4	13
Doc#4	16	8	8	7	4	16
Doc#5	24	12	14	9	6	8
Σ	96	58	68	43	26	56

5. Discussion

In the context of the "knowledge reengineering bottleneck", we introduced a semantic annotation pipeline to semi-automatically streamline the knowledge aggregation from publications within the domain of tribology. The inputs for the pipeline are publications of experimental investigations from the domain of tribology and in particular experiments of the category model test. The output is structured and linked data in form of json-files, which can be visualized as graphs (cf. Section 4.2). The pipeline is built on state-of-the-art language models and NLP techniques and was evaluated on five representative documents. Since NLP is not in common use within the domain of tribology, there are no datasets and standard documents for training and evaluating language models. This limits the significance of the performance test conducted within this contribution since a Gold Standard accepted by the community is missing and the pipeline cannot be compared to similar projects. However, as we work with standard language models, which are approved to be reliable within NLP communities and we conducted a first evaluation of our fine-tuned models by manually annotating five representative documents, some assertions can still be made about the current performance. Thereby, the document extraction (module 1) has shown reliable performance on different structured and formatted publications under the premise that the provided PDFs are not defect. This was substantiated by one tested PDF document, which contains invisible overlays and therefore shows high deviations from the GT in comparison to the other documents. The PDF extraction is always a critical step within NLP processes as it depends on the quality of the PDF and accessibility of the textual and other entities within the PDF. This is one reason for the modular structure of the pipeline. The PDF extraction is only required if the input publications are in the form of PDF format (which is a common format for textual documents). Since nowadays publications are frequently available online as well, the accessibility of textual data from HTML-Websites via an API is easier when the access is provided by publishers. Therefore,

PDF extraction is a pragmatic approach to access the textual information from publications. The annotation process (module 2) is performed using the SpanBERT language model, which shows remarkable high F1 scores. The NER model introduced within this publication is currently limited to publications on model tests (without claiming completeness), since those are well structured and mostly standardized. To our best knowledge, NER tagsets or language models itself as available for example in the domain of biomedicine (e.g., BioBERT) so far do not exist for the domain of tribology. In the future, the development and training of tribological language models can therefore improve the performance of applications in NLP within the domain of tribology. Furthermore, knowledge object generation is only a first step in named entity linking.

We discussed the role of knowledge objects within semantic knowledge bases within Section 2.3. The knowledge objects here are an aggregation of annotations from the document extracted by the pipeline. However, successful semantic knowledge sharing is usually community driven within a domain (e.g., OBO foundry). An established knowledge graph within the domain of tribology containing knowledge objects can therefore be extended with aggregated objects from the annotation module. Further established knowledge objects can also be enriched by the annotations. Thus, information about entities of interest in the domain of tribology can be semi-automatically acquired. Within the last module, we exploited QA to generate structured output from the unstructured and annotated data. The templates contain questions referring to tribAIn-ontology (e.g., questions about input parameters, the tribological system structure and output parameters). Overall, the QA system showed plausible answers to the tested question templates. During evaluation, we recognized a frequently appearing misconduct of the decision maker, which often could not differentiate the properties of the body and the counterbody. On the one hand, this can be attributed to an insufficient differentiation within the textual description and on the other hand to the question generation process within the QA system. The experiences with the QA module further led to two major perceptions in the context of extracting information from tribological publications. First, analyzing publications by a QA system can be exploited for a quality check and improvement of standardization of the description of experimental studies and outcomes. Thereby, question templates can be specified as a check list, what a sufficient description of experimental studies and results should contain to enable understanding and reproduction of the results. Second and relating thereto, the question templates itself have to be carefully designed to gain an answer and aggregate structured data from texts. Therefore, analyzing the publication practices and further the research practices of tribologists can give interesting insights for improving knowledge and data aggregation within the domain. However, the pipeline is intended to be human supervised, since trust is a critical issue especially within neural NLP processes which generate output without explanations of the process itself. This is the second reason for the modular architecture. The output from every module can be checked and adapted before continuing with the pipeline. This is especially important if automatic extraction is used to extend semantic knowledge bases or aggregate structured data for further processing. Besides the quality and trust of the results from the pipeline, another important issue is the computational costs. As mentioned within Section 4.3, the training of the language models took less than an hour (20 to 30 min) for each model. The low computational costs are due to the currently available pre-trained language models, which merely must be fine-tuned to be tailored to a specific domain. The execution of the annotation further only takes a few seconds. Therefore, the pipeline can be considered as very efficient compared to manual annotation.

6. Conclusions

Sharing knowledge in publications has a long tradition in scientific research since this is the elemental way of consuming and communicating information and knowledge by human scientists. Within the domain of tribology, the vast amount of available information overcomes the cognitive capacity of humans in terms of efficient aggregation and processing.

Therefore, AI provides a lot of potential to support scientists by handling this flood of information. Nevertheless, the way of knowledge sharing within tribology is still mainly based on publications and thus human focused. Descriptions in natural language are vague and insufficient from a formalization perspective. This phenomenon is due to an intended human consumer, for whom the provision of formal sufficient information would result in far too long publications and re-reading the same information over and over again. In contrast, machines need a formal and explicit model for aggregating and processing information. Therefore, if the available amount of information overwhelms the capacity of human processability, the question arises if we better should create representations for sharing information with an AI instead of humans in the future? The answer is: Not necessarily. As pointed out by Gruber [74], "the purpose of AI is to empower humans with machine intelligence". This is referred to as "humanistic AI", an artificial intelligence designed to meet human needs by collaborating with and augmenting people. In terms of an AI empowered tribological knowledge sharing, we introduced a semantic annotation pipeline towards generating knowledge graphs from natural language publications to bridge the gap between a human-understandable and a machine-processable knowledge representation. The pipeline is built upon state-of-the-art NLP methods and is inspired by similar challenges from the biomedical domain. Although we demonstrate the potential of the approach (NER and QA show reliable computational performance scores), further validation of the approach to ensure practical usability is recommended. This includes especially the definition of the specific objective of the extracted information, e.g., for trend studies or identifying research gaps and contradictions within the domain of tribology. Furthermore, the annotation model is currently limited to model tests and is not validated to suit the practical information needs of tribologists. Therefore, user studies for analyzing the capability of information extraction compared with human experts provide possibilities to improve the performance of the approach.

Author Contributions: Conceptualization, P.K. and M.M.; methodology, P.K. and R.D.; software, R.D.; validation, R.D. and P.K.; writing—original draft preparation, P.K. and M.M.; writing—review and editing, P.K., M.M., R.D., B.S. and S.W.; visualization, P.K. and M.M.; supervision, B.S. and S.W.; funding acquisition, S.W. All authors have read and agreed to the published version of the manuscript.

Funding: This research received no external funding.

Institutional Review Board Statement: Not applicable.

Informed Consent Statement: Not applicable.

Data Availability Statement: The code for the Semantic Annotation Pipeline is available under https://github.com/snow0815/TriboAnnotation.git, accessed on 14 December 2021.

Acknowledgments: P. Kügler, B. Schleich and S. Wartzack gratefully acknowledge the financial support of project WA 2913/22-2 within the Priority Program 1921 by the German Research Foundation (DFG). M. Marian acknowledges the support from Pontificia Universidad Católica de Chile.

Conflicts of Interest: The authors declare no conflict of interest.

References

1. Holmberg, K.; Erdemir, A. Influence of tribology on global energy consumption, costs and emissions. *Friction* **2017**, *3*, 263–284. [CrossRef]
2. Zhang, Z.; Yin, N.; Chen, S.; Liu, C. Tribo-informatics: Concept, architecture, and case study. *Friction* **2021**, *3*, 642–655. [CrossRef]
3. Rosenkranz, A.; Marian, M.; Profito, F.J.; Aragon, N.; Shah, R. The Use of Artificial Intelligence in Tribology—A Perspective. *Lubricants* **2021**, *1*, 2. [CrossRef]
4. Hasan, M.S.; Kordijazi, A.; Rohatgi, P.K.; Nosonovsky, M. Triboinformatics Approach for Friction and Wear Prediction of Al-Graphite Composites Using Machine Learning Methods. *J. Tribol.* **2022**, *144*, 011701. [CrossRef]
5. Subrahmanyam, M.; Sujatha, C. Using neural networks for the diagnosis of localized defects in ball bearings. *Tribol. Int.* **1997**, *10*, 739–752. [CrossRef]
6. Prost, J.; Cihak-Bayr, U.; Neacşu, I.A.; Grundtner, R.; Pirker, F.; Vorlaufer, G. Semi-Supervised Classification of the State of Operation in Self-Lubricating Journal Bearings Using a Random Forest Classifier. *Lubricants* **2021**, *5*, 50. [CrossRef]

7.   Sathiya, P.; Aravindan, S.; Haq, A.N.; Paneerselvam, K. Optimization of friction welding parameters using evolutionary computational techniques. *J. Mater. Process. Technol.* **2009**, *5*, 2576–2584. [CrossRef]

8.   Cetinel, H. The artificial neural network based prediction of friction properties of Al$_2$O$_3$-TiO$_2$ coatings. *Ind. Lubr. Tribol.* **2012**, *5*, 288–293. [CrossRef]

9.   Boidi, G.; Da Silva, M.R.; Profito, F.J.; Machado, I.F. Using Machine Learning Radial Basis Function (RBF) Method for Predicting Lubricated Friction on Textured and Porous Surfaces. *Surf. Topogr. Metrol. Prop.* **2020**, *4*, 44002. [CrossRef]

10.   Bhaumik, S.; Pathak, S.D.; Dey, S.; Datta, S. Artificial intelligence based design of multiple friction modifiers dispersed castor oil and evaluating its tribological properties. *Tribol. Int.* **2019**, *140*, 105813. [CrossRef]

11.   Sattari Baboukani, B.; Ye, Z.; Reyes, K.G.; Nalam, P.C. Prediction of Nanoscale Friction for Two-Dimensional Materials Using a Machine Learning Approach. *Tribol. Lett.* **2020**, *68*, 1–14. [CrossRef]

12.   Marian, M.; Tremmel, S. Current Trends and Applications of Machine Learning in Tribology—A Review. *Lubricants* **2021**, *9*, 86. [CrossRef]

13.   Stipčević, M.; Bowers, J.E. Spatio-temporal optical random number generator. *Opt. Express* **2015**, *9*, 11619–11631. [CrossRef]

14.   Vinoth, A.; Datta, S. Design of the ultrahigh molecular weight polyethylene composites with multiple nanoparticles: An artificial intelligence approach. *J. Compos. Mater.* **2020**, *2*, 179–192. [CrossRef]

15.   Bhaumik, S.; Mathew, B.R.; Datta, S. Computational intelligence-based design of lubricant with vegetable oil blend and various nano friction modifiers. *Fuel* **2019**, *241*, 733–743. [CrossRef]

16.   Feigenbaum, E.A. Knowledge Engineering: The Applied Side of Artificial Intelligence. *Ann. N. Y. Acad. Sci.* **1984**, *426*, 91–107. [CrossRef]

17.   Cullen, J.; Bryman, A. The knowledge acquisition bottleneck: Time for reassessment? *Expert Syst.* **1988**, *3*, 216–225. [CrossRef]

18.   Tallian, T.E. Tribological Design Decisions Using Computerized Databases. *J. Tribol.* **1987**, *3*, 381–386. [CrossRef]

19.   Tallian, T.E. A Computerized Expert System for Tribological Failure Diagnosis. *J. Tribol.* **1989**, *2*, 238–244. [CrossRef]

20.   Studer, R.; Benjamins, V.R.; Fensel, D. Knowledge engineering: Principles and methods. *Data Knowl. Eng.* **1998**, *1–2*, 161–197. [CrossRef]

21.   Morik, K. Underlying assumptions of knowledge acquisition and machine learning. *Knowl. Acquis.* **1991**, *2*, 137–156. [CrossRef]

22.   Hoekstra, R. The knowledge reengineering bottleneck. *Semant. Web* **2010**, *2*, 111–115. [CrossRef]

23.   Chandrasegaran, S.K.; Ramani, K.; Sriram, R.D.; Horváth, I.; Bernard, A.; Harik, R.F.; Gao, W. The evolution, challenges, and future of knowledge representation in product design systems. *Comput.-Aided Des.* **2013**, *2*, 204–228. [CrossRef]

24.   Verhagen, W.J.; Bermell-Garcia, P.; van Dijk, R.E.; Curran, R. A critical review of Knowledge-Based Engineering: An identification of research challenges. *Adv. Eng. Inform.* **2012**, *1*, 5–15. [CrossRef]

25.   Gruber, T.R. A translation approach to portable ontology specifications. *Knowl. Acquis.* **1993**, *2*, 199–220. [CrossRef]

26.   Bodenreider, O. Biomedical Ontologies in Action: Role in Knowledge Management, Data Integration and Decision Support. *Yearb. Med. Inform.* **2008**, *1*, 67–79. [CrossRef]

27.   The Gene Ontology Consortium. Creating the gene ontology resource: Design and implementation. *Genome Res.* **2001**, *8*, 1425–1433. [CrossRef]

28.   Esnaola Gonzalez, I.; Fernandez, I. Materials' Tribological Characterisation: An OntoCommons Use Case. In Proceedings of the ESWC Workshop DORIC-MM, Online, 7 June 2021.

29.   Kügler, P.; Marian, M.; Schleich, B.; Tremmel, S.; Wartzack, S. tribAIn—Towards an Explicit Specification of Shared Tribological Understanding. *Appl. Sci.* **2020**, *13*, 4421. [CrossRef]

30.   Auer, S.; Bizer, C.; Kobilarov, G.; Lehmann, J.; Cyganiak, R.; Ives, Z. *DBpedia: A Nucleus for a Web of Open Data*; ISWC 2007, ASWC 2007; Springer: Berlin/Heidelberg, Germany, 2007; pp. 722–735. [CrossRef]

31.   Giorgi, J.M.; Bader, G.D. Transfer learning for biomedical named entity recognition with neural networks. *Bioinformatics* **2018**, *23*, 4087–4094. [CrossRef]

32.   Lee, J.; Yoon, W.; Kim, S.; Kim, D.; Kim, S.; So, C.H.; Kang, J. BioBERT: A pre-trained biomedical language representation model for biomedical text mining. *Bioinformatics* **2020**, *4*, 1234–1240. [CrossRef]

33.   Devlin, J.; Chang, M.-W.; Lee, K.; Toutanova, K. *BERT: Pre-Training of Deep Bidirectional Transformers for Language Understanding*; NAACL-HLT; Association for Computational Linguistics: Stroudsburg, PA, USA, 2019; pp. 4171–4186. [CrossRef]

34.   Czichos, H.; Habig, K.-H. *Tribologie-Handbuch*, 4th ed.; Springer: Wiesbaden, Germany, 2015.

35.   Marian, M. *Numerische Auslegung von Oberflächenmikrotexturen für Geschmierte Tribologische Kontakte*; FAU University Press: Erlangen, Germany, 2021. [CrossRef]

36.   Vakis, A.I.; Yastrebov, V.A.; Scheibert, J.; Nicola, L.; Dini, D.; Minfray, C.; Almqvist, A.; Paggi, M.; Lee, S.; Limbert, G.; et al. Modeling and simulation in tribology across scales: An overview. *Tribol. Int.* **2018**, *125*, 169–199. [CrossRef]

37.   Soldatova, L.N.; King, R.D. An ontology of scientific experiments. *J. R. Soc. Interface* **2006**, *11*, 795–803. [CrossRef] [PubMed]

38.   W3C Recommendation. OWL 2 Web Ontology Language Overview. Available online: http://www.w3.org/TR/owl2-overview/ (accessed on 13 October 2021).

39.   Baader, F. *The description logic handbook: Theory, Implementation, and Applications*; Cambridge University Press: Cambridge, UK, 2003.

40.   W3C Recommendation. RDF 1.1 Turtle: Terse RDF Triple Language. Available online: https://www.w3.org/TR/turtle/ (accessed on 14 December 2021).

41. Gruber, T.R. Toward principles for the design of ontologies used for knowledge sharing? *Int. J. Hum.-Comput. Stud.* **1995**, 5–6, 907–928. [CrossRef]
42. Navigli, R.; Velardi, P. From Glossaries to Ontologies: Extracting Semantic Structure from Textual Definitions. In *Ontology Learning and Population. Bridging the Gap between Text and Knowledge*; Buitelaar, P., Cimiano, P., Eds.; IOS Press: Amsterdam, The Netherlands, 2008.
43. Marian, M.; Feile, K.; Rothammer, B.; Bartz, M.; Wartzack, S.; Seynstahl, A.; Tremmel, S.; Krauß, S.; Merle, B.; Böhm, T.; et al. Ti$_3$C$_2$T$_x$ solid lubricant coatings in rolling bearings with remarkable performance beyond state-of-the-art materials. *Appl. Mater. Today* **2021**, 25, 101202. [CrossRef]
44. Wyatt, B.C.; Rosenkranz, A.; Anasori, B. 2D MXenes: Tunable Mechanical and Tribological Properties. *Adv. Mater.* **2021**, 17, e2007973. [CrossRef] [PubMed]
45. Marian, M.; Berman, D.; Rota, A.; Jackson, R.L.; Rosenkranz, A. Layered 2D Nanomaterials to Tailor Friction and Wear in Machine Elements—A Review. *Adv. Mater. Interfaces* **2021**, 9, 2101622. [CrossRef]
46. Marian, M.; Tremmel, S.; Wartzack, S.; Song, G.; Wang, B.; Yu, J.; Rosenkranz, A. Mxene nanosheets as an emerging solid lubricant for machine elements–Towards increased energy efficiency and service life. *Appl. Surf. Sci.* **2020**, 523, 146503. [CrossRef]
47. Domingue, J. *Handbook of Semantic Web Technologies*; Springer: Berlin, Germany, 2011.
48. Goldberg, Y. Neural Network Methods for Natural Language Processing. *Synth. Lect. Hum. Lang. Technol.* **2017**, 1, 1–309. [CrossRef]
49. Jurafsky, D.; Martin, J.H. Speech and Language Processing (Draft-Version from Dec 2020). Available online: https://web.stanford.edu/~jjurafsky/slp3/ (accessed on 11 August 2021).
50. Hu, J. An Overview for Text Representations in NLP. Blogpost: Towards Data Science. Available online: https://towardsdatascience.com/an-overview-for-text-representations-in-nlp-311253730af1?gi=4f92ddbafc7d (accessed on 18 October 2021).
51. Schuster, M.; Nakajima, K. Japanese and Korean voice search. In Proceedings of the ICASSP 2012—2012 IEEE International Conference on Acoustics, Speech and Signal Processing, Kyoto, Japan, 25–30 March 2012; IEEE: New York, NY, USA; pp. 5149–5152. [CrossRef]
52. Mikolov, T.; Chen, K.; Corrado, G.; Dean, J. Efficient Estimation of Word Representations in Vector Space. *arXiv* **2013**, arXiv:1301.3781.
53. Mikolov, T.; Sutskever, I.; Chen, K.; Corrado, G.S.; Dean, J. Distributed Representations of Words and Phrases and Their Compositionality. *arXiv* **2013**, arXiv:1310.4546.
54. Koroteev, M.V. BERT: A Review of Applications in Natural Language Processing and Understanding. *arXiv* **2021**, arXiv:2103.11943.
55. Peters, M.; Neumann, M.; Iyyer, M.; Gardner, M.; Clark, C.; Lee, K.; Zettlemoyer, L. *Deep Contextualized Word Representations*; NAC, Association for Computational Linguistics: Stroudsburg, PA, USA, 2018; pp. 2227–2237. [CrossRef]
56. Radford, A.; Narasimhan, K.; Salimans, T.; Sutskever, I. Improving Language Understanding by Generative Pre-Training. Available online: https://s3-us-west-2.amazonaws.com/openai-assets/research-covers/language-unsupervised/language_understanding_paper.pdf (accessed on 14 December 2021).
57. Vaswani, A.; Shazeer, N.; Parmar, N.; Uszkoreit, J.; Jones, L.; Gomez, A.N.; Kaiser, L.; Polosukhin, I. Attention Is All You Need. *arXiv* **2017**, arXiv:1706.03762.
58. Ghelani, S. From Word Embeddings to Pretrained Language Models—A New Age in NLP-Part 2. Blogpost: Towards Data Science. Available online: https://towardsdatascience.com/from-word-embeddings-to-pretrained-language-models-a-new-age-in-nlp-part-2-e9af9a0bdcd9 (accessed on 7 October 2021).
59. Alammar, J. The Illustrated BERT, ELMo, and co. (How NLP Cracked Transfer Learning). Blogpost. Available online: http://jalammar.github.io/illustrated-bert/ (accessed on 14 December 2021).
60. Alammar, J. The Illustrated Transformer. Blogpost. Available online: http://jalammar.github.io/illustrated-transformer/ (accessed on 20 October 2021).
61. Beltagy, I.; Lo, K.; Cohan, A. SciBERT: A Pretrained Language Model for Scientific Text. In Proceedings of the 2019 Conference on Empirical Methods in Natural Language Processing and the 9th International Joint Conference on Natural Language Processing (EMNLP-IJCNLP), Hong Kong, China, 3–7 November 2019; Association for Computational Linguistics: Stroudsburg, PA, USA, 2019; pp. 3613–3618. [CrossRef]
62. Joshi, M.; Chen, D.; Liu, Y.; Weld, D.S.; Zettlemoyer, L.; Levy, O. SpanBERT: Improving Pre-training by Representing and Predicting Spans. *Trans. Assoc. Comput. Linguist.* **2020**, 8, 64–77. [CrossRef]
63. Russell, S.J.; Norvig, P. *Artificial Intelligence—A Modern Approach, 4. Auflage*; Pearson: Boston, MA, USA, 2021.
64. Cho, H.; Lee, H. Biomedical named entity recognition using deep neural networks with contextual information. *BMC Bioinform.* **2019**, 1, 735. [CrossRef]
65. McKie, J.X. PyMuPDF 1.19.0. GitHub Repository. Available online: https://github.com/pymupdf/PyMuPDF (accessed on 21 October 2021).
66. Akbik, A.; Bergmann, T.; Blythe, D.; Rasul, K.; Schweter, S.; Vollgraf, R. FLAIR: An easy-to-use framework for state-of-the-art NLP. In Proceedings of the 2019 Conference of the North American Chapter of the Association for Computational Linguistics (Demonstrations), Minneapolis, MN, USA, 2–7 June 2019; pp. 54–59. [CrossRef]
67. Marian, M.; Song, G.C.; Wang, B.; Fuenzalida, V.M.; Krauß, S.; Merle, B.; Tremmel, S.; Wartzack, S.; Yu, J.; Rosenkranz, A. Effective usage of 2D MXene nanosheets as solid lubricant–Influence of contact pressure and relative humidity. *Appl. Surf. Sci.* **2020**, 531, 147311. [CrossRef]

68. Mekgwe, G.N.; Akinribide, O.J.; Langa, T.; Obadele, B.A.; Olubambi, P.A.; Lethabane, L.M. Effect of graphite addition on the tribological properties of pure titanium carbonitride prepared by spark plasma sintering. *IOP Conf. Ser. Mater. Sci. Eng.* **2019**, *499*, 12011. [CrossRef]
69. Li, X.; Xu, J. Coordinating influence of multilayer graphene and spherical SnAgCu for improving tribological properties of a 20CrMnTi material. *RSC Adv.* **2018**, *25*, 14129–14137. [CrossRef]
70. Wang, Z.; Hu, S.; Feng, C.; Chen, E. The high temperature and varying temperature tribological performance of TiC coatings. In *IOP Conference Series: Materials Science and Engineering*; IOP Publishing: Bristol, UK, 2018; p. 22032. [CrossRef]
71. Byeong-Choon, G.; In-Sik, C. Microstructural Analysis and Wear Performance of Carbon-Fiber-Reinforced SiC Composite for Brake Pads. *Materials* **2017**, *7*, 701. [CrossRef]
72. Hinton, G.E.; Srivastava, N.; Krizhevsky, A.; Sutskever, I.; Salakhutdinov, R.R. Improving neural networks by preventing co-adaptation of feature detectors. *arXiv* **2012**, arXiv:1207.0580.
73. Srivastava, N.; Hinton, G.; Krizhevsky, A.; Sutskever, I.; Salakhutdinov, R. Dropout: A Simple Way to Prevent Neural Networks from Overfitting. *J. Mach. Learn. Res.* **2014**, *1*, 1929–1958.
74. Gruber, T.R. How AI Can Enhance Our Memory, Work and Social Lives. TED Talk. 2017. Available online: https://www.ted.com/talks/tom_gruber_how_ai_can_enhance_our_memory_work_and_social_lives (accessed on 14 December 2021).

 *lubricants*

Article

Design of Amorphous Carbon Coatings Using Gaussian Processes and Advanced Data Visualization

Christopher Sauer *, Benedict Rothammer, Nicolai Pottin, Marcel Bartz, Benjamin Schleich and Sandro Wartzack

Engineering Design, Friedrich-Alexander-Universität Erlangen-Nürnberg, Martensstraße 9,
91058 Erlangen, Germany; rothammer@mfk.fau.de (B.R.); nicolai.pottin@fau.de (N.P.); bartz@mfk.fau.de (M.B.);
schleich@mfk.fau.de (B.S.); wartzack@mfk.fau.de (S.W.)
* Correspondence: sauer@mfk.fau.de

Abstract: In recent years, an increasing number of machine learning applications in tribology and coating design have been reported. Motivated by this, this contribution highlights the use of Gaussian processes for the prediction of the resulting coating characteristics to enhance the design of amorphous carbon coatings. In this regard, by using Gaussian process regression (GPR) models, a visualization of the process map of available coating design is created. The training of the GPR models is based on the experimental results of a centrally composed full factorial 2^3 experimental design for the deposition of a-C:H coatings on medical UHMWPE. In addition, different supervised machine learning (ML) models, such as Polynomial Regression (PR), Support Vector Machines (SVM) and Neural Networks (NN) are trained. All models are then used to predict the resulting indentation hardness of a complete statistical experimental design using the Box–Behnken design. The results are finally compared, with the GPR being of superior performance. The performance of the overall approach, in terms of quality and quantity of predictions as well as in terms of usage in visualization, is demonstrated using an initial dataset of 10 characterized amorphous carbon coatings on UHMWPE.

Keywords: machine learning; amorphous carbon coatings; UHWMPE; total knee replacement; Gaussian processes

Citation: Sauer, C.; Rothammer, B.; Pottin, N.; Bartz, M.; Schleich, B.; Wartzack, S. Design of Amorphous Carbon Coatings Using Gaussian Processes and Advanced Data Visualization. *Lubricants* **2022**, *10*, 22. https://doi.org/10.3390/lubricants10020022

Received: 14 December 2021
Accepted: 3 February 2022
Published: 7 February 2022

Publisher's Note: MDPI stays neutral with regard to jurisdictional claims in published maps and institutional affiliations.

Copyright: © 2022 by the authors. Licensee MDPI, Basel, Switzerland. This article is an open access article distributed under the terms and conditions of the Creative Commons Attribution (CC BY) license (https://creativecommons.org/licenses/by/4.0/).

1. Introduction

Machine Learning (ML) as a subfield of artificial intelligence (AI) has become an integral part of many areas of public life and research in recent years. ML is used to create learning systems that are considerably more powerful than rule-based algorithms and are thus predestined for problems with unclear solution strategies and a high number of variants. ML algorithms are used from product development and production [1] to patient diagnosis and therapy [2]. ML algorithms are also playing an increasingly important role in the field of medical technology, for example, in coatings for joint replacements.

Particularly in coating technology and design, the use of ML algorithms enables the identification of complex relationships between several deposition process parameters on the process itself as well as on the properties of the resulting coatings [3,4]. From this view on the complex relationships between the deposition process parameters, coating designers can base their experiments and obtain valuable insights on their coating designs and the necessary parameter settings for coating deposition.

This contribution looks into the application of a possible ML algorithm in the coating design of amorphous carbon coatings. It first provides an overview of the necessary experimental setup for data generation and the concept of machine learning and its algorithms. Likewise, the deposition of amorphous carbon coatings and their properties are presented. Subsequently, the capabilities of the selected supervised ML algorithms: Polynomial Regression (PR), Support Vector Machines (SVM), Neural Networks (NN), Gaussian Process Regression (GPR) are explained and the resulting data visualization is shown. Afterwards, the obtained results are discussed, with the GPR being the superior prediction model. Finally, the main findings are summarized and an outlook is given as well as further potentials and applications are identified.

2. Related Work and Main Research Questions

2.1. Amorphous Carbon Coating Design

An example of a complex process is the coating of metal and plastic parts, as used for joint replacements, with amorphous carbon coatings [5]. In the field of machine elements [6,7], engine components [8,9] and tools [10,11], amorphous carbon coatings are commonly used. In contrast, amorphous carbon coatings are rarely used for load-bearing, tribologically stressed implants [12,13]. The coating of engine and machine elements has so far been used with the primary aim of reducing friction, whereas the coating of forming tools has been used to adjust friction while increasing the service life of the tools. Therefore, the application of tribologically effective coating systems on the articulating implant surfaces is a promising approach to reduce wear and friction [14–16].

The coating process depends on many different coating process parameters, such as the bias voltage [17], the target power [18], the gas flow [19] or the temperature, which influence the chemical and mechanical properties as well as the tribological behavior of the resulting coatings [20]. Therefore, it is vital to ensure both the required coating properties and a robust and reproducible coating process to meet the high requirements for medical devices. Compared to experience-based parameter settings, which are often based on trial-and-error, ML algorithms provide clearer and more structured correlations.

However, several experimental investigations focus on improving the tribological effectiveness of joint replacements [21–23] and lubrication conditions in prostheses [24–26], some experimental investigations are complemented with computer-aided or computational methods to improve the prediction and findings [27–29]. Nevertheless, the exact interactions of coating process parameters and resulting properties are mostly qualitative and only valid for certain coating plants and in certain parameter ranges.

2.2. Coating Process and Design Parameters

The use of ML algorithms is a promising approach [30] to not only qualitatively describe such interactions, which have to be determined in elaborate experiments, but also to quantify them [21]. Using ML, the aim is to generate reproducible, robust coating processes with appropriate, required coating properties. For this purpose, the main coating properties, such as coating thickness, roughness, adhesion, hardness and indentation modulus, of the coating parameter variations have to be analyzed and trained with suitable ML algorithms [31].

Within this contribution, the indentation modulus and the coating hardness are examined in more detail, since these parameters can be determined and reproduced with high accuracy and have a relatively high predictive value for the subsequent tribological behavior, such as the resistance to abrasive wear [32,33].

2.3. Research Questions

Resulting from the above-mentioned considerations it was found that existing solutions are solely based on a trial-and-error approach. ML was not considered in the specific coating design in joint replacements. So, in brief, this contribution wants to answer the following central questions. The first one is can ML algorithms predict resulting properties in amorphous carbon coatings? Based on this, the second one is how good is the resulting prediction of resulting properties in terms of quality and quantity? And lastly, can ML support in visualizing the coating properties results and the coating deposition parameters leading to those results? When ML can be used in these cases, the main advantages would be a more efficient approach to coating design with fewer to none trial-and-error steps and, lastly, the co-design of coating experts and ML. The following sections are to present the materials and methods used in trying to answer the stated research questions and provide an outlook on what would be possible via ML.

3. Materials and Methods

First, the studied materials and methods will be described briefly. In this context, the application of the amorphous carbon coating to the materials used (UHMWPE) as well as the setup and procedure of the experimental tests to determine the mechanical properties (hardness and elasticity) are described. Secondly, the pipeline for ML and the used methods are explained. Finally, the programming language Python and the deployed toolkits are described.

3.1. Experimental Setup

3.1.1. Materials

The investigated substrate was medical UHMWPE [34] (Chirulen® GUR 1020, Mitsubishi Chemical Advanced Materials, Vreden, Germany). The specimens to be coated were flat disks, which have been used for mechanical characterization (see [35]). The UHMWPE disks had a diameter of 45 mm and a height of 8 mm. Before coating, the specimens were mirror-polished in a multistage polishing process (Saphir 550-Rubin 520, ATM Qness, Mammelzen, Germany) and cleaned with ultrasound (Sonorex Super RK 255 H 160 W 35 Hz, Bandelin electronic, Berlin, Germany) in isopropyl alcohol.

3.1.2. Coating Deposition

Monolayer a-C:H coatings were prepared on UHMWPE under two-fold rotation using an industrial-scale coating equipment (TT 300 K4, H-O-T Härte- und Oberflächentechnik, Nuremberg, Germany) for physical vapor deposition and plasma-enhanced chemical vapor deposition (PVD/PECVD). The recipient was evacuated to a base pressure of at least 5.0×10^{-4} Pa before actual deposition. The recipient was not preheated before deposition on UHMWPE to avoid the deposition-related heat flux into UHMWPE. The specimens were then cleaned and activated for 2 min in an argon (Ar, purity 99.999%)$^+$-ion plasma with a bipolar pulsed bias of -350 V and an Ar flow of 450 sccm. The deposition time of 290 min was set to achieve a resulting a-C:H coating thickness of approximately 1.5 to 2.0 μm. Using reactive PVD, the a-C:H coating was deposited by medium frequency (MF)-unbalanced magnetron (UBM) sputtering of a graphite (C, purity 99.998%) target under Ar–ethyne (C_2H_2) atmosphere (C_2H_2, purity 99.5%). During this process, the cathode (dimensions 170×267.5 mm) was operated with bipolar pulsed voltages. The negative pulse amplitudes correspond to the voltage setpoints, whereas the positive pulses were represented by 15% of the voltage setpoints. The pulse frequency f of 75 kHz was set with a reverse recovery time RRT of 3 μs. A negative direct current (DC) bias voltage was used for all deposition processes. The process temperature was kept below 65 °C during the deposition of a-C:H functional coatings on UHMWPE. In Table 1, the main, varied deposition process parameters are summarized. Besides the reference coating (Ref), the different coating variations (C1 to C9) of a centrally composed full factorial 2^3 experimental design are presented in randomized run order. In this context, the deposition process parameters shown here for the generation of different coatings represent the basis for the machine learning process.

Table 1. Summary of the main deposition process parameters for a-C:H on UHMWPE.

Designation	Coating	Sputtering Power/kW	Bias Voltage/V	Combined Ar and C_2H_2 Flow/sccm
Ref		0.6	−130	187
C1		0.6	−90	187
C2		2.0	−170	91
C3		1.3	−130	133
C4		2.0	−90	187
C5	a-C:H	0.6	−170	91
C6		2.0	−170	187
C7		0.6	−170	187
C8		0.6	−90	91
C9		2.0	−90	91

3.1.3. Mechanical Characterization

According to [36,37], the indentation modulus E_{IT} and the indentation hardness H_{IT} were determined by nanoindentation with Vickers tips (Picodentor HM500 and WinHCU, Helmut Fischer, Sindelfingen, Germany). For minimizing substrate influences, care was taken to ensure that the maximum indentation depth was considerably less than 10% of the coating thicknesses [38,39]. Considering the surface roughness, lower forces also proved suitable to obtain reproducible results. Appropriate distances of more than 40 µm were maintained between individual indentations. For statistical reasons, 10 indentations per specimen were performed and evaluated. A value for Poisson's ratio typical for amorphous carbon coatings was assumed to determine the elastic–plastic parameters [40,41]. The corresponding settings and parameters are shown in Table 2. In Section 3, the results of nanoindentation are presented and discussed.

Table 2. Settings for determining the indentation modulus E_{IT} and the indentation hardness H_{IT}.

Parameters	Settings for a-C:H Coatings
Maximum load/mN	0.05
Application time/s	3
Delay time after lowering/s	30
Poisson's ratio ν	0.3

3.2. Machine Learning and Used Models

3.2.1. Supervised Learning

The goal of machine learning is to derive relationships, patterns and regularities from data sets [42]. These relationships can then be applied to new, unknown data and problems to make predictions. ML algorithms can be divided into three subclasses: supervised, unsupervised and reinforced learning. In the following, only the class of supervised learning will be discussed in more detail, since algorithms from this subcategory were used in this paper, namely Gaussian process regression (GPR). Supervised ML was used because of the available labelled data.

In supervised learning, the system is fed classified training examples. In this data, the input values are already associated with known output data values. This can be done, for example, by an already performed series of measurements with certain input parameters (input) and the respective measured values (output). The goal of supervised learning is to train the model or the algorithms using the known data in such a way that statements and predictions can also be made about unknown test data [42]. Due to the already classified data, supervised learning represents the safest form of machine learning and is therefore very well suited for optimization tasks [42].

In the field of supervised learning, one can distinguish between the two problem types of classification and regression. In a classification problem, the algorithm must divide the data into discrete classes or categories. In contrast, in a regression problem, the model is to estimate the parameters of pre-defined functional relationships between multiple features in the data sets [42,43].

A fundamental danger with supervised learning methods is that the model learns the training data by role and thus learns the pure data points rather than the correlations in data. As a result, the model can no longer react adequately to new, unknown data values. This phenomenon is called overfitting and must be avoided by choosing appropriate training parameters [31]. In the following, basic algorithms of supervised learning are presented, ranging from PR and SVM to NN and GPR.

3.2.2. Polynomial Regression

At first, we want to introduce polynomial regression (PR) for supervised learning. PR is a special case of linear regression and tries to predict data with a polynomial regression curve. The parameters of the model are often fitted using a least square estimator and the overall approach is applied to various problems, especially in the engineering domain. A basic PR model can lead to the following equation [44]:

$$y_i = \beta_0 + \beta_1 x_{i1} + \beta_2 x_{i2} + \ldots + \beta_k x_{ik} + e_i \text{ for } i = 1, 2, \ldots, n \tag{1}$$

with β being the regression parameters and e being the error values. The prediction targets are formulated as y_i and the features used for prediction are described as x_i. A more sophisticated technique based on regression models are support vector machines, which are described in the next section.

3.2.3. Support Vector Machines

Originally, support vector machines (SVM) are a model commonly used for classification tasks, but the ideas of SVM can be extended to regression as well. SVM try to find higher order planes within the parameter space to describe the underlying data [45]. Thereby, SVM are very effective in higher dimensional spaces and make use of kernel functions for prediction. SVM are widely used and can be applied to a variety of problems. In this regard, SVM can also be applied nonlinear problems. For a more detailed theoretical insight, we refer to [45].

3.2.4. Neural Networks

Another supervised ML technique is neural networks (NN), which rely on the concept of the human brain to build interconnected multilayer perceptrons (MLP) capable of predicting arbitrary feature–target correlations. The basic building block of such MLP are neurons based on activation functions which allow the neuron to fire when different threshold values are reached [46]. When training a NN, the connections and the parameters of those activation functions are optimized to minimize training errors; this process is called backpropagation [31].

3.2.5. Gaussian Process Regression

The Gaussian processes are supervised generic learning methods, which were developed to solve regression and classification problems [43]. While classical regression algorithms apply a polynomial with a given degree or special models like the ones mentioned above, GPR uses input data more subtly [47]. Here, the Gaussian process theoretically generates an infinite number of approximation curves to approximate the training data points as accurately as possible. These curves are assigned probabilities and Gaussian normal distributions, respectively. Finally, the curve which fits its probability distribution best to that of the training data is selected. In this way, the input data gain significantly more influence on the model, since in the GPR altogether fewer parameters are fixed in advance than in the classical regression algorithms [47]. However, the behavior of the different GPR models can be defined via kernels. This can be used, for example, to influence how the model should handle outliers and how finely the data should be approximated.

In Figure 1, two different GPR models have been used to approximate a sinusoid. The input data points are sinusoidal but contain some outliers. The model with the lightblue approximation curve has an additional kernel extension for noise suppression compared to the darkblue model. Therefore, the lightblue model is less sensitive to outliers and has a smoother approximation curve. This is also the main advantage when using GPR compared to other regression models like linear or polynomial regression. GPR are more robust to outliers or messy data and are also relatively stable on small datasets [47] like the one used for this contribution. That is why they were mainly selected for the later-described use case.

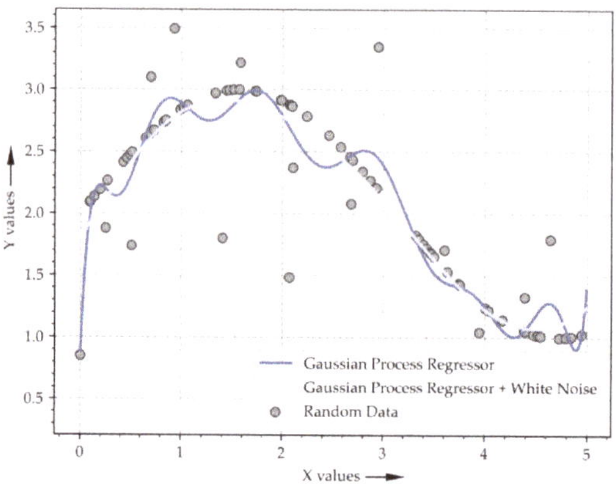

Figure 1. Gaussian process regression for the regression of a sinusoid.

3.2.6. Python

The Python programming language was chosen for the present work, as it is the de-facto standard language for ML and Data Science. This programming environment is particularly suitable in the field of machine learning, as it allows the easy integration of external libraries. In order to use machine learning algorithms in practice, many libraries and environments have been developed in the meantime. One of them is the open-source Python library scikit-learn [48]. For the above-described methods, the following scikit-learn libraries were used: the scikit-learn module Polynomial Features for the modeling of the PR models, which was combined with the Linear Regression module to facilitate a PR model for prediction of coating parameters. For modeling via SVM, the SVR or support vector regressor module of scikit-learn was used. The NN were modeled via the MLP Regressor module and lastly the GPR were implemented using the Gaussian Process Regressor module of scikit-learn. All models were trained using the standard parameters, and only for the GPR model was the kernel function smoothed via adding some white noise; this was necessary because the GPR of scikit-learn has no real standard parameters.

4. Use Case with Practical Example in a-C:H Coating Design

4.1. Data Generation

The average indentation modulus and indentation hardness values are presented in Figure 2 Obviously, elasticity and hardness differed significantly between the various coated groups. A considerable influence of the sputtering power on the achieved E_{IT} and H_{IT} values was revealed. For example, C2, C4, C6 and C9, which were produced with a sputtering power of 2.0 kW, had indentation modulus between 13.3 and 16.4 GPa and indentation hardness between 3.7 and 5.1 GPa. In contrast, specimens Ref, C1, C5, C7 and C8 exhibited significantly lower E_{IT} and H_{IT} values, ranging from 3.6 to 4.9 GPa and 1.2 to 1.5 GPa, respectively. Compared to the latter, the central point represented by C3 did not indicate significantly higher elastic–plastic values. The variation of the bias voltage or the combined gas flow did not allow us to derive a distinct trend, especially concerning the standard deviation. In general, increased sputtering power could increase E_{IT} and H_{IT} by more than a factor of three. Accordingly, the higher coating hardness is expected to shield the substrates from adhesive and abrasive wear and also to shift the cracking towards higher stresses [28,35]. At the same time, the relatively lower indentation modulus leads to an increased ability of the coatings to sag without flowing [33]. As a result, the pressures induced by tribological loading may be reduced by increasing the contact dimensions [49]. Thus, it can be considered that the developed a-C:H coatings enable a very advantageous wear behavior [28,50].

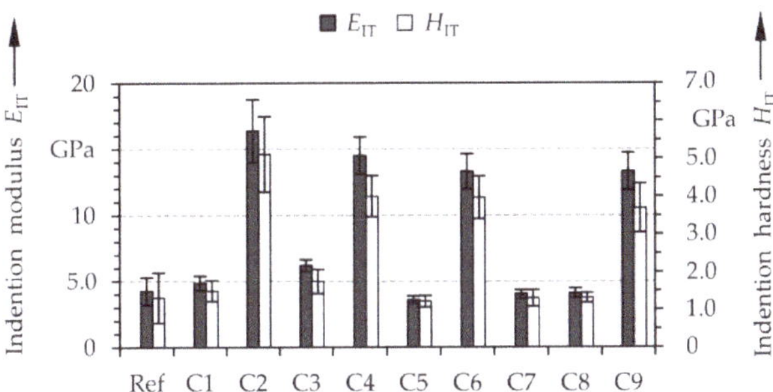

Figure 2. Averaged values of indentation modulus E_{IT} and indentation hardness H_{IT} and standard deviation of the different a-C:H coatings ($n = 10$).

4.2. Data Processing

4.2.1. Reading in and Preparing Data

After the coating characterization, the measured values were available in a standardized Excel dataset, which contains the plant parameters and the resulting coating characteristics for each sample. It could also be possible that the relevant measurements are already in a machine-readable format, for example the tribAIn ontology [51], but for our case we focused on the data handling via Excel and Python. To facilitate the import of the data into Python, the dataset had to be modified in such a way that a column-by-column import of the data was possible. Afterwards, the dataset needed to be imported into our Python program via the pandas library [52]. To facilitate further data processing, the plant parameters sputtering power, bias voltage and combined Ar and C_2H_2 were combined in an array of features and the coating characteristic such as the indentation hardness as a target for prediction.

4.2.2. Model Instantiation

The class Gaussian Process Regressor (GPR) of the scikit-learn package class allows the implementation of Gaussian process models. For the instantiation in particular, a definition of a kernel was needed. This kernel is also called covariance function in connection with Gaussian processes and influences the probability distributions of the Gaussian processes decisively. The main task of the kernel is to calculate the covariance of the Gaussian process between the individual data points. Two GPR objects were instantiated with two different kernels. The first one was created with a standard kernel and the second one was additionally linked with a white noise kernel. During the later model training, the hyperparameters of the kernel were optimized. Due to possibly occurring local maxima, the passing parameter n_restarts_optimizer can be used to determine how often this optimization process should be run. In the case of GPR, a standardization of the data was carried out. This standardization was achieved by scaling the data mean to 0 and the standard deviation to 1.

4.2.3. Training the Model

As described before, one of the main tasks of machine learning algorithms was the training of the model. The scikit-learn environment offers the function fit(X,y), with the input variables X and y. Here, X was the feature vector, which contains the feature data of the test data set (the control variables of the coating plant). The variable y was defined as the target vector and contains the target data of the test data set (the characteristic values of the coating characterization). By calling the method reg_model.fit(X,y) with the available data and the selected regression model (GPR, in general reg_model) the model was trained and fitted on the available data.

Particularly with small datasets, there was the problem that the dataset shrank even further when the data was divided into training and test data. For this reason, the k-fold cross-validation approach could be used [31]. Here, the training data set was split into k smaller sets, with one set being retained as a test data set per training run. In the following runs, the set distributions change. This approach can be used to obtain more training datasets despite small datasets, thus significantly improving the training performance of the model.

4.2.4. Model Predictions

After the models were trained on the available data, the models can compute or predict corresponding target values for the feature variables that were previously unknown to the model. Unknown feature values are equally distributed data points from a specified interval as well as the features of a test data set. For the former, the minima and maxima of the feature values of the training data set were extracted. Afterwards, equally distributed data points were generated for each feature in this min-max interval.

For predicting the targets, the scikit-learn library provides the method predict(x), where the feature variables are passed as a vector x to the function. Calling the method reg_model.predict(x) then returns the corresponding predicted target values. The predictions for the test data were further evaluated in terms of the root mean squared error, the mean absolute error and the coefficient of prognosis (CoP) [53] and showed good quality, especially for the GPR model (see Table 3).

Table 3. Prediction quality of the models based on the initial dataset.

Model	Root Mean Squared Error	Mean Absolute Error	Coefficient of Prognosis
Gaussian Process Regressor	540 MPa	474 MPa	91%
Polynomial Regression	699 MPa	653 MPa	45%
Support Vector Machine	955 MPa	677 MPa	29%
Neural Network	3405 MPa	3307 MPa	16%

From Table 3, it follows that the GPR model is the most suitable model for further evaluation in our test case since it shows the highest coefficient of prognosis. Therefore, we selected the GPR model for the demonstration and visualization of our use case.

4.2.5. Visualization

The Python library matplotlib was used to visualize the data in Python. This allowed an uncomplicated presentation of numerical data in 2D or 3D. Since the feature vector contained three variables (sputter power P sputter, gas flow φ and bias voltage U_{bias}), a three-dimensional presentation of the feature space was particularly suitable. Here, the three variables were plotted on the x-, y- and z-axis and the measurement points were placed in this coordinate system. For the presentation of the corresponding numerical target value, color-coding serves as the fourth dimension. The target value of the measuring point could then be inferred from a color bar.

This presentation was especially suitable for small data sets, e.g., to get an overview of the actual position of the training data points. For large data sets with several thousand data points, a pure 3D visualization is too confusing, since measurement points inside the feature space were no longer visible. For this reason, a different visualization method was used to display the results of ML prediction of uniformly distributed data.

This visualization method is based on the visualization of computer tomography (CT) data set using a slice-based data view. Here, the 3D images of the body are skipped through layer-by-layer to gain insights into the interior of the workings level-by-level. Similar to this principle, the feature space was also traversed layer-by-layer.

Two feature variables span a 2D coordinate system. The measured values were again colored and displayed in the x–y plane analogous to the 3D display.

The third feature vector served as a run variable in the z-axis, i.e., into the plane. Employing a slider, the z-axis can be traversed, and the view of the feature space was then obtained layer-by-layer.

5. Results and Discussion

5.1. Gaussian Process Regression and Visualization

For the above-described initial dataset created from a design of experiments approach, different GPR models were trained. Before training the different models, the dataset was scaled to only contain values between 0 and 1. This was especially useful for GPR, to reduce training effort and stabilize the optimization of the model parameters. The main difference between the different GPR models was the used kernel function for the gaussian processes. The used GPR supports a variety of different kernel functions which were optimized during the training of the GPR model. It was found that with a dot product kernel with some additional white noise the prediction capabilities of the model were enhanced to reach a mean absolute error of around 440 MPa. Moreover, the root mean squared error was around 387 MPa. This results in an CoP of around 90%, which means that the prediction quality and quantity is acceptable to classify this model for a prediction model. For model training, a train-test-split of 80–20% was used and the training data was shuffled before training. The overall prediction quality is a notable finding since the dataset used for training is relatively small. Here also GPR with little white noise show their strengths on sparse datasets. However, model performance can further benefit from more data. This prediction model is also capable of visualizing the prediction space, see Figure 3.

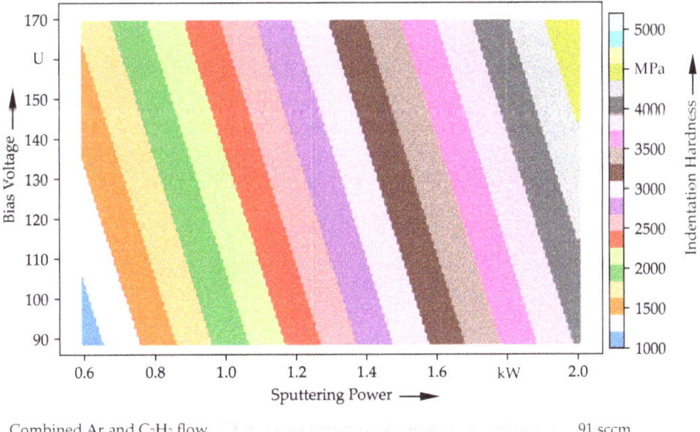

Figure 3. Predicted space in a 20-color colormap for better differentiation between the different areas of resulting hardness for minimum combined gas flow.

The striped pattern emerges from the usage of a 20-color-based colormap for drawing. This is done to further show the different sections of the predicted data. The whole plot can be viewed as a process map. In order to find the ideal coating properties, the tribology experts need to look for their color in indentation hardness and then easily see the bias voltage and sputtering power needed. For tuning purposes, the gas flow can be changed via the slider at the bottom. The plot for the maximum combined gas flow is depicted in Figure 4.

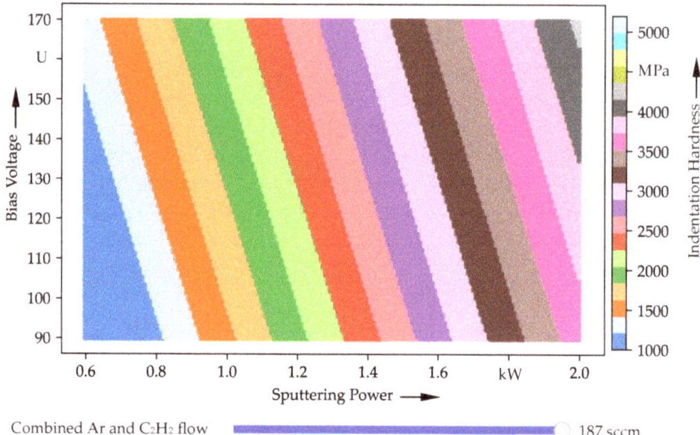

Figure 4. Predicted space for maximum combined gas flow.

The space for lower indentation hardness is getting bigger and the highest indentation hardness of around 4.2 GPa vanished. This correlates with the experience made from initial experimental studies. It was expected that the gas flow—especially the C_2H_2 gas flow [15]—influenced the hydrogen content and thus the mechanical properties and further affected the tribologically effective behavior. Based on these visualizations, it can be easily seen which parameters lead to the desired indentation hardness. This visualization technique benefits the process of where to look for promising parameter sets for ideal indentation hardness.

For validation of our model, we performed another experimental design study based on a Box–Behnken design with 3 factors and two stages (see Table 4). Initially, the indentation hardness was predicted using our GPR model. Subsequently, the GPR model was evaluated—after coating the specimens—by determining the indentation hardness experimentally. For illustrative purposes, the prediction of the central point, which was deposited at a sputtering power of 3 kW, a bias voltage of 200 V, and a combined gas flow of 108 sccm, is shown in Figure 5. In this context, it should be noted that the prediction space included a significant extension of the training space and thus could be influenced by many factors.

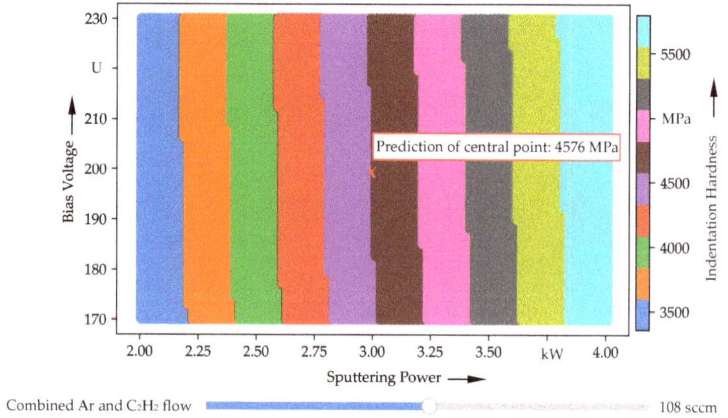

Figure 5. Predicted extended space for probe points.

Table 4. Summary of main deposition process parameters and predictions for a-C:H on UHMWPE, prediction of H_{IT} by the GPR model as well as experimental determination of H_{IT} based on the average values and standard deviations of the different a-C:H coatings ($n = 10$).

Designation	Coating	Sputtering Power/kW	Bias Voltage/V	Combined Ar and C_2H_2 Flow/Sccm	GPR Model Prediction of H_{IT}/MPa	Experimentally Determined H_{IT}/MPa
P1		2	−230	108	3397	3040 ± 223
P2		2	−170	108	3355	2441 ± 537
P3		2	−200	125	3363	3069 ± 401
P4		2	−200	91	3389	2965 ± 328
P5.1		3	−200	108	4576	4699 ± 557
P5.2		3	−200	108	4576	4577 ± 731
P5.3		3	−200	108	4576	4837 ± 634
P6	a-C:H	3	−170	125	4542	4180 ± 399
P7		3	−170	91	4568	4256 ± 622
P8		3	−230	125	4584	4627 ± 1055
P9		3	−230	91	4610	4415 ± 675
P10		4	−170	108	5755	5081 ± 1361
P11		4	−200	91	5789	5476 ± 1637
P12		4	−230	108	5797	4313 ± 1513
P13		4	−200	125	5763	6224 ± 1159

As shown in Figure 5 and Table 4, the H_{IT} values of the previously performed prediction of the GPR model largely coincided with the experimentally determined H_{IT} values. Especially with regard to the standard deviation of the experimentally determined H_{IT} values, all values were in a well-usable range for further usage and processing of the data. Despite a similar training space, the prediction for the coating variations P1–P4 showed a slightly lower accuracy than for the coating variations beyond the training space, but this could be attributed to the difficulty of determining the substrate-corrected coating hardness. Thus, during the indentation tests, the distinct influence of the softer UHMWPE substrate [54,55] was more pronounced for the softer coatings (P1–P4), which were coated with lower target power than for the harder coatings (P5–P13). However, the standard deviation of the hardness values increased with hardness, which could be attributed to increasing coating defects locations and roughness. In brief, the predictions match with the implicit knowledge of the coating experts. This is the only physical conceivable conceptual model that can be considered when looking at the results presented, as the coating deposition is a complex and multi-scale process.

Though the visualization of the prediction space in Figure 5 differed slightly from the prediction spaces in Figure 3 and in Figure 4 due to steeper dividing lines, the prediction space in Figure 5 spanned larger coating process parameter dimensions.

Generally, the prediction quality and especially the quantity of the model was very good, so the model can be used for further coating development processes and adjustments of the corresponding coating process parameters. An extension of the GPR model to other coating types, such as ceramic coatings, e.g., CrN, or solid lubricants, e.g., MoS_2, or different coating systems on various substrates is conceivable.

5.2. Comparison to Polynomial Regression, Support Vector Machines and Neural Network Models

For the purpose of comparing our results and trained models with the other models described previously, Table 5 shows the different predictions generated by the models for the previously unknown dataset in our test study.

Table 5. Comparison of the predictions of the different models used in this contribution.

Designation	Experimentally Determined H_{IT}/MPa	GPR Model Prediction of H_{IT}/MPa	PR Model Prediction of H_{IT}/MPa	SVM Model Prediction of H_{IT}/MPa	NN Model Prediction of H_{IT}/MPa
P1	3040 ± 223	3397	3861	4220	1402
P2	2441 ± 537	3355	3566	4218	1153
P3	3069 ± 401	3363	3567	4219	1346
P4	2965 ± 328	3389	4101	4219	1209
P5.1	4699 ± 557	4576	4863	4219	1281
P5.2	4577 ± 731	4576	4863	4219	1281
P5.3	4837 ± 634	4576	4863	4219	1281
P6	4180 ± 399	4542	4548	4217	1225
P7	4256 ± 622	4568	4850	4218	1088
P8	4627 ± 1055	4584	5063	4219	1475
P9	4415 ± 675	4610	5376	4220	1336
P10	5081 ± 1361	5755	4912	4218	1160
P11	5476 ± 1637	5789	5447	4219	1216
P12	4313 ± 1513	5797	5659	4220	1409
P13	6224 ± 1159	5763	5366	4219	1354

It is shown that only the GPR model is capable of producing meaningful outputs, while the other models are not able to achieve a prediction quality close to the GPR model. When comparing the training results on root mean squared error, mean absolute error and coefficient of prognosis set, the story becomes even more clearer (see Table 6).

Table 6. Comparison of the prediction qualities of the models on the unknown data set.

Model	Root Mean Squared Error	Mean Absolute Error	Coefficient of Prognosis
Gaussian Process Regressor	551 MPa	415 MPa	78%
Polynomial Regression	720 MPa	587 MPa	71%
Support Vector Machine	991 MPa	781 MPa	0.1%
Neural Network	3156 MPa	2999 MPa	1%

The results show that the GPR model was the best model compared to PR, SVM and NN. It is worth noting that we have used polynomial degree of 2 for the PR models, as this produced the best prediction results, a higher polynomial degree of 3 to 5 led to a decrease in RMSE, MAE and CoP. This also shows that especially the SVM and NN are not capable of producing meaningful prediction output. The PR fitting overall shows acceptable prediction quality of around 70%, however the GPR has better RMSE and MAE values, so it would be selected for further consideration. Furthermore, GPR provided better results on the training dataset. It is important to always evaluate RMSE, MAE and CoP together, as all three values allow a thorough evaluation of the prediction model. In brief, RMSE and MAE characterize the spread predictions better than the CoP, the CoP returns an overall performance score of the model. The weak performance of SVM can possibly be explained by the small dataset used for training, since SVM need way more training data, as the model only scores around 30% CoP on the training dataset. For extrapolation on the test dataset the trained SVM model was not feasible. The same could be the case for the NN, as NN rely on big datasets for training and show weaker extrapolation capabilities.

6. Conclusions

This contribution evaluated the use of Gaussian processes and advanced data visualization in the design of amorphous carbon coatings on UHMWPE. This study focused on elaborating an overview of the required experimental setup for data generation and the concepts of ML, and also provided the corresponding ML algorithms. Afterwards, the deposition and characterization of amorphous carbon coatings were presented.

The use of ML in coating technology and tribology represents a very promising approach for the selective optimization of coating process parameters and coating properties. In particular, this could be demonstrated by the GPR models used to optimize the mechanical properties of the coatings and, consequently, the tribological behavior, by increasing the hardness and thus the abrasive wear resistance. However, further experimental studies and parameter tuning are needed to obtain better predictive models and better process maps. The initial results of these visualizations and the GPR models provide a good basis for further studies. For our approach the following conclusions could be drawn:

- The GPR models and the materials used showed the potentials of the selected ML algorithms. One data visualization method using the GPR was detailed;
- The usage of ML looked very promising in this case, which can benefit the area of ML in coating technology and tribology. The prediction accuracy of the hardness values with our approach showed a high agreement with the experimentally determined hardness values;
- The used data visualization (see Figures 3 and 4) is a neat feature for coating process experts to tune their parameters into the desired parameter space. The plotted process maps can further enhance the coating design or other coating types.

For our use case we implemented a four-step process, mainly consisting of data generation via design of experiments to create the initial dataset. This initial dataset was then analyzed via Python-based scripting tools, to create meaningful prediction models via GPR. Those GPR models are then used for the presented visualization approach. To put it all together one Python script was created to lead through the process. This Python script can be configured to look into different values, however we focused on indentation hardness.

Based on this work, further experimental studies will be conducted, and the proposed models will then be re-trained using the available data. The dataset generated for this article was considered as a starting point for the ML algorithms used and will be supplemented with future experimental data and thus grow. When more data is available, maybe different ML models like neural networks will come into perspective.

Author Contributions: Conceptualization, C.S. and B.R.; methodology, C.S. and B.R.; specimen preparation D.R., coating deposition, B.R.; mechanical experiments, B.R.; data analysis, C.S., B.R. and N.P.; writing—original draft preparation, C.S. and B.R.; writing—review and editing, N.P., M.B., B.S. and S.W.; visualization, C.S. and B.R.; supervision, M.B., B.S. and S.W. All authors have read and agreed to the published version of the manuscript.

Funding: We acknowledge financial support by Deutsche Forschungsgemeinschaft and Friedrich-Alexander-Universität Erlangen-Nürnberg within the funding programme "Open Access Publication Funding".

Institutional Review Board Statement: Not applicable.

Informed Consent Statement: Not applicable.

Data Availability Statement: For tinkering with the visualization a similar model can be found under http://csmfk.pythonanywhere.com/ (accessed on 1 February 2022). The related data generated and analyzed for the contribution is available from the corresponding author on request.

Conflicts of Interest: The authors declare no conflict of interest.

References

1. Bertolini, M.; Mezzogori, D.; Neroni, M.; Zammori, F. Machine Learning for industrial applications: A comprehensive literature review. *Expert Syst. Appl.* **2021**, *175*, 114820. [CrossRef]
2. Lynch, C.J.; Liston, C. New machine-learning technologies for computer-aided diagnosis. *Nat. Med.* **2018**, *24*, 1304–1305. [CrossRef]
3. Marian, M.; Tremmel, S. Current Trends and Applications of Machine Learning in Tribology—A Review. *Lubricants* **2021**, *9*, 86. [CrossRef]

4. Caro, M.A.; Csányi, G.; Laurila, T.; Deringer, V.L. Machine learning driven simulated deposition of carbon films: From low-density to diamondlike amorphous carbon. *Phys. Rev. B* **2020**, *102*, 174201. [CrossRef]

5. Shah, R.; Gashi, B.; Hoque, S.; Marian, M.; Rosenkranz, A. Enhancing mechanical and biomedical properties of protheses—Surface and material design. *Surf. Interfaces* **2021**, *27*, 101498. [CrossRef]

6. Kröner, J.; Kursawe, S.; Musayev, Y.; Tremmel, S. Analysing the Tribological Behaviour of DLC-Coated Dry-Running Deep Groove Ball Bearings with Regard to the Ball Material. *Appl. Mech. Mater.* **2016**, *856*, 143–150. [CrossRef]

7. Khadem, M.; Penkov, O.V.; Yang, H.-K.; Kim, D.-E. Tribology of multilayer coatings for wear reduction: A review. *Friction* **2017**, *5*, 248–262. [CrossRef]

8. Marian, M.; Weikert, T.; Tremmel, S. On Friction Reduction by Surface Modifications in the TEHL Cam/Tappet-Contact-Experimental and Numerical Studies. *Coatings* **2019**, *9*, 843. [CrossRef]

9. Liu, K.; Kang, J.; Zhang, G.; Lu, Z.; Yue, W. Effect of temperature and mating pair on tribological properties of DLC and GLC coatings under high pressure lubricated by MoDTC and ZDDP. *Friction* **2020**, *9*, 1390–1405. [CrossRef]

10. Häfner, T.; Rothammer, B.; Tenner, J.; Krachenfels, K.; Merklein, M.; Tremmel, S.; Schmidt, M. Adaption of tribological behavior of a-C:H coatings for application in dry deep drawing. *MATEC Web Conf.* **2018**, *190*, 14002. [CrossRef]

11. Krachenfels, K.; Rothammer, B.; Zhao, R.; Tremmel, S.; Merklein, M. Influence of varying sheet material properties on dry deep drawing process. *IOP Conf. Ser. Mater. Sci. Eng.* **2019**, *651*, 012012. [CrossRef]

12. Hauert, R.; Thorwarth, K.; Thorwarth, G. An overview on diamond-like carbon coatings in medical applications. *Surf. Coat. Technol.* **2013**, *233*, 119–130. [CrossRef]

13. Hauert, R. A review of modified DLC coatings for biological applications. *Diam. Relat. Mater.* **2003**, *12*, 583–589. [CrossRef]

14. McGeough, J.A. *The Engineering of Human Joint Replacements*; John Wiley & Sons Ltd.: Chichester, UK, 2013; ISBN 978-1-118-53684-1.

15. Döring, J.; Crackau, M.; Nestler, C.; Welzel, F.; Bertrand, J.; Lohmann, C.H. Characteristics of different cathodic arc deposition coatings on CoCrMo for biomedical applications. *J. Mech. Behav. Biomed. Mater.* **2019**, *97*, 212–221. [CrossRef] [PubMed]

16. Dorner-Reisel, A.; Gärtner, G.; Reisel, G.; Irmer, G. Diamond-like carbon films for polyethylene femoral parts: Raman and FT-IR spectroscopy before and after incubation in simulated body liquid. *Anal. Bioanal. Chem.* **2017**, *390*, 1487–1493. [CrossRef]

17. Wang, L.; Li, L.; Kuang, X. Effect of substrate bias on microstructure and mechanical properties of WC-DLC coatings deposited by HiPIMS. *Surf. Coat. Technol.* **2018**, *352*, 33–41. [CrossRef]

18. Bociaga, D.; Sobczyk-Guzenda, A.; Szymanski, W.; Jedrzejczak, A.; Jastrzebska, A.; Olejnik, A.; Jastrzębski, K. Mechanical properties, chemical analysis and evaluation of antimicrobial response of Si-DLC coatings fabricated on AISI 316 LVM substrate by a multi-target DC-RF magnetron sputtering method for potential biomedical applications. *Appl. Surf. Sci.* **2017**, *417*, 23–33. [CrossRef]

19. Bobzin, K.; Bagcivan, N.; Theiß, S.; Weiß, R.; Depner, U.; Troßmann, T.; Ellermeier, J.; Oechsner, M. Behavior of DLC coated low-alloy steel under tribological and corrosive load: Effect of top layer and interlayer variation. *Surf. Coat. Technol.* **2013**, *215*, 110–118. [CrossRef]

20. Hetzner, H.; Schmid, C.; Tremmel, S.; Durst, K.; Wartzack, S. Empirical-Statistical Study on the Relationship between Deposition Parameters, Process Variables, Deposition Rate and Mechanical Properties of a-C:H:W Coatings. *Coatings* **2014**, *4*, 772–795. [CrossRef]

21. Kretzer, J.P.; Jakubowitz, E.; Reinders, J.; Lietz, E.; Moradi, B.; Hofmann, K.; Sonntag, R. Wear analysis of unicondylar mobile bearing and fixed bearing knee systems: A knee simulator study. *Acta Biomater.* **2011**, *7*, 710–715. [CrossRef]

22. Polster, V.; Fischer, S.; Steffens, J.; Morlock, M.M.; Kaddick, C. Experimental validation of the abrasive wear stage of the gross taper failure mechanism in total hip arthroplasty. *Med. Eng. Phys.* **2021**, *95*, 25–29. [CrossRef] [PubMed]

23. Ruggiero, A.; Zhang, H. Editorial: Biotribology and Biotribocorrosion Properties of Implantable Biomaterials. *Front. Mech. Eng.* **2020**, *6*, 17. [CrossRef]

24. Rufaqua, R.; Vrbka, M.; Choudhury, D.; Hemzal, D.; Křupka, I.; Hartl, M. A systematic review on correlation between biochemical and mechanical processes of lubricant film formation in joint replacement of the last 10 years. *Lubr. Sci.* **2019**, *31*, 85–101. [CrossRef]

25. Rothammer, B.; Marian, M.; Rummel, F.; Schroeder, S.; Uhler, M.; Kretzer, J.P.; Tremmel, S.; Wartzack, S. Rheological behavior of an artificial synovial fluid—Influence of temperature, shear rate and pressure. *J. Mech. Behav. Biomed. Mater.* **2021**, *115*, 104278. [CrossRef]

26. Nečas, D.; Vrbka, M.; Marian, M.; Rothammer, B.; Tremmel, S.; Wartzack, S.; Galandáková, A.; Gallo, J.; Wimmer, M.A.; Křupka, I.; et al. Towards the understanding of lubrication mechanisms in total knee replacements—Part I: Experimental investigations. *Tribol. Int.* **2021**, *156*, 106874. [CrossRef]

27. Gao, L.; Hua, Z.; Hewson, R.; Andersen, M.S.; Jin, Z. Elastohydrodynamic lubrication and wear modelling of the knee joint replacements with surface topography. *Biosurface Biotribol.* **2018**, *4*, 18–23. [CrossRef]

28. Rothammer, B.; Marian, M.; Neusser, K.; Bartz, M.; Böhm, T.; Krauß, S.; Schroeder, S.; Uhler, M.; Thiele, S.; Merle, B.; et al. Amorphous Carbon Coatings for Total Knee Replacements—Part II: Tribological Behavior. *Polymers* **2021**, *13*, 1880. [CrossRef]

29. Ruggiero, A.; Sicilia, A. A Mixed Elasto-Hydrodynamic Lubrication Model for Wear Calculation in Artificial Hip Joints. *Lubricants* **2020**, *8*, 72. [CrossRef]

30. Rosenkranz, A.; Marian, M.; Profito, F.J.; Aragon, N.; Shah, R. The Use of Artificial Intelligence in Tribology—A Perspective. *Lubricants* **2020**, *9*, 2. [CrossRef]

31. Witten, I.H.; Frank, E.; Hall, M.; Pal, C. *Data Mining: Practical Machine Learning Tools and Techniques*, 4th ed.; Morgan Kaufmann: Burlington, MA, USA, 2017. [CrossRef]

32. Fontaine, J.; Donnet, C.; Erdemir, A. Fundamentals of the Tribology of DLC Coatings. In *Tribology of Diamond-Like Carbon Films*; Springer: Boston, MA, USA, 2020; pp. 139–154. [CrossRef]

33. Leyland, A.; Matthews, A. On the significance of the H/E ratio in wear control: A nanocomposite coating approach to optimised tribological behaviour. *Wear* **2000**, *246*, 1–11. [CrossRef]

34. *ISO 5834-2:2019*; Implants for Surgery—Ultra-High-Molecular-Weight Polyethylene—Part 2: Moulded Forms. ISO: Geneva, Switzerland, 2019.

35. Rothammer, B.; Neusser, K.; Marian, M.; Bartz, M.; Krauß, S.; Böhm, T.; Thiele, S.; Merle, B.; Detsch, R.; Wartzack, S. Amorphous Carbon Coatings for Total Knee Replacements—Part I: Deposition, Cytocompatibility, Chemical and Mechanical Properties. *Polymers* **2021**, *13*, 1952. [CrossRef] [PubMed]

36. Oliver, W.C.; Pharr, G.M. An improved technique for determining hardness and elastic modulus using load and displacement sensing indentation experiments. *J. Mater. Res.* **1992**, *7*, 1564–1583. [CrossRef]

37. Oliver, W.C.; Pharr, G.M. Measurement of hardness and elastic modulus by instrumented indentation: Advances in understanding and refinements to methodology. *J. Mater. Res.* **2004**, *19*, 3–20. [CrossRef]

38. *DIN EN ISO 14577-1:2015-11*; Metallic Materials—Instrumented Indentation Test for Hardness and Materials Parameters—Part 1: Test Method. DIN: Berlin, Germany, 2015.

39. *DIN EN ISO 14577-4:2017-04*; Metallic Materials—Instrumented Indentation Test for Hardness and Materials Parameters—Part 4: Test Method for Metallic and Non-Metallic Coatings. DIN: Berlin, Germany, 2017.

40. Jiang, X.; Reichelt, K.; Stritzker, B. The hardness and Young's modulus of amorphous hydrogenated carbon and silicon films measured with an ultralow load indenter. *J. Appl. Phys.* **1989**, *66*, 5805–5808. [CrossRef]

41. Cho, S.-J.; Lee, K.-R.; Eun, K.Y.; Hahn, J.H.; Ko, D.-H. Determination of elastic modulus and Poisson's ratio of diamond-like carbon films. *Thin Solid Films* **1999**, *341*, 207–210. [CrossRef]

42. Müller, A.C.; Guido, S. *Introduction to Machine Learning with Python: A Guide for Data Scientists*; O'Reilly: Beijing, China, 2017; ISBN 9781449369880.

43. Williams, C.K.; Rasmussen, C.E. *Gaussian Processes for Machine Learning*; Adaptive Computation and Machine Learning; MIT Press: Cambridge, MA, USA, 2003; Volume 2, p. 4, ISBN 026218253X.

44. Ostertagová, E. Modelling using Polynomial Regression. *Procedia Eng.* **2012**, *48*, 500–506. [CrossRef]

45. Smola, A.J.; Schölkopf, B. A tutorial on support vector regression. *Stat. Comput.* **2004**, *14*, 199–222. [CrossRef]

46. Hinton, G.E. 20—Connectionist Learning Procedures. In *Machine Learning: An Artificial Intelligence Approach*; Elsevier: Amsterdam, The Netherlands, 1990; Volume III. [CrossRef]

47. Ebden, M. Gaussian Processes: A Quick Introduction. *arXiv* **2015**, arXiv:1505.02965. Available online: https://arxiv.org/abs/1505.02965 (accessed on 1 February 2022).

48. Pedregosa, F.; Varoquaux, G.; Gramfort, A.; Michel, V.; Thirion, B.; Grisel, O.; Blondel, M.; Prettenhofer, P.; Weiss, R.; Dubourg, V.; et al. Scikit-Learn: Machine Learning in Python. *JMLR* **2011**, *12*, 2825–2830.

49. Weikert, T.; Wartzack, S.; Baloglu, M.V.; Willner, K.; Gabel, S.; Merle, B.; Pineda, F.; Walczak, M.; Marian, M.; Rosenkranz, A.; et al. Evaluation of the surface fatigue behavior of amorphous carbon coatings through cyclic nanoindentation. *Surf. Coat. Technol.* **2021**, *407*, 126769. [CrossRef]

50. Rothammer, B.; Weikert, T.; Tremmel, S.; Wartzack, S. Tribologisches Verhalten amorpher Kohlenstoffschichten auf Metallen für die Knie-Totalendoprothetik. *Tribol. Schmierungstech.* **2019**, *66*, 15–24. [CrossRef]

51. Kügler, P.; Marian, M.; Schleich, B.; Tremmel, S.; Wartzack, S. tribAIn—Towards an Explicit Specification of Shared Tribological Understanding. *Appl. Sci.* **2020**, *10*, 4421. [CrossRef]

52. McKinney, W. Data structures for statistical computing in Python. In Proceedings of the 9th Python in Science Conference (SCIPY 2010), Austin, TX, USA, 28 June–3 July 2010; pp. 51–56.

53. Most, T.; Will, J. Metamodel of Optimal Prognosis—An automatic approach for variable reduction and optimal metamodel selection. *Proc. Weimar. Optim. Stochastiktage* **2008**, *5*, 20–21.

54. Poliakov, V.P.; Siqueira, C.J.d.M.; Veiga, W.; Hümmelgen, I.A.; Lepienski, C.M.; Kirpilenko, G.G.; Dechandt, S.T. Physical and tribological properties of hard amorphous DLC films deposited on different substrates. *Diam. Relat. Mater.* **2004**, *13*, 1511–1515. [CrossRef]

55. Martínez-Nogués, V.; Medel, F.J.; Mariscal, M.D.; Endrino, J.L.; Krzanowski, J.; Yubero, F.; Puértolas, J.A. Tribological performance of DLC coatings on UHMWPE. *J. Phys. Conf. Ser.* **2010**, *252*, 012006. [CrossRef]

Article

Using Machine Learning Methods for Predicting Cage Performance Criteria in an Angular Contact Ball Bearing

Sebastian Schwarz [1,*], Hannes Grillenberger [2], Oliver Graf-Goller [2], Marcel Bartz [1], Stephan Tremmel [3] and Sandro Wartzack [1]

[1] Department of Mechanical Engineering, Friedrich-Alexander-Universität Erlangen-Nürnberg (FAU), Engineering Design, Martensstraße 9, 91058 Erlangen, Germany; bartz@mfk.fau.de (M.B.); wartzack@mfk.fau.de (S.W.)

[2] Schaeffler Technologies AG & Co. KG, Industriestraße 1-3, 91074 Herzogenaurach, Germany; hannes.grillenberger@schaeffler.com (H.G.); Oliver.Graf-Goller@schaeffler.com (O.G.-G.)

[3] Faculty of Engineering, Universität Bayreuth, Engineering Design and CAD, Universitätsstr. 30, 95447 Bayreuth, Germany; stephan.tremmel@uni-bayreuth.de

* Correspondence: schwarz@mfk.fau.de

Abstract: Rolling bearings have to meet the highest requirements in terms of guidance accuracy, energy efficiency, and dynamics. An important factor influencing these performance criteria is the cage, which has different effects on the bearing dynamics depending on the cage's geometry and bearing load. Dynamics simulations can be used to calculate cage dynamics, which exhibit high agreement with the real cage motion, but are time-consuming and complex. In this paper, machine learning algorithms were used for the first time to predict physical cage related performance criteria in an angular contact ball bearing. The time-efficient prediction of the machine learning algorithms enables an estimation of the dynamic behavior of a cage for a given load condition of the bearing within a short time. To create a database for machine learning, a simulation study consisting of 2000 calculations was performed to calculate the dynamics of different cages in a ball bearing for several load conditions. Performance criteria for assessing the cage dynamics and frictional behavior of the bearing were derived from the calculation results. These performance criteria were predicted by machine learning algorithms considering bearing load and cage geometry. The predictions for a total of 10 target variables reached a coefficient of determination of $R^2 \approx 0.94$ for the randomly selected test data sets, demonstrating high accuracy of the models.

Keywords: rolling bearing dynamics; cage instability; regression; machine learning; neural networks; random forest; gradient boosting; evolutionary algorithms

Citation: Schwarz, S.; Grillenberger, H.; Graf-Goller, O.; Bartz, M.; Tremmel, S.; Wartzack, S. Using Machine Learning Methods for Predicting Cage Performance Criteria in an Angular Contact Ball Bearing. *Lubricants* **2022**, *10*, 25. https://doi.org/10.3390/lubricants10020025

Received: 14 December 2021
Accepted: 4 February 2022
Published: 11 February 2022

Publisher's Note: MDPI stays neutral with regard to jurisdictional claims in published maps and institutional affiliations.

Copyright: © 2022 by the authors. Licensee MDPI, Basel, Switzerland. This article is an open access article distributed under the terms and conditions of the Creative Commons Attribution (CC BY) license (https://creativecommons.org/licenses/by/4.0/).

1. Introduction

The use of dynamics and noise behavior as criteria to assess the performance of a rolling bearing are coming into increasing focus besides the lifetime and energy efficiency. In addition to potentially negative health consequences of noise pollution [1], one reason for this is the increasing electrification of passenger cars and the associated sensitivity regarding disturbing and unpleasant noise of all machine elements contained in the technical system [2]. Besides unpleasant noise caused by bearing dynamics, in precision applications such as the bearing assembly of the main spindle of machine tools, vibration of the bearing can lead to a negative influence on manufacturing accuracy [3].

The vibrations emitted by a rolling bearing may have various causes. Due to the rotation of the rolling element set, the force transmitting points between the inner and outer ring differ. This leads to a changing stiffness and to unavoidable vibrations of the rolling bearing caused by the design itself and is known as variable compliance [4]. The characteristics of these vibrations differ depending on the rolling bearing type (geometry, number of rolling elements, and pitch diameter) and load conditions (operating contact angle and load zone). In addition to the geometry-related causes of vibrations in rolling bearings, production-related geometric deviations of the bearing rings or rolling elements, such as roughness, waviness, or surface damage (scratches and inclusions in the material), influence the radial displacement of the rings and can cause undesired vibrations [5]. Thus, depending on the frequency of vibration occurring, isolated surface deviations can be assigned to the inner or outer bearing part based on the respective ball-pass frequency [6].

The cage of the rolling bearing can also be a source of vibrations and noise. An example are highly dynamic cage movements, which are called "cage rattling" or "cage instability" in the literature and are associated with strong noise generation [7–9]. The normal and frictional forces at the guiding surfaces accelerate the cage, so that certain operating conditions lead to a high-frequency motion and severe deformation of the cage [10,11]. These cage dynamics lead to a sharp increase in frictional torque [8,9,12] and temperature in the rolling bearing [8] and can have a negative effect on cage life due to severe deformations and component stresses. Cage dynamics depend on many influencing factors; an overview of previous research papers is provided in Table 1.

Table 1. Influencing parameters on the cage dynamics that have been investigated in research papers.

Group	Parameter
Bearing and cage properties	Internal clearance [13] Rolling element size [13] Rolling element profile [13] Pocket clearance [13–15] Guidance clearance [14,15] Pocket shape [9]
Bearing load	Load ratio [14,16] Rotational speed [7,13–16] External vibrations [8]
Friction	Coefficient cage/rolling elements [7,14,15,17] Coefficient cage/raceway [17] Rolling element/raceway traction [18]
Lubrication	Viscosity [8,19] Temperature [8,19] Oil injection [8]

The dynamic behavior of the cage depends on the bearing and cage properties as well as the operating conditions of the bearing. As the cage is (besides the rib contact) accelerated by the rolling elements contact, the dynamic behavior of the rolling elements has an influence on the cage motion. The kinematics of the rolling elements is affected by various factors, such as the bearing load and speed, the friction in the contact to the raceway, the rolling element geometry and the bearing clearance. However, these parameters are determined depending on the intended application with focus on bearing lifetime and accuracy of shaft guidance. The influence of the bearing design and load on the cage dynamics during the application is not usually in the focus in the bearing selection. Therefore, the cage dynamics must be adjusted by adapting the cage geometry in the available design space of the selected bearing. By varying the cage geometry, properties such as the pocket and guidance clearance, the mass inertia and stiffness, and the shape of the cage pocket are affected. By defining the cage properties, the dynamics can be adjusted, for example, to avoid unstable cage movements or to minimize the friction loss caused by the cage as well as the robustness against shock loads.

The influencing parameters on the resulting cage dynamics can be named in general, but the quantification of their effects is only partially known so far. There are two primary reasons for this. First, the calculation using numerical computer simulations or the measurement of the cage dynamics (motion, forces, or deformation) on a test rig are time consuming and complicated. In particular for experimental tests, the range of influencing parameters that can be investigated is usually limited. Second, the interaction of the influencing variables is complex, so that it is not possible to determine the influence of the individual effects directly on the basis of the observed dynamics. Cage instability, for example, is caused by high frictional forces in the cage contacts and high rotational speeds of the bearing [14]. If one of the two parameters is low, the probability that highly dynamic movements will be excited is reduced. In addition to this example, other interactions can be found, making it more difficult to determine the cage dynamics depending of the influencing factors such as cage geometry and bearing load.

Machine learning methods are suitable for identifying complex patterns and relationships in the data provided. The application of machine learning algorithms in the field of tribological problems is increasing, especially in recent years. A comprehensive overview of the use of machine learning for tribological problems was provided by Marian and Tremmel [20]. Based on experimental test results, calculations, or information collected from the literature, regression methods are used to predict typical tribological behavior in the form of temperature, specific wear, or coefficient of friction. In addition to applications at the nano or micro scale, machine learning methods are also used at the macro scale, such as in bearing technology. Schwarz et al. used an ensemble classification model to determine the dependence between geometric parameters and load of a rolling bearing and the resulting dynamics of a cage. The result of the classification was one of the classes "unstable", "stable", or "circling" that were used to assess the qualitative behavior of the cage [21]. By extending this approach with a regression algorithm, not only the cage motion class but also the resulting forces on the cage or the acceleration of the cage can be estimated.

In previous research investigations [21], it was possible to quantify the dynamics of the cage for different operating conditions, but this was usually completed in isolated cases within the framework of complex numerical calculations or tests. A method for the time-efficient estimation of the quantitative dynamic behavior of rolling bearing cages for certain cage properties and rolling bearing loads is not yet available. The aim of this paper is to present a procedure for predicting the dynamics of a rolling bearing cage in an angular contact ball bearing using dynamics simulations and regression machine learning algorithms. This enables time-efficient estimation of the dynamics for the intended application during the development and selection of rolling bearing cages and also for operating conditions that are not directly included in the training data.

2. Materials and Methods

2.1. Methodology

The application of machine learning regression methods to predict the dynamics of a rolling bearing cage requires data representing the correlation between the varied parameters and the calculated cage dynamics. The starting point was the multi-body simulation model defined in the software Caba3D [22,23]. The calculation parameters of the model such as initial and boundary conditions, friction models, and elastic modeling of the cage are described in Section 2.2. The geometry of the cage as well as the bearing load and rotational speed were modified with the help of a comprehensive simulation plan using the design of experiment, see Section 2.3. A Latin hypercube sampling was used to ensure that the varied parameters are distributed uniformly in the entire mathematical space defined by previously specified boundaries [24]. The limits of the simulation plan were chosen so that the operating conditions prevailing in reality are mainly covered. On the basis of the uniformly distributed parameter values in the simulation plan, the correlations between the parameters can be efficiently learned by the algorithm. After performing the calculations, the simulation results were used to determine the input and output parameters and thus the data sets for machine learning, see Section 2.4. Characteristic values such as the Cage Dynamics Indicator (CDI) defined by Schwarz et al. [21] were derived from the calculated time series, which can be used for the assessment of the cage dynamics and as target values for machine learning. Artificial neural networks (ANN) [25], random forest (RF) [26], and XGBoost [27] were applied to predict the target variables based on the varied calculation parameters, see Section 2.5. The optimization of the hyperparameters of the used algorithms was performed as part of the training process using an evolutionary algorithm (EA) [28]. Finally, the predictions of the optimized models for test data sets were compared so that the most suitable algorithm could be selected. Figure 1 illustrates the procedure for generating a regression model for the prediction of characteristic values representing cage dynamics.

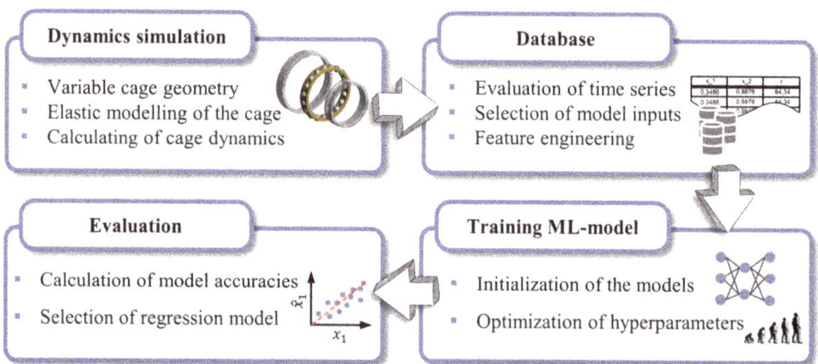

Figure 1. Procedure in this study, starting with the dynamics simulation and the design of experiments, followed by the creation of a database, the training of the machine learning models, and analysis of the predictions.

2.2. Calculation of Bearing Cage Dynamics

The multi-body simulation software Caba3D [29] developed by SCHAEFFLER Technologies AG & Co. KG was used to determine the rolling bearing dynamics. This tool allows the calculation of the dynamics of all rolling bearing components for a previously defined time step and simulation time using a Runge–Kutta-method for performing the numerical time step integration. The results of the multi-body simulation include the kinematics (position, velocity, and acceleration in all degrees of freedom) of the rolling bearing components as well as contact results (pressures, relative velocities, etc.) and node displacements of the elastically modeled cage [23].

The discretization of the contacts rolling element/raceway as well as cage/ring was achieved by means of slices. Contact results such as pressure or forces were calculated for each of the slices and thus resolved locally [22]. For the contact calculation between rolling element and cage pocket, the 'node-to-surface model' was used. This approach determines the contact results using the surface nodes of the finite element (FE) model of the cage and the slices of the rolling element. This allowed the elastic deformations of the cage and their effects on the contact conditions to be determined during the calculation [22]. An elastohydrodynamic model with consideration of mixed friction and the surface roughness was used for the calculation of the friction between rolling elements and raceways. The lubricant film thickness was calculated according to Dowson–Higginson [30]. Coulomb's friction law was used to calculate the frictional force in the contact between the cage and the other bearing elements.

The calculation of the node displacements of a FE model of the rolling bearing cage with several thousand degrees of freedom would be too computationally intensive in the context of a multi-body simulation. Therefore, a model order reduction according to Craig and Bampton [31] was performed to consider the node displacements of the FE model during the dynamics simulation. This allows the number of degrees of freedom to be significantly reduced without a meaningful degradation in accuracy [29]. For the reduction in the FE model, eigenfrequencies up to 20 kHz and a maximum of 100 eigenmodes were considered. The deviation of the eigenfrequencies from the original model and thus the quality of the reduced FE model was verified using various quality criteria (e.g., modal assurance criteria and normalized relative eigenfrequency difference [29,32]).

The modeling considered the angular contact ball bearing without adjacent machine elements consisting of two bearing raceways, the rolling elements, and the outer ring guided cage, see Figure 2.

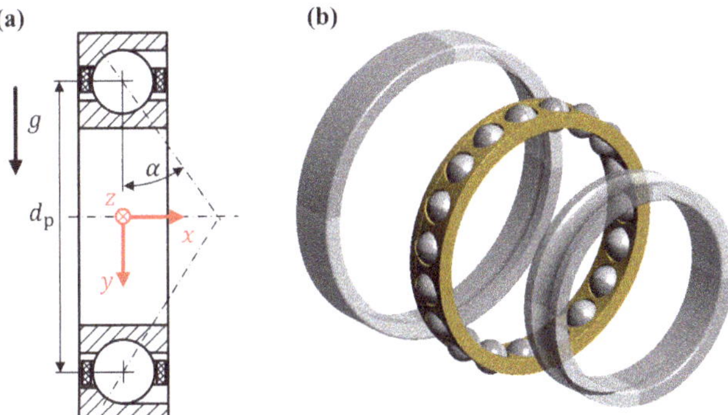

Figure 2. (**a**) Cross-section of the angular contact ball bearing as typically used in machine tools. (**b**) Exploded view of the three-dimensional dynamics simulation model consisting of two raceways, 19 rolling elements, and the outer ring-guided window cage.

The degrees of freedom of the outer ring were disabled, while the other rolling bearing components could move along all six degrees of freedom. The angular contact ball bearing was loaded axially (F_x) and radially (F_y) by a force on the inner ring. Further parameters important for the calculation can be found in Table 2.

Table 2. Calculation information of the simulation model.

Group	Property	Value
Integration	Integration method	Runge–Kutta
	Output time step	0.0001 s
	Calculation time	1.0 s
Cage properties	Cage guidance type	rib
	Cage material	fibre reinforced phenolic resin
Bearing properties	Inner diameter	90 mm
	Outer diameter	140 mm
	Pitch diameter d_p	115 mm
	Rolling elements	19 balls
	Rolling element diameter	15.875 mm
	Ring and ball material	100 Cr6
	Contact angle α	15°
	Static load capacity $C_{0,\mathrm{r}}$	51 kN

The data-driven approach to employ machine learning methods for cage dynamics prediction requires a high-quality set of data. The source of the data is the multi-body simulation software Caba3D, for which a high correlation with the real cage motion has already been found several times [10,14,21] and is therefore considered as a reliable source for the generation of datasets. Schwarz et al. used a test rig specially developed for testing cages of rolling bearings and high-speed cameras for optical measurement of cage dynamics. As in the calculations, cage instability could be observed in the experiment. For the shape, amplitude, and frequency of the cage deformation, high agreement was found with the measurement results [21].

2.3. Simulation Plan for the Cage Geometry and Bearing Load

Twenty cage variants were generated for the angular contact ball bearing, and their dynamic behavior was calculated for 100 different operating conditions using multi-body simulations as described in Section 2.2. In total, 2000 dynamics simulations were performed. Figure 3 illustrates the geometry parameters used to generate different cage designs. The chosen parametrization of the cage geometry enables the shape to be represented as generically as possible by 7 parameters. This allows cages with different properties to be created in the given design space and their dynamic behavior to be investigated. Using the parameter d_g, the clearance between the cage and the outer ring and thus the guidance clearance can be influenced, see Figure 3a. The cross-section of the cage is defined by the height h_c and width of the cage b_c, see Figure 3a. Both parameters affect important properties such as mass, moment of inertia, and stiffness of the cage. The shape of the cage pocket was varied using the parameters c_0, c_1, c_2, and c_3, which represent the pocket clearance along the circumference, see Figure 3b.

By choosing the pocket shape parameters, the pocket clearance of the cage on the one hand and the contact point between cage and the rolling element on the other hand can be influenced. The pocket clearance has a significant effect on the cage dynamics, as the number of contacts to the rolling elements increases with decreasing pocket clearance and can cause highly dynamic cage movements [21]. The contact point between the rolling element and the cage defines the direction of the normal and frictional force vector in the contact and finally the direction of the cage acceleration.

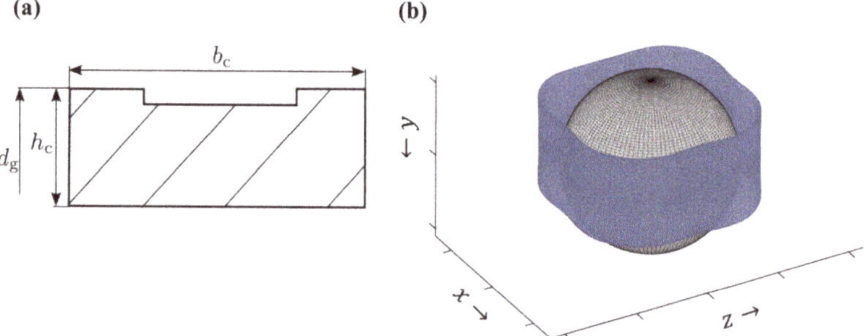

Figure 3. (**a**) Cross-section of the angular contact ball bearing cage. (**b**) Three-dimensional view of cage pocket and a rolling element. The blue area represents the geometry of the cage pocket and shows an exemplary shape defined by four parameters c_0, c_1, c_2, and c_3.

Using the geometry parameters, a total of 20 different cage variations were created using Latin hypercube sampling. The boundaries for the sampling shown in Table 3 were chosen in such a way that there are no dependencies between the cage design parameters. For the smallest guidance diameter d_g and largest cage height h_c, the clearance cage/inner ring is greater than the clearance cage/outer ring, and the same guidance type is provided.

Besides the modifications of the cage geometry, the load on the rolling bearing was also modified using an additional Latin hypercube sampling. The forces acting on the inner ring were varied using the load ratio R and the equivalent dynamic bearing load P. Based on the two parameters in Equations (1) and (2), the forces F_x and F_y to be defined in the simulation can be calculated. In addition to the forces, the inner ring was also loaded by the torque T_z acting around the z-axis, see Figure 2. The frictional force in the rolling element/cage contact was varied via the coefficient of friction μ_c.

$$P = X \cdot F_x + Y \cdot F_y \tag{1}$$

$$R = \frac{F_x}{F_y} \tag{2}$$

The speed of the inner ring n_i was also taken into account in the sampling. The kinematic speed of the rolling elements n_r and the cage n_c at the beginning of the simulation were determined for the initial time step by the Equations (3) and (4) depending on the defined inner ring rotational speed [33].

$$n_r = -\frac{n_i}{2} \cdot \left(\frac{d_p}{d_{re}} - \frac{d_{re} \cdot \cos^2(\alpha)}{d_p} \right) \tag{3}$$

$$n_c = \frac{n_i}{2} \cdot \left(1 + \frac{d_{re} \cdot \cos(\alpha)}{d_p} \right) \tag{4}$$

A simulation plan consisting of a total of 100 operating conditions (inner ring rotational speed, force and torque on the inner ring, and friction coefficient in the rolling element/cage contact) was created using the boundary values in Table 3 and Latin hypercube sampling. Simulation models were generated for each of the 20 cage variants according to the same operating conditions defined by the created simulation plan, so that a total of 2000 dynamics simulations were performed.

Table 3. Minimum and maximum values of the parameters for the Latin hypercube sampling.

Parameter	Minimum	Maximum
Load ratio R	0.25	10
Equivalent dynamic bearing load P in N	1000	10,000
Torque on inner ring T_z in Nm	10	50
Rotational speed inner ring n_i in rpm	1000	9000
Friction coefficient rolling element/cage μ_c	0.05	0.35
Pocket shape parameters c_0, c_1, c_2, c_3 in mm	0.1	0.35
Cage width b_c in mm	19	23
Cage height h_c in mm	4	8
Guidance diameter d_g	122	124.2

2.4. Features and Targets for Machine Learning

The input and output parameters for machine learning were derived from the calculation models and results and formed the database. The input parameters were structured by mechanical and geometrical properties of the cage as well as the loading parameters and the resulting class of the cage motion (according to Schwarz et al. [21]), see Table 4. Stiffness as a mechanical property is defined using a weighted area moment of inertia and cross-sectional area of the cage as input parameters. The weighting of the cross-section properties in the pocket and in the bar is based on a nonlinear function that provides a disproportionate amount of the area moment of inertia and the cross-sectional area in the cage pocket according to Schwarz et al. [21]. The cage mass and the mass moments of inertia complement the mechanical properties. The geometrical properties consist of the pocket shape parameters and the pocket and guidance clearance of the cage. The mechanical and geometrical parameters represent the essential properties that can be derived from a given cage geometry. The axial and radial loads, as well as the torque acting on the inner ring, were defined as relative quantities in relation to the basic static load rating $C_{0,r}$ and the pitch diameter d_p as input parameters. Thus, the database can be supplemented by calculation results of other bearing sizes in the future. The cage motion class is represented by one of the basic observable cage motion types, "unstable", "stable", or "circling", and was determined using Quadratic Discriminant Analysis based on the simulation results. The movement types differ in their dynamic behavior and can be classified based on their kinematics [34]. The cage motion type can also be predicted with high reliability by the classification algorithm AdaBoostM1 using the input parameters of the simulation [21].

However, the cage motion class provides information about the dynamics of the cage in a qualitative level. By extending the prediction using a regression algorithm, the relevant kinematic results can be specified more precisely.

Table 4. Features for the machine learning model.

Group	Property	Symbol
Mechanical and material properties	Weighted area moment of inertia	$\tilde{I}$
	Weighted cross-sectional area	$\tilde{A}$
	Cage mass	m
	Cage moment of inertia	J_x, J_y
Geometric properties	Pocket shape parameters	c_0, c_1, c_2, c_3
	Pocket clearance circumferential	c_c
	Pocket clearance axial	c_a
	Guidance clearance	c_g
Load parameter	Axial Force	$F_x / C_{0,r}$
	Radial Force	$F_y / C_{0,r}$
	Torque	$T_z / (C_{0,r} \cdot d_p)$
	Rotational speed inner ring	n_i
	Coefficient of friction	μ_c
Predicted cage motion	Cage motion class	C

The Cage Dynamics Indicator (CDI) defined by Schwarz et al. [21] contains all necessary parameters for the assessment of the cage dynamics and was used as the target of the regression task. The median (med) and the quantile distance (qd) indicate the distribution of the motion quantities contained in the CDI and were determined from the calculated time series. For the evaluation of the cage motion, the Ω-ratio, the cage coordinates normalized to the guidance or pocket clearance $\tilde{x}_c$, $\tilde{y}_c$, and $\tilde{z}_c$, the rotational ratio $\tilde{n}_c$, and the equivalent deformation force F_e were used.

In addition to the CDI, the output parameters include the median of the frictional torque T_f, the median of the contact forces on the cage $|F_c|$ and the median of the translational acceleration $|a_c|$ of the cage. In total, the output parameters for the regression algorithm consist of 10 parameters, which can be used to assess the cage dynamics as well as the energy efficiency of the bearing. In previous research papers, the CDI has been used as a key figure to assess the cage motion calculated by the dynamics simulation [14,21,34]. In this contribution, machine learning methods will be used to predict the CDI in order to accurately assess cage dynamics.

A strong scatter of the target variables reduces the prediction accuracy of the algorithms. Therefore, an anomaly detection for each motion class identified outliers of the target variables and removed them from the database. A density-based approach developed by Breunig et al. was used for anomaly detection. The local outlier factor (LOF) determines the degree of isolation of a data set compared to the immediately neighboring data sets [35].

2.5. Regression Algorithms and Hyperparameter Optimization

In this paper, the prediction accuracy of three different regression algorithms (Random Forest, XGBoost, and Artificial Neural Networks) used to estimate rolling bearing cage dynamics are compared. The hyperparameters of the models were determined by an EA as part of an optimization of the prediction accuracy [28].

RF is an ensemble method based on the 'wisdom of the crowd' paradigm. According to this, a prediction made by a large number of different persons/models achieves better results than the prediction of a single person/model. Accordingly, an RF regressor contains multiple regression trees that learn the regression problem using different sub-sets of the original training data. These sub-sets are regenerated by bagging for each regression tree. The degree of randomness is further increased by using only a random selection of features for training the decision trees. Random components (bagging or feature selection) reduce the model's tendency to overfit the training data [36].

Gradient boosting is another ensemble method developed by Friedman [37]. In an iterative process, multiple regression trees are trained. The training process of a decision tree depends on predictions and loss of the already trained decision trees in the ensemble. One implementation of gradient boosting is XGBoost (extreme gradient boosting) [27], which was used for predicting cage dynamics in the present case. As the regression algorithm in XGBoost is designed to predict only a single value, one model was trained for each output parameter. However, this allows interactions of the targets to be represented less effectively than with the random forest regressor.

ANNs are widely used algorithms for classification and regression in the field of machine learning. The input value of a neuron results from the weighted sum of the output values of the neurons of the previous layer and a so-called constant bias value. The neuron's input value is converted into the output by a nonlinear activation function. During the training of the ANN, the weights as well as the bias values are optimized so that the relationship in the training data between the input values and the output values can be predicted as accurately as possible [38]. For the prediction of the cage dynamics in this paper, an ANN consisting of a total of five layers was trained using the training algorithm Adam [39]. The target variable of the optimization procedure is the mean square error (MSE) between the ANN's predictions and the target values contained in the training data.

For the ML algorithms, hyperparameters such as the ANN's number of neurons per layer need to be specified. With the help of an EA, the hyperparameters were determined so that the prediction accuracy of the models were optimized. The remaining parameters of the models are listed in Appendix A. The EA uses mechanisms of biological evolution such as selection, recombination and mutation to improve the fitness (metric for assessing regression results, e.g., coefficient of determination R^2) of the individuals (set of hyperparameters) contained in a population (amount of individuals) for a predefined number of generations, see Figure 4. Starting from an initial population generated by Latin hypercube sampling, the fitness of each individual is determined. The fitness of the individuals and target value of the EA was represented by the mean $\overline{R^2}$ according to Equation (5). Using a K-fold ($K = 5$) cross-validation, a total of K validation data sets were generated from the training data for fitness evaluation. The data set was randomly split so that 85% is used for hyperparameter optimization as well as the cross-validation contained within the loop and 15% for subsequent testing of model predictions. $\overline{R^2}$ was calculated by evaluating the arithmetic mean of the R^2 for each validation data set and target variable. The prediction accuracy for the validation data is an indicator of the generalization capability of the model, which can be finally evaluated after training by the test data sets.

$$\overline{R^2} = \frac{1}{K} \cdot \frac{1}{N} \cdot \sum_{i}^{K} \sum_{j}^{N} R_{ij}^2 \tag{5}$$

The R^2 of each output parameter was calculated by Equation (6) using the predictions of the algorithm $\hat{y}_i$, the target parameter according to the test data y_i, and its arithmetic mean $\overline{y}$. Thus, R^2 can reach a maximum value of 1 in case of an error-free prediction of the algorithm.

$$R^2 = 1 - \frac{\sum_i (y_i - \hat{y}_i)^2}{\sum_i (y_i - \overline{y})^2} \tag{6}$$

After calculating the fitness of the initial population, the evolutionary process consisting of selection, recombination, mutation, and evaluation of fitness was repeated in a given number of generations. Individuals for recombination were selected by the fitness proportional method stochastic universal sampling. Each individual received an area proportionate to its fitness value on a wheel. By spinning the wheel once and with n arrows equally distributed around the circumference, n individuals were selected by the pointers. Recombination was performed in pairs for the selected individuals. The list of hyperparameters of two individuals for mating were separated at two points and the new resulting individuals were defined by alternating the combination of the sections, see Figure 4. After recombination, mutation was performed for each parameter contained in the individual by a uniformly distributed random variable. Mutation served to generate new parameter specifications in the population and was performed with a previously defined probability. The individuals produced by recombination and mutation, as well as the best individual from the previous population (elite), formed the new population for the following generation.

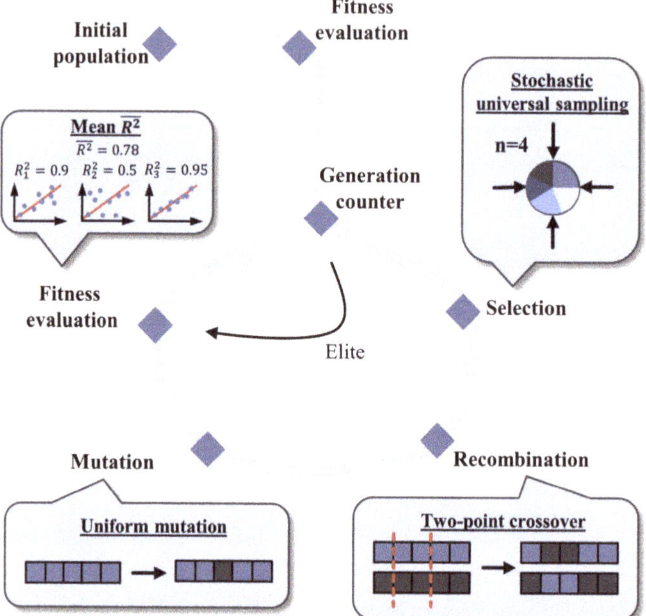

Figure 4. Steps of the EA used for hyperparameter optimization of the regression models.

After a predetermined number of generations, the model with the highest fitness and best prediction accuracy for the test data was returned by the EA. The parameters controlling the behavior of the EA can be taken from Table 5.

Table 5. Parameters of the EA for the optimization of the hyperparameters of the regression algorithms.

Parameter	Value	Parameter	Value
Number of generations	30	Population size	50
Elite individuals	1	Crossover probability	0.8
Mutation probability	0.15		

Table 6 shows the hyperparameters of the algorithms and components of the individuals as well as the range of the parameters considered during optimization. The ranges of the hyperparameters were chosen to be comparatively large in order to provide as many parameter combinations as possible. Large ranges of the hyperparameters increase the risk of overfitting (e.g., large number of neurons contained in the ANN). However, overfitting was avoided, including in the training methods of XGBoost and ANN, by using evaluation datasets. Based on the predictions for the evaluation datasets that were not used directly for training, it is determined whether overfitting is present in the current state of the training process. No evaluation dataset was used for Random Forest, because the algorithm generally has a low tendency to overfit the training data [36].

Table 6. Hyperparameters of the regression models optimized using the EA.

Model	Parameter	Minimum	Maximum
	Max depth	20	200
	Number of estimators	100	1500
XGBoost	Learning rate	0.0001	0.01
	L1 regularization	0.0001	0.9
	L2 regularization	0.0001	0.9
	Minimum loss reduction for tree split	0.00001	0.2
	Max depth	20	200
	Number of estimators	100	1500
Random Forest	Minimum samples required for a leaf	2	10
	Maximum number of features for a split	10	17
	Maximum number of leaf nodes	10	500
	Minimal cost-complexity pruning	0	0.5
	Number of neurons in layer 1	100	600
	Number of neurons in layer 2	100	600
ANN	Number of neurons in layer 3	100	600
	Number of neurons in layer 4	100	600
	Learning rate	0.0001	0.9
	Activation function	ELU, RELU, Leaky_RELU	

3. Results

3.1. Dynamics Simulation Results

The results of the dynamics simulations contain time series that include dynamics of the cage as well as the rolling elements. Figure 5 shows an example of the dynamic behavior of a cage for different operating conditions of the bearing. In the qualitative assessment of cage dynamics, a fundamental differentiation is made between "unstable", "stable", and "circling" cage motions [21,34]. These types of movements could also be observed for the cages investigated. Figure 5a–c illustrates an example of an unstable cage motion (loading conditions $\mu_c = 0.21$, $F_x = -8058$ N, $F_y = 1077$ N, $n_i = 8263$ rpm $T_z = 48$ Nm), that is characterized by high dynamics as well as severe and high-frequency cage deformations. The cage was pressed against the outer ring and strongly deformed. This led to the diameter of the circular center of gravity trajectory being significantly larger than in the other two calculations. In addition, high contact forces caused frictional losses, which significantly impair the energy efficiency of the rolling bearing. In the case of stable cage motion (loading conditions $\mu_c = 0.26$, $F_x = -15,126$ N, $F_y = 1636$ N, $n_i = 4407$ rpm $T_z = 14$ Nm), no significant deformations occurred and the dynamics of the cage were generally low, see Figure 5d–f. The contact forces between the cage and the rolling element and outer ring were also significantly reduced compared to an unstable motion, and therefore the frictional losses were also lower. The circling cage motion (loading conditions $\mu_c = 0.12$, $F_x = -3030$ N, $F_y = 377$ N, $n_i = 6844$ rpm $T_z = 12$ Nm) is characterized by a circular motion of the cage center of mass that exhibits small variations in the rotational speed. The rotational speed of the cage center of mass corresponds to the speed of the rolling element set. The cage is pressed in a radial direction due to the centrifugal force acting, so that the number of contacts to the guidance rib and the contact force acting in the contact increase.

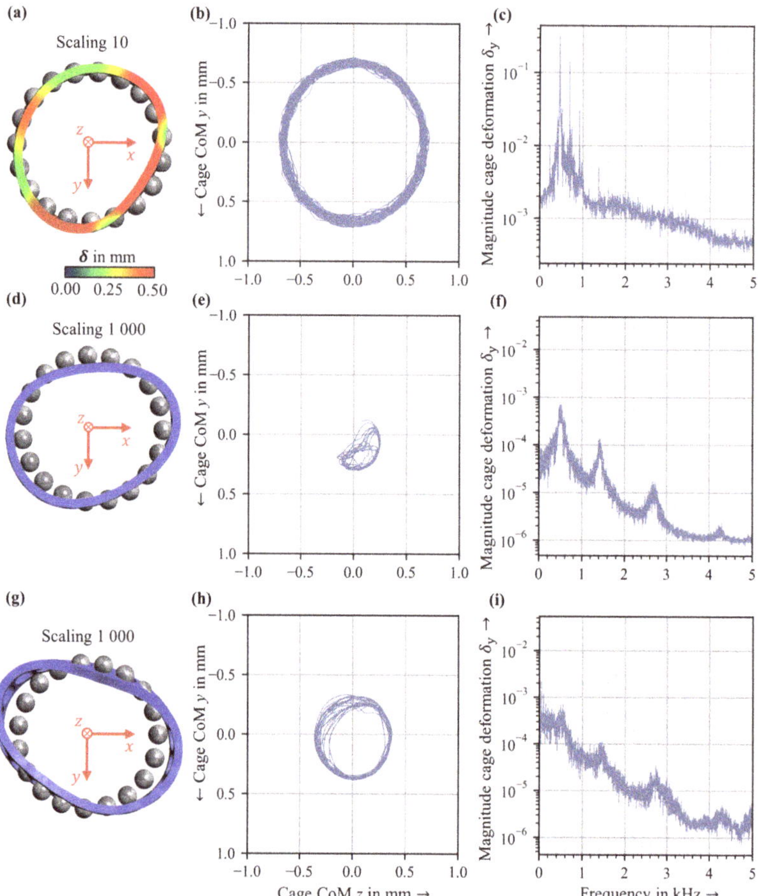

Figure 5. Dynamic behavior of a cage for different operating conditions: an unstable (**a–c**), stable (**d–f**), and circling (**g–i**) cage motion. The three-dimensional deformation of the cages, the center of gravity trajectory, and the amplitude spectrum of the node displacement are illustrated.

In addition to the load on the bearing, the geometry of the cage can also influence the dynamic response of the bearing. Figure 6 shows the cage dynamics for a load situation ($\mu_c = 0.16$, $F_x = -9655$ N, $F_y = 3718$ N, $n_i = 6844$ rpm $T_z = 16$ Nm) and three different cage geometries. The first cage variant performed a highly dynamic cage motion with severe deformations and a high rotational speed of the cage center of mass, see Figure 6a–c. A modification of the cage geometry (cross-section and shape of the cage pocket) for the other two variants and the same operating conditions led to circling cage motions in both cases. The amplitudes of the deformations were significantly smaller compared to the first cage variant and the larger amplitudes were shifted to the low frequency range, see Figure 6d–i.

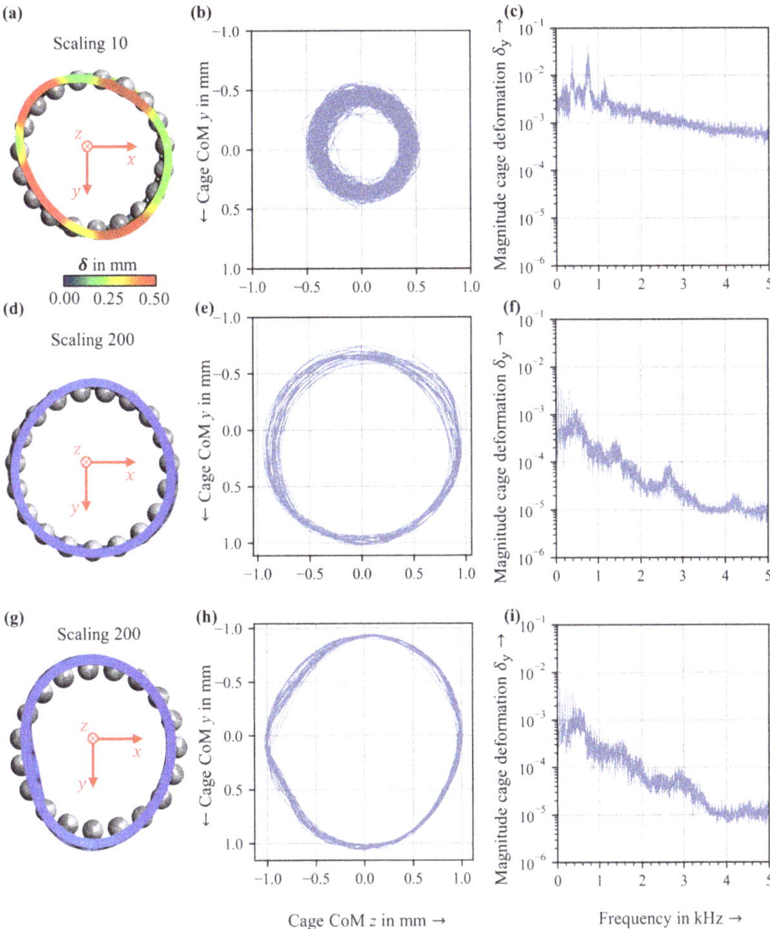

Figure 6. Dynamic behavior of three different cage variants for the same loading conditions. The three-dimensional deformation of the cages (**a,d,g**), the center of gravity trajectory (**b,e,h**), and the amplitude spectrum of the node displacement (**c,f,i**) are illustrated.

An overview of the simulations performed and the resulting cage motion types is shown in Figure 7. Certain cage geometries (ID 02 or 05) had a high proportion of unstable cage motions, while other cage variants exhibited a much lower tendency to unstable cage motions (ID 14 or 10). In addition, differences in the proportion of circumferential and stable cage movements were also evident for the different cage variants. The dynamic behavior of the cage variants illustrates the potential of the geometry parameters to positively influence the dynamics of the cage. A clear influence could also be identified in the loading conditions, as was found, for example, by Schwarz et al. [14]. However, as the operating conditions often cannot be influenced, these serve only as a reference for comparing the dynamic behavior of the cage geometries.

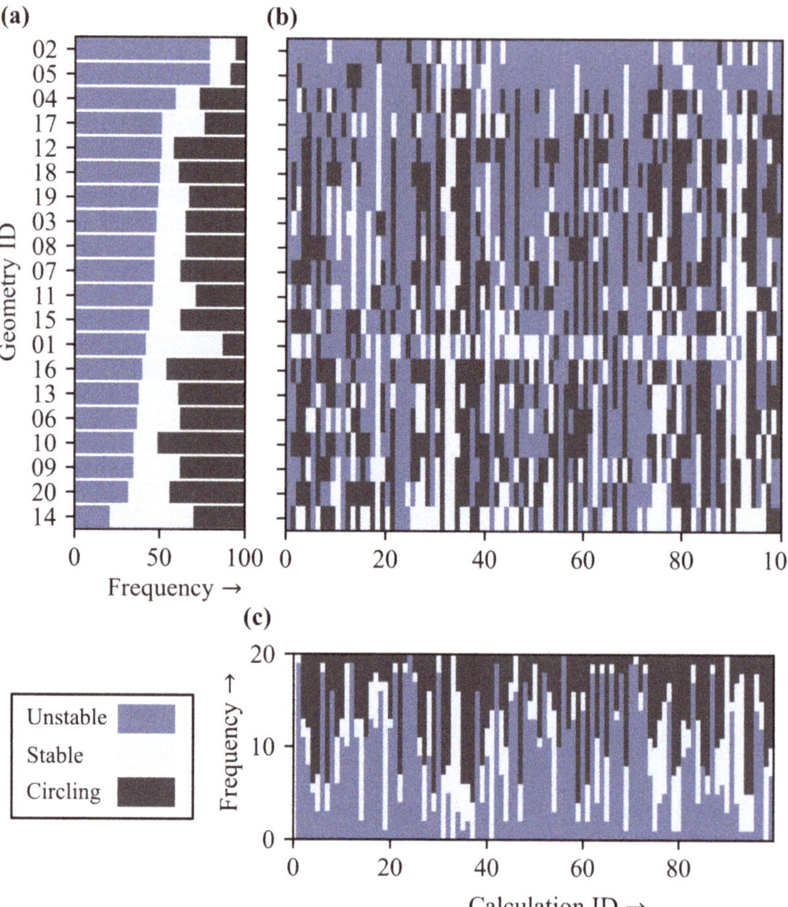

Figure 7. Overview of the results of the dynamics simulation. (**a**) Number of motion types "unstable", "stable", and "circling" for each cage variant. (**b**) Motion type as a function of cage geometry and bearing load. (**c**) Number of motion types for each bearing load in the experimental design.

The simulation results were further processed so that the influence of cage geometry and bearing load was represented by a database consisting of input and target variables and could be used for machine learning.

3.2. Preprocessing of Calculation Results and Data Analysis

The calculated time series were the starting point for determining the targets for machine learning. For the evaluation of the cage dynamics, the time range $t = 0.5...1$ s was analyzed to avoid unrepresentative cage motions due to the initial conditions at the beginning of the calculation.

In addition it was checked whether the simulation results are suitable to be integrated into the database. Especially for simulations with high friction coefficients, a severe deformation of the cage occurred, which led to a termination of the simulation. Nonphysical results as the automatically generated inputs are out of a reasonable range for this application and were removed from the database. Using the density-based LOF approach, outliers in the database could be identified and removed. The LOF approach was applied to each of the classes "unstable", "stable" and "circling". Outliers with respect to the dynamic behavior typical for the respective classes were thereby identified. Figure 8 illustrates the outliers (red) and the remaining datasets (blue). Outlier detection reliably removed atypical cage movements, ensuring a high-quality database for machine learning. After preparing the simulation results, the database for machine learning contained a total of 1362 data sets.

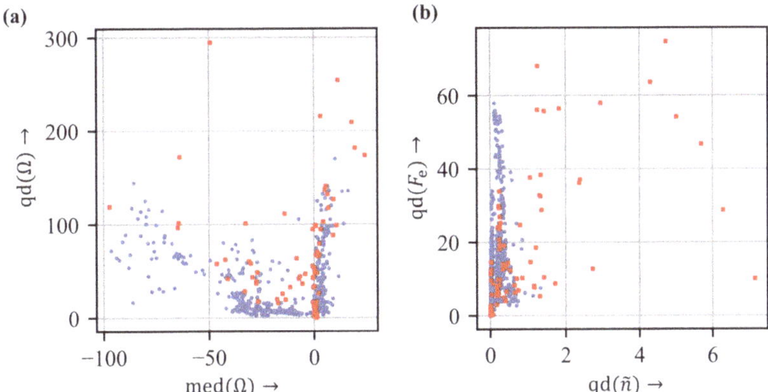

Figure 8. Distribution of target regression variables (**a**) med(Ω) and qd(Ω) as well as (**b**) qd(F_e) and qd($\tilde{n}$) in the database. The data sets marked in red are identified as outliers using the LOF approach and not considered for training the regression models.

Figure 9 shows the correlation matrix for determining the qualitative relationship between input and output parameters. The mechanical properties of the cage (area moment of inertia $\tilde{I}$, mass m, area cross section $\tilde{A}$, and moment of inertia J) had similar values for the correlation coefficient and thus a related influence on the target parameters, see Figure 9a. A mathematical negative correlation existed between the mechanical properties and the center of mass acceleration of the cage $|a_c|$. Accordingly, lower accelerations occur at higher masses of the cage, which can be justified by the inertia of the geometry. There is also a positive correlation between the cage mass and the equivalent force F_e representing the deformation of the cage. Thus, for the cages with larger masses, the equivalent deformation force tend to be larger. With respect to the bearing speed n_i and friction coefficient μ_c, a mathematical positive correlation to cage acceleration, contact forces, and finally a highly-dynamic cage movement could be clearly determined. This is due to the increased relative velocity and frictional force in the contact between the cage and the other components, which leads to a stronger excitation of the cage and an increased tendency to highly dynamic movements.

Based on the matrix in Figure 9b, a mutual correlation of the output parameters was also evident. Highly dynamic cage movements are characterized by strong deformations of the cage, high accelerations, and a high frictional torque, for which reason these parameters exhibited a strong correlation. Due to the opposite movement of the center of mass in the case of unstable cage dynamics, there is a mathematical negative relationship between the median of the Ω-ratio and the other parameters. The weak relationship of the normalized $\tilde{x}_c$-coordinate of the cage to the other target quantities is also noticeable. The contact forces between the cage and the rolling element/rib point primarily in radial direction, which is the direction of the resulting acceleration. Therefore, the correlation between the quantile distance of the two non-axial coordinates is more significant, especially in the case of an unstable cage motion. The quantile distance of the Ω-ratio also indicates a slightly lower correlation to the other parameters, but still stronger than the quantile distance of the $\tilde{x}_c$-coordinate of the cage center of mass.

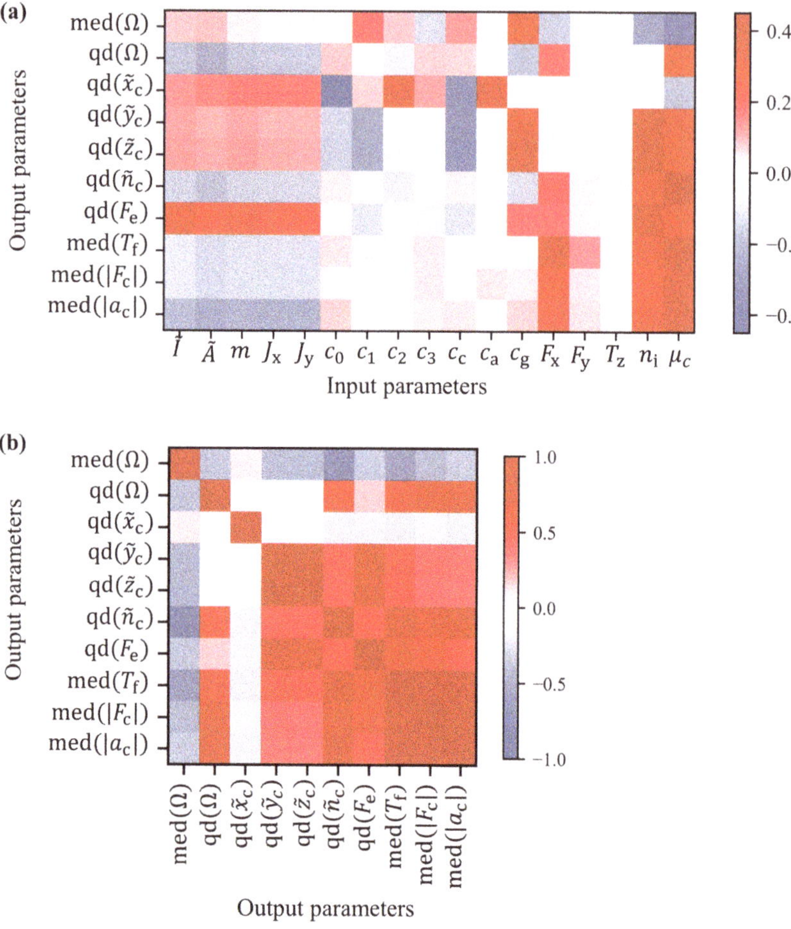

Figure 9. Matrix with correlation coefficients for determining the relationship between (**a**) the input and output parameters and (**b**) the output parameters among each other.

Although there were trends based on the correlation matrix that suggest the resulting dynamic behavior of the cage, the relationship is highly nonlinear due to interactions between the parameters. Therefore, the regression algorithms are trained in the following to learn the relationship between input and output parameters.

3.3. Evaluating Optimization and Regression Results

The EA determined the hyperparameters of the models to maximize the average coefficient of determination for the validation data sets. The best individuals or parameter combinations are shown in Table 7. A large number of neurons, or many estimators in the ensemble methods, increase the adjustable model parameters, the risk of overfitting to the training data, and poor prediction accuracy for test data. However, the hyperparameters causing overfitting were not chosen by the optimization to maximize the number of model parameters to reach high values for the prediction accuracy based on the training data. In general, this is a first indication that a generalization capable model was created by the training and optimization.

Table 7. Optimized parameters of the regression models.

Model	Parameter	Minimum	Maximum	Value
XGBoost	Max depth	20	200	52
	Number of estimators	100	1500	832
	Learning rate	0.0001	0.01	0.045
	L1 regularization	0.0001	0.9	0.865
	L2 regularization	0.0001	0.9	0.791
	Minimum loss reduction for tree split	0.00001	0.2	0.021
Random Forest	Max depth	20	200	46
	Number of estimators	100	1500	812
	Minimum samples required for a leaf	2	10	2
	Maximum number of features for a split	10	17	17
	Maximum number of leaf nodes	10	500	441
	Minimal cost-complexity pruning	0	0.5	0.0009
ANN	Number of neurons in layer 1	100	600	447
	Number of neurons in layer 2	100	600	568
	Number of neurons in layer 3	100	600	577
	Number of neurons in layer 4	100	600	455
	Learning rate	0.0001	0.9	0.003
	Activation function	ELU, RELU, Leaky_RELU		Leaky_RELU

The hyperparameters optimized by the EA were used for training the algorithms. Afterwards, the models were evaluated using the coefficient of determination R^2 for test and training data, see Figure 10. For the training data, acceptable values for R^2 were obtained for all algorithms. The quantile distance for the normalized $\tilde{x}_c$-coordinate of the cage reached $R^2 \approx 0.41$ in the case of the random forest regressor, which is to be assessed as a medium correlation. The excitation of the cage, as well as the translational center of mass movement, occurs both for the contact of the cage to the rolling elements and to the rib in the radial direction. The relationship between the geometry and load parameters as well as the axial center of mass movement and finally the R^2 of the predictions were therefore lower compared to the other center of mass coordinates. Random forest regressor predicted very well for all target values, but reached a slightly lower R^2 compared to XGboost and the neural network for training data. The random components in the random forest algorithm (e.g., feature selection) prevent possible overfitting to the training data and led to slightly inferior prediction. The coefficients of determination $R^2 \approx 1$ for XGBoost were very high and indicate a significant fit to the data sets.

The test datasets generally showed a lower coefficient of determination than the training datasets but were within an acceptable range apart from the quantile distance of the normalized $\tilde{x}_c$ coordinate of the cage. qd($\tilde{x}_c$) exhibited the worst values of $R^2 \approx 0.41$ for the random forest and $R^2 \approx 0.6$ the ANN. Thus, while qd($\tilde{x}_c$) is suitable for assessing cage dynamics when derived from calculated time series, there is no strong correlation to bearing load or cage geometry. The difference in prediction accuracy for training and test data was lowest for random forest, which indicates a generalization of the model. However, the difference for XGBoost and the ANN was also in an acceptable range, which is also a sufficient generalization capability. All models reached comparable values for the coefficient of determination R^2 based on the test data sets and thus can be used equally for the prediction of cage dynamics. The best prediction values for R^2 based on the test data sets were obtained for the quantile distance of the equivalent deformation force F_e, the median of the Ω-ratio and the median of the friction torque T_f in the range of $R^2 \in [0.90 \ldots 0.94]$. For the remaining target parameters, with the exception of qd($\tilde{x}_c$), at least one of the models investigated achieved a coefficient of determination $R^2 > 0.8$ and sufficient prediction accuracy.

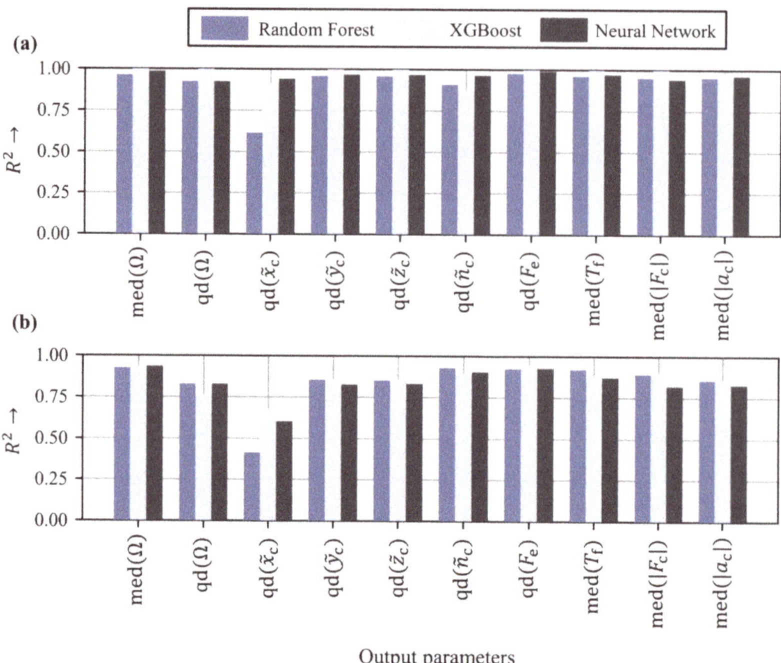

Figure 10. Coefficient of determination R^2 for the target variables of the regression algorithms Random Forest, XGBoost, and Neural Network for training (**a**) and test data (**b**).

The scatter plot in Figure 11 shows, representative of the trained models, the predictions of the ANN compared to the true values in the training (blue) and test (red) data. As can be seen from the correlation matrix and the coefficient of determination, the predictions for the quantile distance of the axial coordinate of the cage $\mathrm{qd}(\tilde{x}_c)$ were considerably more scattered than the other target variables. For the quantile distance of the omega ratio, the deviation of the predictions from the test data sets was smaller, but a stronger, though still acceptable, scatter was also present here. For the remaining parameters, a good correlation was present, analogous to the R^2. The deviations are within a tolerable range, as can be seen by the intervals containing 90% of the errors determined for the test data (blue area).

The hyperparameters obtained from the optimization by the EA were used to perform a 10-fold cross-validation. This allowed us to determine how strong the predictions of the algorithm differ depending on the used training and test data set, see Figure 12. Based on this, the sensitivity of the prediction results for different training and test data sets could be investigated. Figure 12 exhibits the distribution of the average prediction of the target values for the training data and a 10-fold cross-validation including (a) and excluding (b) the quantile distance of the cage coordinate $\tilde{x}_c$ as regression target. For all three models, omitting the normalized coordinate improves the average prediction quality, as lower R^2 values are obtained for $\mathrm{qd}(\tilde{x}_c)$ than for the other values in all iterations of the cross-validation. The minima and maxima of $\overline{R^2}$ for the three models without considering $\tilde{x}_c$ in the cross-validation were very similar and differ only slightly. As no obvious favorite could be identified based on the prediction accuracies, all three algorithms were suitable for predicting the cage dynamics with a comparable error tolerance.

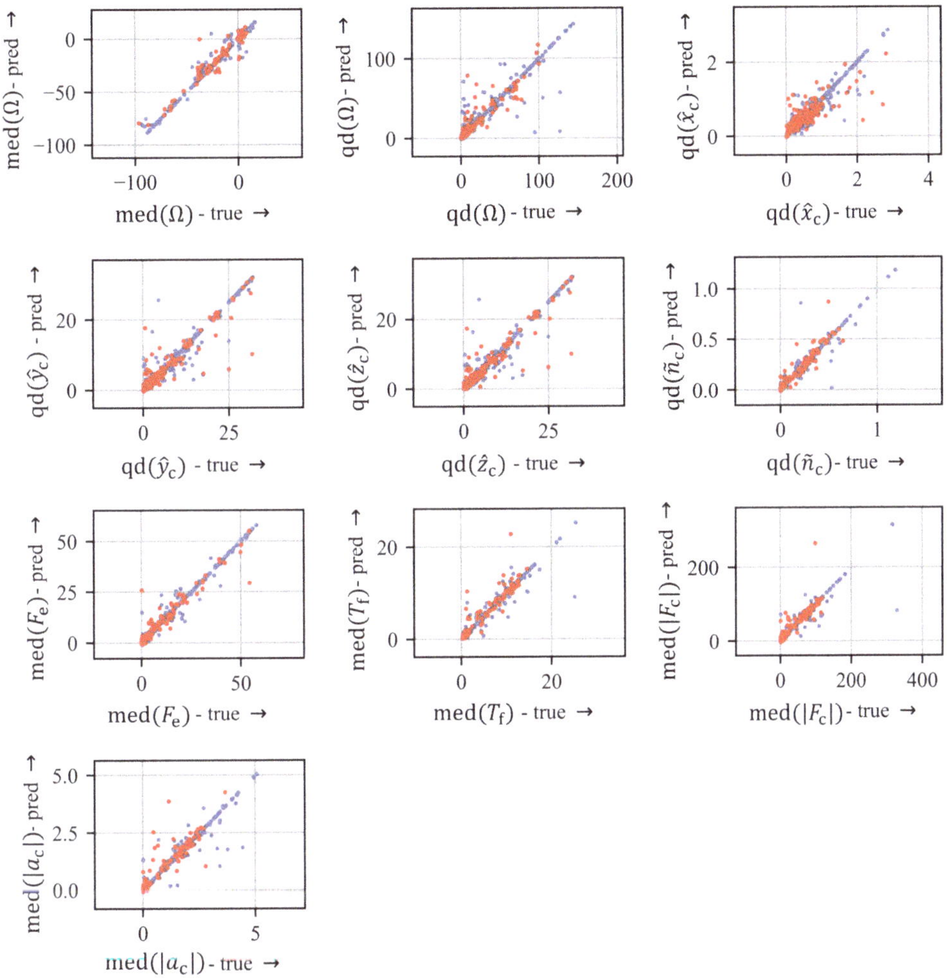

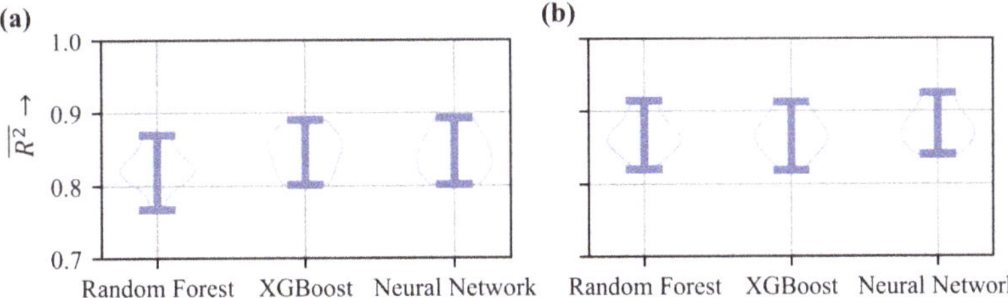

Figure 11. Scatter plot of the target parameters for training (**blue**) and test (**red**) data sets and the predicted values by the neural network. The colored area represents the range where 90% of the errors for the test data sets are located.

Figure 12. Distribution of $\overline{R^2}$ values for all target variables (**a**) and without the normalized $\tilde{x}_c$-coordinate (**b**) for a 10-fold cross-validation. Besides the minimum and maximum, the distribution of the values is also illustrated.

4. Discussion

The results of the dynamics simulation illustrate the strong influence of the bearing load and cage geometry on the resulting cage dynamics, see Figure 7. Depending on the geometry, the tendency of a cage to highly dynamic and unstable cage movements varies significantly. However, the relationship between the geometry parameters and the resulting cage dynamics is very complex and difficult to determine using conventional methods of descriptive statistics, as can be seen from the covariance matrix in Figure 9. The complex relationship between the input and output parameters can basically be determined with the help of the investigated algorithms. Analyzing the prediction results for test data, it can be seen that for the quantile distance of the normalized center of mass coordinate $\tilde{x}_c$ of the cage, mediocre prediction values could be obtained. As the frictional forces acting in contact between the cage and the other components accelerate the cage primarily in the bearing plane, the physical relationship between the input parameters of the model and the resulting axial cage motion is less than for the other parameters. The normalized $\tilde{x}_c$ coordinate of the cage is thus less suitable for predicting the cage dynamics. Though, as a component of the multivariate metric CDI, which can be derived from calculated time series representing cage dynamics, $\tilde{x}_c$ is a contribution to improve the classification performance.

The algorithms Random Forest, XGBoost, and ANN achieved similar values for the R^2 of the different target variables for the test data sets, see Figure 10. A 10-fold cross-validation exhibited that the differences between the models are small, and thus all algorithms are suitable for the prediction of the cage dynamics. The robustness of the predicted targets for a given cage geometry with respect to deviations from the true values can be improved by a large number of predictions by the regression algorithm with a subsequent statistical analysis. This reduces the influence of single incorrect predictions and improves the comparability of the dynamic behavior of different cage variants.

A transfer of the predictions to other rolling bearing sizes is possible in general. For this purpose, new training data must be generated and the existing database expanded. However, a similar effect on the cage dynamics can be expected, especially for the load conditions as shown, for example, by Schwarz et al. for various bearings [14,21]. Therefore, the amount of training data for the same bearing type and similar cage shapes can probably be lower than for the investigated angular contact ball bearing. In addition to the extension to other bearing types, other parameters can also be added as input variables, so that depending on the existing application, the model can also be designed flexibly. As with the geometry parameters, new data sets must be created for the training, but the database established so far serves as an initial starting point for further investigations.

5. Summary and Conclusions

The aim of this paper was to present a procedure for predicting the dynamics of cages in an angular contact ball bearing using dynamics simulations and machine learning regression methods. To achieve this aim, the approach in this paper is structured as follows: starting with a comprehensive simulation study, a database was created to represent the relationship between the input (cage geometry and bearing load) and output (cage dynamics and bearing friction) parameters for the regression models. As part of the training, the hyperparameters of the random forest, XGBoost, and artificial neural network models were optimized using an evolutionary algorithm. The optimized hyperparameters were used to train the regression models. The prediction accuracy of the models was compared using the coefficient of determination R^2 and regression plots. Based on the models and their predictions, the dynamics of the cage represented by the target variables can be predicted with high accuracy. The following conclusions can be drawn from the results of this paper:

- The cage geometry has a significant influence on the resulting cage dynamics. The occurrence of unstable cage movements can be significantly reduced by changing the geometry of the cage.

- The influence of the geoemtric parameters is non-linear and characterized by strong alternating effects and can therefore hardly be assigned to single parameters.
- There is a low correlation between the axial movement of the cage and the influencing factors such as bearing load and cage geometry. The reason for this is that the contact forces acting on the cage point mostly in radial or circumferential direction. The forces acting on the cage are influenced by the parameters such as cage mass, cage speed, etc.
- In this study, all regression algorithms achieved acceptable values for the coefficient of determination in the range of $R^2 \in [0.75 \ldots 0.94]$ for the target variables except for the quantile distance of the normalized axial center of mass coordinate of the cage. Therefore, the models appear to be suitable to compare the performance (dynamics, friction) of different cages.
- The use of machine learning algorithms allows prediction even for new data sets of the analyzed bearing for which no dynamics simulation has been performed. The duration of the prediction is less than one second, while the computation time for a simulation is about 10 h.

Author Contributions: Conceptualization, S.S., M.B., H.G., O.G.-G., S.T. and S.W.; methodology, S.S., H.G., M.B., O.G.-G., S.T. and S.W.; software, S.S.; validation, S.S., H.G., M.B., S.T. and S.W.; formal analysis, S.S., H.G. and M.B.; investigation, S.S., H.G. and M.B.; resources, O.G.-G. and S.W.; data curation, S.S.; writing—original draft preparation, S.S., H.G. and M.B.; writing—review and editing, O.G.-G., S.T. and S.W.; visualization, S.S.; supervision, O.G.-G., S.T. and S.W.; project administration, O.G.-G., M.B. and S.W.; and funding acquisition, H.G., S.S., S.T., O.G.-G. and S.W. All authors have read and agreed to the published version of the manuscript.

Funding: This research was executed within the project Machine Learned Dynamics (MeLD), which is funded by Bayerische Forschungsstiftung (BFS) grant number AZ-1398-19.

Institutional Review Board Statement: Not applicable.

Informed Consent Statement: Not applicable.

Data Availability Statement: The data presented in this work are available on request from the corresponding author.

Acknowledgments: We acknowledge financial support by Deutsche Forschungsgemeinschaft (DFG) and Friedrich-Alexander-Universität Erlangen-Nürnberg (FAU) within the funding programme "Open Access Publication Funding".

Conflicts of Interest: The authors declare no conflicts of interest.

Abbreviations

The following abbreviations are used in this manuscript:

ANN	Artificial neural network
CDI	Cage Dynamics Indicator
EA	Evolutionary Algorithm
FE	Finite element
MSE	Mean squared error
RF	Random Forest
med	Median
qd	Quantile Distance

Appendix A

Table A1. Parameters of the machine learning models, that were not optimized. The literature reference for each model is the implementation of the algorithms.

Random Forest [26]		XGBoost [27]		ANN [25]	
Parameter	Value	Parameter	Value	Parameter	Value
Split criterion	MSE	Importance type	gain	Weight initializer	HeNormal/ Glorot
Minimum samples for split	2	Objective	MSE	Slope coefficient for Leaky ReLu activation	0.01
Use out of bag score	False	Subsampling	False	Bias initializer	None
Bootstraping when building trees	True	Booster	gbtree	Regularizer	None

References

1. Goines, L.; Hagler, L. Noise pollution: A modem plague. *South Med. J.* **2007**, *100*, 287–294. [CrossRef] [PubMed]
2. Migal, V.; Lebedev, A.; Shuliak, M.; Kalinin, E.; Arhun, S.; Korohodskyi, V. Reducing the vibration of bearing units of electric vehicle asynchronous traction motors. *J. Vib. Control* **2021**, *27*, 1123–1131. [CrossRef]
3. Alfares, M.A.; Elsharkawy, A.A. Effects of axial preloading of angular contact ball bearings on the dynamics of a grinding machine spindle system. *J. Mater. Process. Technol.* **2003**, *136*, 48–59. [CrossRef]
4. Harris, T.A.; Kotzalas, M.N. *Essential Concepts of Bearing Technology*, 5th ed.; Harris, T.A., Kotzalas, M.N., Eds.; Rolling Bearing Analysis; CRC Press: Boca Raton, FL, USA, 2007.
5. Aschenbrenner, A.; Schleich, B.; Tremmel, S.; Wartzack, S. A variational simulation framework for the analysis of load distribution and radial displacement of cylindrical roller bearings. *Mech. Mach. Theory* **2020**, *147*, 103769. [CrossRef]
6. Thomson, W.T. *Vibration Monitoring of Induction Motors: Practical Diagnosis of Faults via Industrial Case Studies*; Cambridge University Press: Cambridge, UK, 2020. [CrossRef]
7. Grillenberger, H.; Hahn, B.; Koch, O. Elastic Cage Instability in Rolling Element Bearings. Simulation and Test. In Proceedings of the 70th STLE Annual Meeting and Exhibition, Dallas, TX, USA, 17–21 May 2015; Society of Tribologists and Lubrication Engineers: Park Ridge, IL, USA, 2015.
8. Boesiger, E.A.; Donley, A.D.; Loewenthal, S. An Analytical and Experimental Investigation of Ball Bearing Retainer Instabilities. *J. Tribol.* **1992**, *114*, 530–538. [CrossRef]
9. Sathyan, K.; Gopinath, K.; Lee, S.H.; Hsu, H.Y. Bearing Retainer Designs and Retainer Instability Failures in Spacecraft Moving Mechanical Systems. *Tribol. Trans.* **2012**, *55*, 503–511. [CrossRef]
10. Schwarz, S.; Grillenberger, H.; Tremmel, S. Investigations on cage dynamics in rolling bearings by test and simulation. In Proceedings of the 74th STLE Annual Meeting and Exhibition, Nashville, TN, USA, 19–23 May 2019; Society of Tribologists and Lubrication Engineers: Park Ridge, IL, USA, 2019.
11. Kingsbury, E.; Walker, R. Motions of an Unstable Retainer in an Instrument Ball Bearing. *J. Tribol.* **1994**, *116*, 202–208. [CrossRef]
12. Kingsbury, E.P. Torque Variations in Instrument Ball Bearings. *Tribol. Trans.* **1965**, *8*, 435–441. [CrossRef]
13. Ghaisas, N.; Wassgren, C.R.; Sadeghi, F. Cage Instabilities in Cylindrical Roller Bearings. *J. Tribol.* **2004**, *126*, 681–689. [CrossRef]
14. Schwarz, S.; Grillenberger, H.; Tremmel, S.; Wartzack, S. Investigations on the rolling bearing cage dynamics with regard to different operating conditions and cage properties. In Proceedings of the IOP Conference Series: Materials Science and Engineering, Sanya, China, 12–14 November 2021; Volume 1097, p. 012009. [CrossRef]
15. Nogi, T.; Maniwa, K.; Matsuoka, N. A Dynamic Analysis of Cage Instability in Ball Bearings. *J. Tribol.* **2017**, *140*, 011101. [CrossRef]
16. Sakaguchi, T.; Harada, K. Dynamic Analysis of Cage Behavior in a Tapered Roller Bearing. *J. Tribol.* **2006**, *128*, 604. [CrossRef]
17. Boesiger, E.; Warner, M. Spin bearing retainer design optimization. In *The 25th Aerospace Mechanisms Symposium*; Aeronautics, U.S.N., Administration, S., Technology, C.I.O., Missiles, L., Company, S., Laboratory, J.P., Eds.; NASA Office of Management, Scientific and Technical Information Division: Washington, DC, USA, 1991; Volume 25, pp. 161–178.
18. Gupta, P.K. Frictional Instabilities in Ball Bearings. *Tribol. Trans.* **1987**, *31*, 258–268. [CrossRef]
19. Wenhu, Z.; Sier, D.; Guoding, C.; Yongcun, C. Impact of lubricant traction coefficient on cage's dynamic characteristics in high-speed angular contact ball bearing. *Chin. J. Aeronaut.* **2016**, *30*, 827–835. [CrossRef]
20. Marian, M.; Tremmel, S. Current Trends and Applications of Machine Learning in Tribology—A Review. *Lubricants* **2021**, *9*, 86. [CrossRef]
21. Schwarz, S.; Grillenberger, H.; Tremmel, S.; Wartzack, S. Prediction of Rolling Bearing Cage Dynamics Using Dynamics Simulations and Machine Learning Algorithms. *Tribol. Trans.* **2021**. [CrossRef]

22. Hahn, B.; Vlasenko, D.; Gaile, S. Detailed Cage Analysis Using CABA3D. In Proceedings of the 73rd STLE Annual Meeting and Exhibition, Minneapolis, MI, USA, 20–24 May 2018.
23. Vesselinov, V. Dreidimensionale Smulation der Dynamik von Wälzlagern. Ph.D. Thesis, Karlsruher Institut für Technologie (KIT), Karlsruhe, Germany, 2003.
24. McKay, M.D.; Beckman, R.J.; Conover, W.J. A Comparison of Three Methods for Selecting Values of Input Variables in the Analysis of Output from a Computer Code. *Technometrics* **1979**, *21*, 239–245.
25. Abadi, M.; Agarwal, A.; Barham, P.; Brevdo, E.; Chen, Z.; Citro, C.; Corrado, G.S.; Davis, A.; Dean, J.; Devin, M.; et al. TensorFlow: Large-Scale Machine Learning on Heterogeneous Systems. 2015. Available online: https://arxiv.org/abs/1603.04467 (accessed on 11 December 2021).
26. Pedregosa, F.; Varoquaux, G.; Gramfort, A.; Michel, V.; Thirion, B.; Grisel, O.; Blondel, M.; Prettenhofer, P.; Weiss, R.; Dubourg, V.; et al. Scikit-learn: Machine Learning in Python. *J. Mach. Learn. Res.* **2011**, *12*, 2825–2830.
27. Chen, T.; Guestrin, C. XGBoost: A Scalable Tree Boosting System. In Proceedings of the 22nd ACM SIGKDD International Conference on Knowledge Discovery and Data Mining, San Francisco, CA, USA, 13–17 August 2016; ACM: New York, NY, USA, 2016; pp. 785–794. [CrossRef]
28. Fortin, F.A.; De Rainville, F.M.; Gardner, M.A.; Parizeau, M.; Gagné, C. DEAP: Evolutionary Algorithms Made Easy. *J. Mach. Learn. Res.* **2012**, *13*, 2171–2175.
29. Vlasenko, D.; Hahn, B. Modeling of Elastic Cages in the Rolling Bearing Multi-Body Tool CABA3D. In *Multibody Dynamics 2019*; Kecskeméthy, A., Geu Flores, F., Eds.; Springer: Cham, Switzerland, 2020; Volume 53, pp. 96–103. [CrossRef]
30. Dowson, D.; Higginson, G.R. A Numerical Solution to the Elasto-Hydrodynamic Problem. *J. Mech. Eng. Sci.* **1959**, *1*, 6–15. [CrossRef]
31. Craig, J.R.; Bampton, M. Coupling of substructures for dynamic analyses. *AIAA J.* **1968**, *6*, 1313–1319. [CrossRef]
32. Makhavikou, V. Line-Fitting Method of Model Order Reduction in a Context of Elastic Multibody Simulation. Ph.D. Thesis, Otto von Guericke University Magdeburg, Magdeburg, Germany, 2015.
33. Schaeffler Technologies AG & Co. KG. *Wälzlagerpraxis: Handbuch zur Gestaltung und Berechnung von Wälzlagerungen*, 4th ed.; Antriebstechnik, Vereinigte Fachverlage: Mainz, Germany, 2015.
34. Schwarz, S.; Grillenberger, H.; Tremmel, S. Kennzahl zur Identifikation der Wälzlagerkäfigdynamik. In *VDI-Fachtagung. Gleit- und Wälzlagerungen*; Einsatz, G.-B., Ed.; VDI Verlag: Düsseldorf, Germany, 2019; Volume 2348.
35. Breunig, M.M.; Kriegel, H.P.; Ng, R.T.; Sander, J. LOF: Identifying density-based local outliers. *ACM Sigmod Rec.* **2000**, *29*, 93–104. [CrossRef]
36. Breiman, L. Random Forests. *Mach. Learn.* **2001**, *45*, 5–32. [CrossRef]
37. Friedman, J.H. Greedy Function Approximation: A Gradient Boosting Machine. *Ann. Stat.* **2001**, *29*, 1189–1232. [CrossRef]
38. Bishop, C.M. *Pattern Recognition and Machine Learning*, 8th ed.; Springer: New York, NY, USA, 2009.
39. Kingma, D.P.; Ba, J. Adam: A Method for Stochastic Optimization. In Proceedings of the 3rd International Conference on Learning Representations, San Diego, CA, USA, 7–9 May 2015.

Article

On the Importance of Temporal Information for Remaining Useful Life Prediction of Rolling Bearings Using a Random Forest Regressor

Christoph Bienefeld [1,2,*], Eckhard Kirchner [2], Andreas Vogt [1] and Marian Kacmar [1]

[1] Corporate Research, Robert Bosch GmbH, Robert-Bosch-Campus 1, 71272 Renningen, Germany; andreas.vogt1@de.bosch.com (A.V.); marian.kacmar@de.bosch.com (M.K.)

[2] Institute for Product Development and Machine Elements, Technical University of Darmstadt, Otto-Berndt-Straße 2, 64287 Darmstadt, Germany; office@pmd.tu-darmstadt.de

* Correspondence: christoph.bienefeld@de.bosch.com

Abstract: Rolling bearings are frequently subjected to high stresses within modern machines. To prevent bearing failures, the topics of condition monitoring and predictive maintenance have become increasingly relevant. In order to efficiently and reliably maintain rolling bearings in a predictive manner, an estimate of the remaining useful life (RUL) is of great interest. The RUL prediction quality achieved when using machine learning depends not only on the selection of the sensor data used for condition monitoring, but also on its preprocessing. In particular, the execution of so-called feature engineering has a major impact on prediction quality. Therefore, in this paper, various methods of feature engineering are presented based on rolling–bearing endurance tests and recorded structure-borne sound signals. The performance of these methods is evaluated in the context of a regression-based RUL model. Furthermore, the way in which the quality of RUL prediction can be significantly improved is demonstrated, by adding further processed, time-considering features.

Keywords: rolling bearings; remaining useful life; machine learning; feature engineering; condition monitoring; structure-borne sound; random forest; regression

Citation: Bienefeld, C.; Kirchner, E.; Vogt, A.; Kacmar, M. On the Importance of Temporal Information for Remaining Useful Life Prediction of Rolling Bearings Using a Random Forest Regressor. *Lubricants* **2022**, *10*, 67. https://doi.org/10.3390/lubricants10040067

Received: 14 February 2022
Accepted: 7 March 2022
Published: 14 April 2022

Publisher's Note: MDPI stays neutral with regard to jurisdictional claims in published maps and institutional affiliations.

Copyright: © 2022 by the authors. Licensee MDPI, Basel, Switzerland. This article is an open access article distributed under the terms and conditions of the Creative Commons Attribution (CC BY) license (https://creativecommons.org/licenses/by/4.0/).

1. Introduction

Modern machines with rotating components tend to use rolling bearings for their bearing arrangements. For reasons of energy efficiency and limited design space, the bearings are laid out as small as possible, which can lead to them being operated at the limits of their durability. An unforeseen failure of a bearing can cause considerable damage to the entire machine and its environment. Especially in the case of safety-relevant systems, an unforeseen failure must be avoided in any case. In order to prevent such unforeseen failures, condition monitoring and predictive maintenance are becoming increasingly important [1]. Condition monitoring involves using suitable sensors to record measurement data during operation, which is then processed to draw conclusions about the condition of the component [2]. If the condition is judged to be critical in this process, corrective actions such as maintenance can be planned. To be able to carry out such planning with as little risk as possible, it is essential to estimate the remaining useful life (RUL) of components [3].

Rolling bearing damage can occur in various ways. The damage can be caused by lack of lubrication, short-term overload or material fatigue due to long-term stress. Material fatigue usually manifests itself in the form of propagating pitting within the raceway surfaces [4]. Recently, for bearing damage detection, traditional condition monitoring methods have been increasingly combined with Artificial Intelligence (AI). Machine learning (ML), as a subfield of AI, plays an essential role here. ML algorithms can be used to recognize complex structures in data and to evaluate these structures [5]. This offers the possibility of automated inference from the data. Applied to the challenge of RUL prediction, these are approaches to automatically draw conclusions about the RUL from the data measured at the component. Among the machine learning algorithms used for RUL predictions there are different variants of neural networks, such as convolutional neural networks (CNN) [6], recurrent neural networks (RNN) [7], long short-term memory (LSTM) [8], and generative adversarial networks (GAN) [9]. Furthermore, there are contributions to state detection using random forest algorithms [10]. Machine learning is therefore becoming increasingly relevant, not least in the field of tribology [11].

When using machine learning, the achievable prediction quality is highly dependent on the type and quality of the data as well as the preprocessing used. Targeted data preprocessing has a significant impact on both the achievable prediction accuracy and the computational speed of the implemented algorithms [12,13]. In the context of rotating machinery, the measurement of structure-borne sound has proven particularly useful for drawing conclusions about the components' condition [14–16]. Therefore, the present article will also use structure-borne sound measurements to investigate the condition of rolling bearings and to predict their RUL.

Recent approaches for predictive maintenance based on electrical impedance measurements of rolling bearings can complement or even replace structure-borne sound measurements with in situ information [17]. The quality of the underlying model is continuously increased by considering unloaded rolling elements and modeling the detailed rolling contact geometry [18]. ML approaches are used to further enhance the predictive capabilities [19].

In a previous paper presented by the present authors, the influence of feature engineering on condition monitoring of rolling bearings was shown using a random forest regressor [20]. A feature engineering approach is presented in the previous work, which, compared to features from Lei et al. [21], achieves particularly good results in structure-borne sound-based condition detection. Based on these results, the feature-engineering approach is optimized and extended in this study regarding the prediction of remaining useful life. The aim is to develop a methodology that leads to a RUL prediction model with high accuracy and good traceability. Therefore, the investigations are focused on feature engineering and the consideration of information from the temporal past. In order to predict the RUL of rolling bearings, a methodology based on a random forest condition regression is presented.

2. Materials and Methods

To evaluate the developed feature engineering methods in the context of RUL predictions, a methodology in which all other model components and their parameters remain constant as boundary conditions is used. The approach used for this purpose is illustrated in Figure 1. The individual model parts are described in more detail within the subsequent sections.

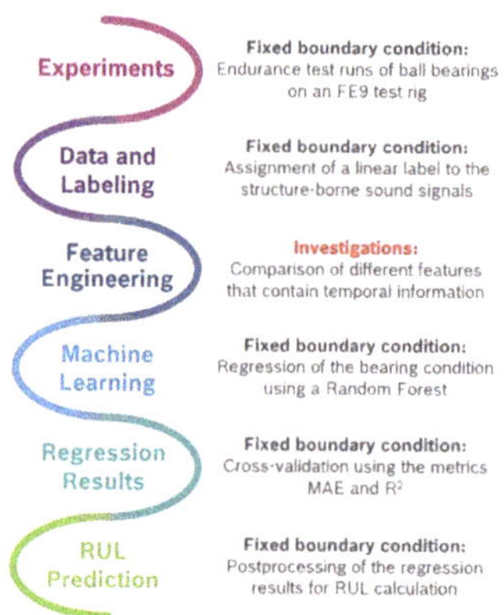

Figure 1. Overview of the methodology used.

2.1. Experiments

The investigations are based on structure-borne sound measurement data, which is recorded on a rolling bearing test rig. The concept of the FE9 test rig used was originally designed for testing rolling bearing greases. An electric motor drives the test head shaft via a belt. On one side of the shaft, an ancillary bearing is mounted, which is provided with circulating oil lubrication. The grease-lubricated test bearing, the wear of which is to be examined, is located on the other side of the shaft. In order to accelerate grease aging and thus, its wear, the test bearing is heated. An axial load is applied with the aid of a spring preload. The test head of the FE9 test rig can be seen in Figure 2.

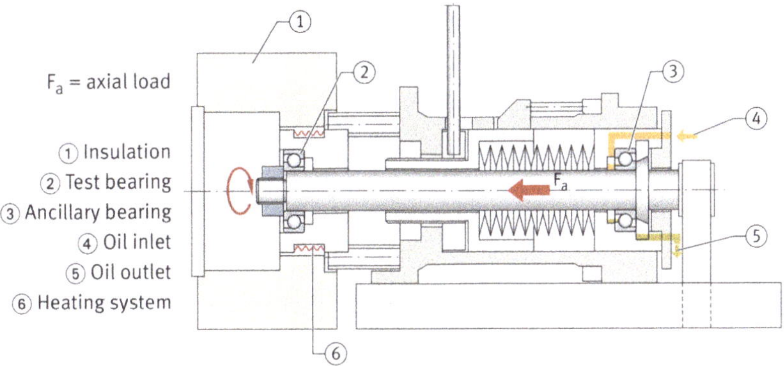

Figure 2. Test head of the FE9 test bench, adapted with permission from [22].

In the case of the tests evaluated in this work, the test bearings used are of type 6206-C-C3 (Schaeffler AG, Herzogenaurach, Germany) and lubricated with a low-temperature grease. The grease is used beyond the limits of its specification due to the applied thermal load, which is why the operating life is greatly reduced. A constant speed of 6000 rpm is present at the test head shaft. The axial load is 1500 N and the temperature of the heater on the test bearing is set to 140 °C. The sensor used is a three-axis piezo accelerometer of type PCB-356A15 (PCB Piezotronics, Depew, NY, USA). The sensor is mounted close to the test bearing, as shown in Figure 3.

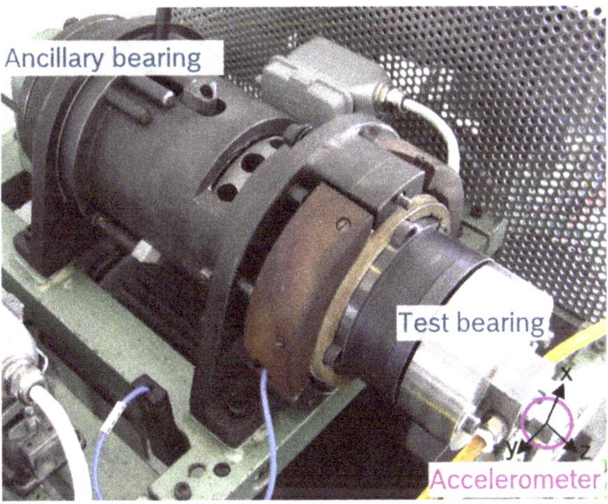

Figure 3. Placement of the accelerometer on the test bench, adapted with permission from ref. [20].

For the investigations carried out here, only data from the sensor's *X*-axis, which is aligned radially to the bearing, is analyzed. The data is acquired at a sampling rate of 20 kHz. An imc CRONOSflex (imc Test & Measurement GmbH, Berlin, Germany) data acquisition system is used in the measuring chain with an 8th order Cauer LP anti-aliasing filter having a cut-off frequency of 8 kHz. The amplitude is resolved with 24 bits. The measurement data is recorded at intervals of 1 s with intervening pauses of 59 s. A total of nine endurance runs are investigated. A threshold value in the power consumption of the driving electric motor is defined as a termination criterion for the experiments. This leads to test run times ranging from 10 h to 20 h. At the end of the tests, the test bearings show very similar damage patterns in the form of pitting. Figure 4 shows an example of the inner ring of one bearing after endurance testing.

Figure 4. Pitting on a ball bearing inner ring after its test run, reprinted with permission from ref. [20].

2.2. Data and Labeling

The aim of the procedure used here is to directly infer the bearings remaining useful life from the trained ML model. Therefore, a supervised learning approach in terms of a regression is used. The label must represent the progressive bearing damage. As already shown in [20], a label that linearly increases from 0 to 1 is used for this purpose. A similar labeling approach has also been presented in [23]. Figure 5 shows the label based on the structure-borne sound signals of a single test run. From a mathematical point of view, the label can be described as the normalized test run duration. The value 0 represents the original condition of the bearing, while 1 indicates the end of its useful life.

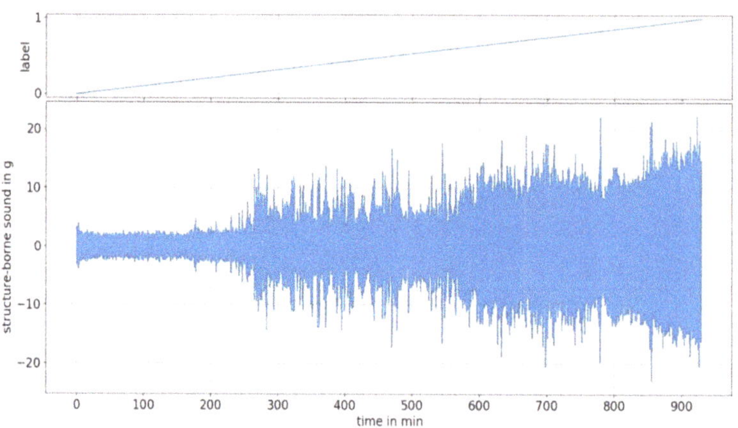

Figure 5. Measurement and assigned label for an exemplary test run.

2.3. Feature Engineering

A wide variety of feature-engineering methods have already been described in the literature [13,24]. The focus of the research conducted here is on the comparison of different feature-engineering methods that consider the temporal past in the context of feature generation. As a basis, the so-called averaged-frequency-band (afb) features are used, which have already been shown in [20] to be particularly performant and computationally efficient compared to the features proposed by Lei et al. [21]. The studies in [20] were based on the same data set also used for the present work. Starting from the afb features, additional features are now to be generated, which contain the information of the temporal past. The influence of these processed features on the RUL prediction is to be investigated.

2.3.1. Averaged-Frequency-Band Features

As base features, the so-called averaged-frequency-band features are used, the calculation method of which is visualized in Figure 6. To calculate the afb features for each 1 s measurement interval, the data of the interval is first transformed into the frequency domain by means of an FFT. The resulting amplitude spectrum is divided into frequency bands of equal width. Finally, the average values of the amplitudes within the formed frequency bands are used as features. Thus, an afb feature describes the average value of the amplitudes within a frequency band.

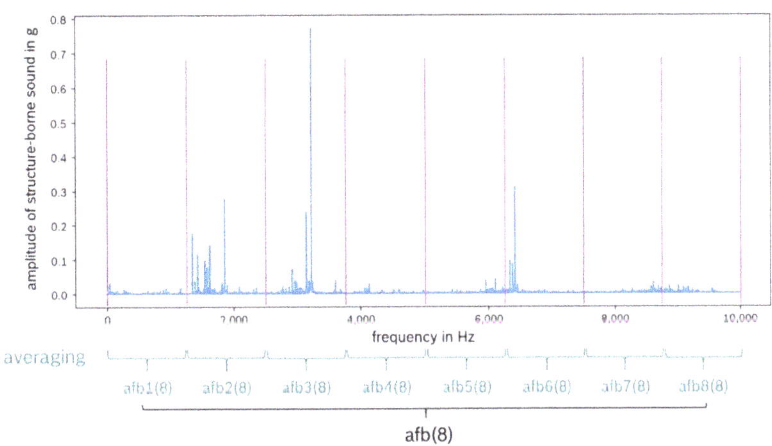

Figure 6. Determination of the averaged-frequency-band features, adapted with permission from ref. [20].

Based on preliminary investigations and in order to keep the total number of features and thus the model complexity at a moderate level, the number of frequency bands in this case is set to 8.

2.3.2. Rolling Mean Features

In order to utilize information from the temporal past, rolling means can be used. In the case presented here, these rolling means were calculated from the afb features presented previously. To be able to represent the short-term dynamics as well as the long-term behavior, several averages are formed over different time spans. Progressively increasing time spans seem to make sense for this use, which was confirmed in preliminary studies. The progressive staggering of rolling means is shown in Figure 7 for three rolling mean durations using the time course of afb1(8).

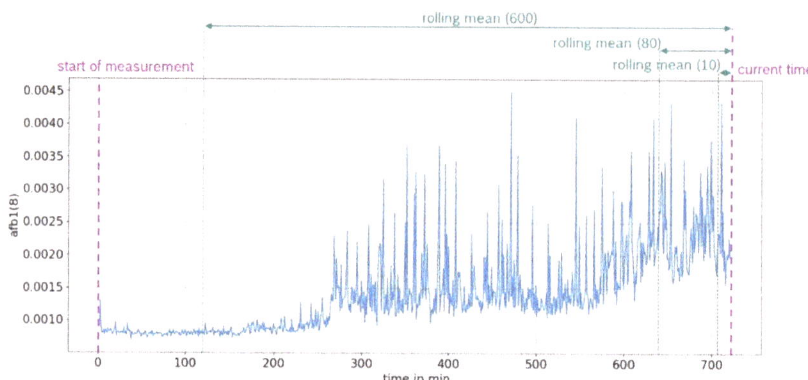

Figure 7. Progression of an averaged-frequency-band and three associated rolling means.

2.3.3. Cumulative Features

Another way to account for temporal information is to use accumulated quantities. Already in [25], the cumulative sum of values was proposed to generate features with monotonic behavior. These accumulated features provide long-term trends, which helps the ML algorithm in its decision making. In the case presented here, the afb features are used for accumulation. Each afb feature is summed up cumulatively from the beginning of the experiment.

2.4. Machine Learning

The goal of the machine learning is to approximate an unknown function, which maps the input features to the label. Since the label is defined in the form of a continuous variable, this is supervised learning in terms of a regression [26]. In the field of machine learning, there is a wide variety of regression algorithms [5]. A comprehensive overview of the available Deep Learning methods can be found in [27]. In [10], a random forest approach has already been used to detect the state of journal bearings. The aim of the present work is to show the influence of targeted feature engineering on RUL prediction performance. Therefore, traceability shall be as good as possible. For this reason, deep learning algorithms are not used here, instead, a random forest regressor is chosen. A random forest is an ensemble method based on decision trees [28]. It is considered to be very robust and to provide continuously good results compared to other regression algorithms.

In the workflow used here to investigate feature engineering, the machine learning algorithm is considered as a constant boundary condition. Therefore, the parameters of the random forest are kept fixed. Based on preliminary studies, the number of trees is set to 500, and the maximum tree depth is limited to 20. The models used in this work are implemented in Python using the numpy, pandas, scipy, and matplotlib libraries. Additionally, the library Scikit-learn is used for the implementation of the random forest and metrics for result evaluation.

To evaluate the models built with the different feature engineering methods, a 9-fold cross-validation is used. Out of the total nine endurance test runs available, eight endurance tests are used for training. The remaining test run is used for the test data set, which means that the test data is always completely separated from the training data. This is repeated a total of nine times so that the data from each test is used as independent test data once.

The quality of the prediction is evaluated using metrics. For this purpose, the MAE and the R^2 are chosen. The MAE (Mean Absolute Error) provides a directly interpretable result of the regression quality in the context of the label used here. For example, an MAE of 0.05 means that the prediction of the current bearing condition is on average 5 % from the true value. Consequently, the MAE tends to 0 in case of a perfect model. In addition, the R^2, which is called the coefficient of determination, provides a general measure of the quality of a regression. It tends towards a maximum value of 1 for optimal predictions. The smaller the value, the worse the prediction [29]. For the overall evaluation in the results section below, the average value of the nine metrics calculated during cross-validation is considered. This ensures an evaluation of the model quality based on the entire data set.

2.5. RUL Prediction

To infer the predicted remaining useful life RUL_{pred} based on the predictions of the label within the regression, the following equation is used:

$$RUL_{pred} = \frac{t}{y_{pred}} \cdot \left(1 - y_{pred}\right) \tag{1}$$

Here, t is the current operating time and y_{pred} is the label predicted at the corresponding time. The described mathematical relationship results from the background of the selected label, which corresponds to the normalized test-run time. At the beginning of the measurement, where the predicted label y_{pred} is close to zero, RUL prediction is not practical due to large inaccuracies, which can be directly justified by Equation (1). Dividing by small y_{pred} then leads to very large fluctuations in the RUL prediction, caused by only slight variations in the predicted label. For this reason, RUL prediction is evaluated exclusively for the second half of the test runs. The result evaluation by means of the RUL-based MAE is also performed exclusively on the second half of the test runs.

In order to compare the predicted with the true remaining useful life RUL_{true}, the latter must also be calculated. This is performed using the total operating time until bearing failure T and the true label at the respective time y_{true}:

$$RUL_{true} = T \cdot (1 - y_{true}) \tag{2}$$

3. Results

The previously presented feature engineering approaches are now compared to each other. In detail, the three approaches listed in Table 1 are considered.

Table 1. Feature engineering approaches used.

Abbreviation	Description	No. of Features
afb(8)	Averaged-frequency-band features using 8 frequency bands	8
rollingmeans(10, 80, 600)	Time domain rolling means calculated based on the afb(8) with window sizes of 10, 80 and 600 min	24
cumsum	Cumulative sum of the afb(8)	8

The first feature set is denoted as afb(8). No temporal past information is used with this feature set. It therefore serves as a reference. For the second feature set, the afb(8) approach is combined with the rollingmeans(10, 80, 600) approach. The third feature set is a combination of the reference features afb(8) and the cumsum approach.

To compare the three feature sets mentioned above, the workflow shown in Figure 1 is used, keeping the boundary conditions constant. The resulting regression and RUL predictions are shown in Figure 8 using a single test data set. While the plots on the left show the regression results of the trained machine learning algorithm, the plots on the right visualize the RUL prediction derived from it. The metrics MAE and R^2 of the visualized results are entered within Figure 8.

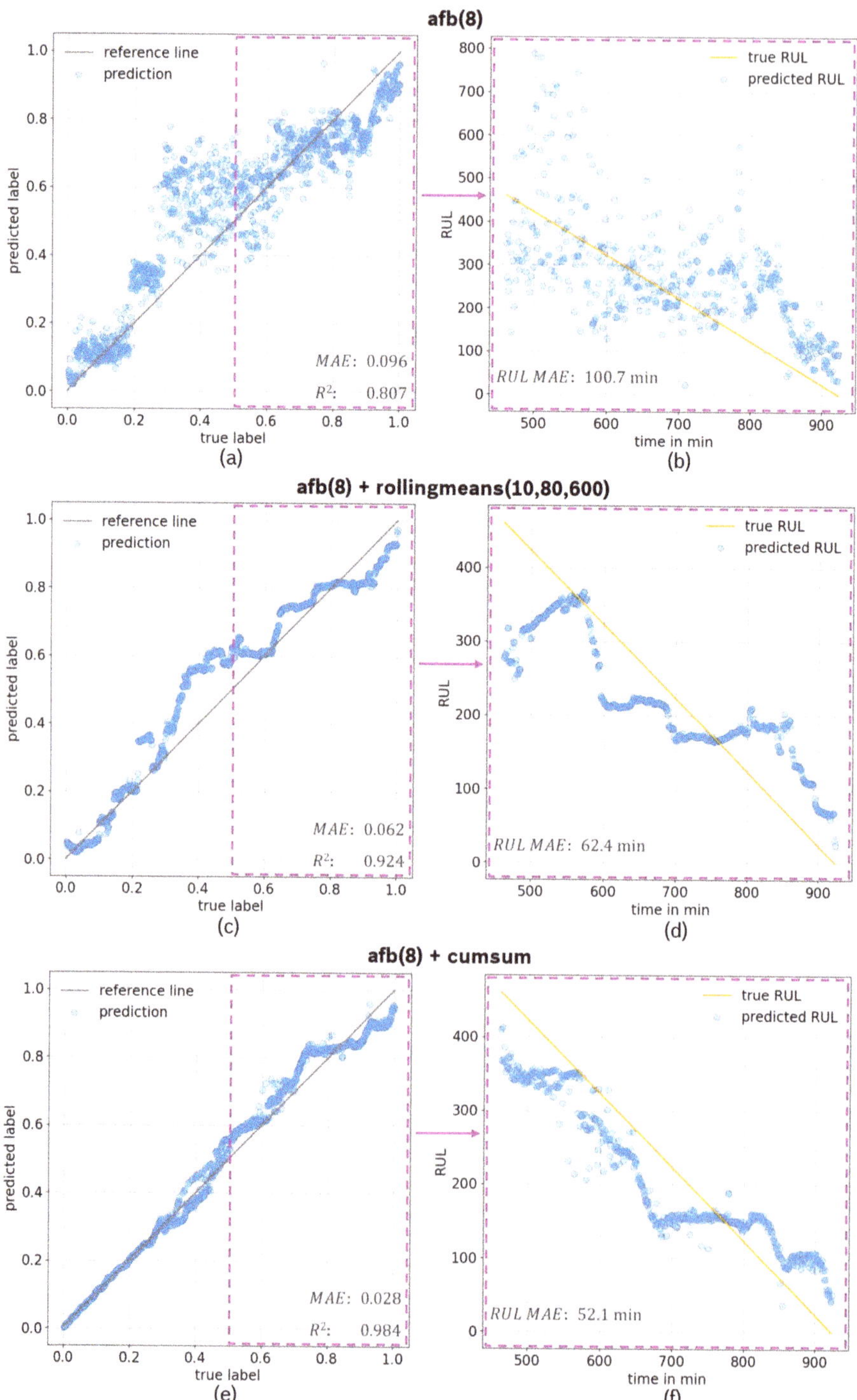

Figure 8. Results of regression (**a,c,e**) and RUL prediction (**b,d,f**) visualized based on the cross-validation run of test bearing No. 1 for the three different feature sets.

The prediction scatters strongly when using only afb(8) features, see Figure 8a. Several states can be identified in the predicted label, between which the prediction changes quite abruptly. In the case of an ideal prediction, all prediction points (blue) would be exactly on top of the reference line. Thus, the vertical deviation of the predictions from the reference line visualizes how inaccurate the prediction is. The same applies to the mapping of the RUL. Here, with optimal prediction, the test data points would align with the orange line, representing the true RUL. The corresponding RUL prediction using the afb(8) features is very inaccurate due to the large prediction spread of the regression results and poorly represents the true RUL, as can be seen in Figure 8b. A significant improvement in prediction quality is achieved by adding the rolling-means, as shown in Figure 8c,d. On average, the forecast shows similar trends, but is much less scattered. This is evident not only in the predicted label, but also within the resulting RUL prediction. Further improvement of the results is achieved with the combination of the afb(8) and cumsum features, which is visualized in Figure 8e,f. The steps visible with the other two feature sets disappear almost completely here. These improvements of the results can be determined not only visually, but also based on the metrics calculated. Smaller MAEs and larger R^2s represent the prediction improvements.

Since Figure 8 only illustrates one of the total of nine cross-validation runs, the overall cross-validation results are summarized in Table 2. For this purpose, the average of the regression MAE and the regression R^2 calculated via cross-validation are entered. Additionally, the averaged MAE of the RUL prediction as well as the relative deviation of the MAE with respect to the test run times are evaluated in the last two rows.

Table 2. Results of cross-validated regression and RUL prediction.

Cross-Validated Metrics of Regression	afb(8)	afb(8) + Rollingmeans (10, 80, 600)	afb(8) + Cumsum
∅ MAE of Regression	0.0892	0.0753	0.0506
∅ R^2 of Regression	0.841	0.875	0.95

Test Bearing No.	Experiment Runtime in min	MAE of RUL Prediction in min		
		afb(8)	afb(8) + Rollingmeans (10, 80, 600)	afb(8) + Cumsum
1	927	100.7	62.4	52.1
2	1073	197.6	75.7	86.5
3	808	91.1	68.1	47.8
4	824	155.3	224.1	107.6
5	1011	123	94.9	48.2
6	882	73.5	70.9	48.7
7	1191	146.7	79	117.9
8	746	92.4	47.3	60
9	668	132.9	154.5	133.8
∅	903.3	123.7	97.4	78.1
∅ relative error of RUL prediction		13.8 %	11.4 %	8.9 %

Looking at the averaged metrics from cross-validation, the results already obtained in Figure 8 are supported. Adding the rolling mean features to the afb(8) yields a significant improvement, with the cumulative features performing even better compared to the rolling mean features. In the MAE of the individual test bearings' RUL, it is noticeable that this sequence of model performance does not apply quantitatively in the same way for each test bearing. Consequently, there is a non-negligible dispersion of the individual test data sets. A possible explanation for this dispersion is the different physical wear behavior of the various bearing endurance test runs used.

4. Discussion

Based on the experimental data used, the results presented show that a clear improvement in RUL prediction is possible with the help of temporal information, implemented by means of time-considering features. By using a well-defined workflow where only the feature sets are changed, the impact of the different features on the RUL prediction performance is evaluated. For the RUL prediction, a random forest regression approach is used. Comparing the two presented ways of incorporating temporal past information in the form of extended feature sets, the approach of cumulatively generated features performs particularly well. By using this extended feature set, the averaged MAE of RUL predictions can be reduced by 37% in comparison to the use of base features only. Calculating rolling means with progressively staggered window widths also adds value in terms of predictive accuracy, although the results are slightly worse than those obtained with the cumulative approach. In the case presented here, the base features are formed from the so-called averaged-frequency-bands, which have already been shown to perform particularly well on the data used in [20]. The authors assume that the methodology presented here will lead to improved RUL predictions for other base features in an analogous manner. A validation of this hypothesis is still pending at this point.

It should be noted that the evaluations carried out here are based on test data obtained on a rolling bearing test rig under constant operating conditions. Limitations are to be expected when implementing the methodology proposed here in a real application, with varying boundary conditions such as speeds, loads or temperatures. In particular, the formation of accumulated features could be error-prone, since each individual point in time has an influence on the entirety of the following time span. Thus, continuous and reliable measurement data acquisition is indispensable for the correct determination of accumulatively formed features.

Future work can investigate further approaches of feature engineering and the possibilities of considering temporal information. The implementation of further RUL prediction methods and the possibilities of deep learning algorithms have been omitted here in order to focus on the integration of temporal information via extended feature engineering approaches. For comparison purposes, it seems reasonable to also consider deep learning methods, such as CNNs, RNNs or LSTMs, which natively offer the possibility to take temporal information into account. However, with these methods, the comprehensibility of decision making is lost. Furthermore, with regard to hybrid models, it seems promising to motivate the development of novel features by physical backgrounds. The investigations should also be extended to additional data that are recorded at non-constant bearing operating conditions. In order to achieve satisfactory RUL prediction results even at non-constant operating conditions, the methods may have to be extended.

Author Contributions: Conceptualization, C.B.; methodology, C.B., A.V. and M.K.; software, C.B.; validation, E.K. and A.V.; formal analysis, E.K. and A.V.; investigation, C.B.; resources, C.B., E.K., A.V. and M.K.; data curation, C.B., A.V. and M.K.; writing—original draft preparation, C.B.; writing—review and editing, E.K., A.V. and M.K.; visualization, C.B.; supervision, E.K.; project administration, M.K. All authors have read and agreed to the published version of the manuscript.

Funding: We acknowledge support by the Deutsche Forschungsgemeinschaft (DFG–German Research Foundation) and the Open Access Publishing Fund of Technical University of Darmstadt.

Data Availability Statement: The measurement data presented in this paper are available on request from the corresponding author after prior approval by Robert Bosch GmbH.

Acknowledgments: The authors would like to thank the Robert Bosch GmbH for enabling the studies presented in this paper, including the measurement data used, to be elaborated in the Corporate Research of the Robert Bosch GmbH.

Conflicts of Interest: The authors declare no conflict of interest.

References

1. Pech, M.; Vrchota, J.; Bednář, J. Predictive Maintenance and Intelligent Sensors in Smart Factory: Review. *Sensors* **2021**, *21*, 1470. [CrossRef] [PubMed]
2. Barszcz, T. *Vibration-Based Condition Monitoring of Wind Turbines*; Springer: New York, NY, USA, 2019; ISBN 978-3-030-05969-9.
3. Kim, N.-H.; An, D.; Choi, J.-H. *Prognostics and Health Management of Engineering Systems*; Springer: New York, NY, USA, 2016; ISBN 978-3-319-44740-7.
4. Schaeffler Monitoring Services GmbH. *Condition Monitoring Praxis*; Vereinigte Fachverlage GmbH: Mainz, Germany, 2019; ISBN 978-3-7830-0419-9.
5. Berry, M.W.; Mohamed, A.; Yap, B.W. *Supervised and Unsupervised Learning for Data Science*; Springer Nature: Cham, Switzerland, 2020; ISBN 978-3-030-22475-2.
6. Li, X.; Ding, Q.; Sun, J.-Q. Remaining Useful Life Estimation in Prognostics Using Deep Convolution Neural Networks. *Reliab. Eng. Syst. Saf.* **2018**, *172*, 1–11. [CrossRef]
7. Han, T.; Pang, J.; Tan, A.C.C. Remaining Useful Life Prediction of Bearing Based on Stacked Autoencoder and Recurrent Neural Network. *J. Manuf. Syst.* **2021**, *61*, 576–591. [CrossRef]
8. Xia, J.; Feng, Y.; Lu, C.; Fei, C.; Xue, X. LSTM-Based Multi-Layer Self-Attention Method for Remaining Useful Life Estimation of Mechanical Systems. *Eng. Fail. Anal.* **2021**, *125*, 105385. [CrossRef]
9. Suh, S.; Lukowicz, P.; Lee, Y.O. Generalized Multiscale Feature Extraction for Remaining Useful Life Prediction of Bearings with Generative Adversarial Networks. *Knowl. Based Syst.* **2022**, *237*, 107866. [CrossRef]
10. Prost, J.; Cihak-Bayr, U.; Neacşu, I.A.; Grundtner, R.; Pirker, F.; Vorlaufer, G. Semi-Supervised Classification of the State of Operation in Self-Lubricating Journal Bearings Using a Random Forest Classifier. *Lubricants* **2021**, *9*, 50. [CrossRef]
11. Marian, M.; Tremmel, S. Current Trends and Applications of Machine Learning in Tribology—A Review. *Lubricants* **2021**, *9*, 86. [CrossRef]
12. Lei, Y. *Intelligent Fault Diagnosis and Remaining Useful Life Prediction of Rotating Machinery*; Butterworth-Heinemann Ltd.: Oxford, UK; Xi'an Jiaotong University Press: Oxford, UK, 2017; ISBN 978-0-12-811534-3.
13. Lei, Y.; Li, N.; Guo, L.; Li, N.; Yan, T.; Lin, J. Machinery Health Prognostics: A Systematic Review from Data Acquisition to RUL Prediction. *Mech. Syst. Signal Process.* **2018**, *104*, 799–834. [CrossRef]
14. Randall, R.B. *Vibration-Based Condition Monitoring: Industrial, Aerospace, and Automotive Applications*; Wiley: Chichester, West Sussex, UK; Hoboken, NJ, USA, 2011; ISBN 978-0-470-74785-8.
15. Wang, X.; Mao, D.; Li, X. Bearing Fault Diagnosis Based on Vibro-Acoustic Data Fusion and 1D-CNN Network. *Measurement* **2021**, *173*, 108518. [CrossRef]
16. Song, Q.; Zhao, S.; Wang, M. On the Accuracy of Fault Diagnosis for Rolling Element Bearings Using Improved DFA and Multi-Sensor Data Fusion Method. *Sensors* **2020**, *20*, 6465. [CrossRef] [PubMed]
17. Schirra, T.; Martin, G.; Vogel, S.; Kirchner, E. Ball Bearings as Sensors for Systematical Combination of Load and Failure Monitoring. In *International Design Conference—Design 2018*; Universität Zagreb, The Design Society: Glasgow, UK, 2018; pp. 3011–3022. [CrossRef]
18. Schirra, T.; Martin, G.; Puchtler, S.; Kirchner, E. Electric Impedance of Rolling Bearings—Consideration of Unloaded Rolling Elements. *Tribol. Int.* **2021**, *158*, 106927. [CrossRef]
19. Kirchner, E.; Bienefeld, C.; Schirra, T.; Moltschanov, A. Predicting the Electrical Impedance of Rolling Bearings Using Machine Learning Methods. *Machines* **2022**, *10*, 156. [CrossRef]
20. Bienefeld, C.; Vogt, A.; Kacmar, M.; Kirchner, E. Feature-Engineering für die Zustandsüberwachung von Wälzlagern mittels maschinellen Lernens. *Tribol. Schmier.* **2021**, *68*, 5–11. [CrossRef]
21. Lei, Y.; He, Z.; Zi, Y. A New Approach to Intelligent Fault Diagnosis of Rotating Machinery. *Expert Syst. Appl.* **2008**, *35*, 1593–1600. [CrossRef]
22. Schaeffler Technologies AG & Co. KG. Lubrication of Rolling Bearings, 2013, p. 160. Available online: https://www.schaeffler.com/remotemedien/media/_shared_media/08_media_library/01_publications/schaeffler_2/tpi/downloads_8/tpi_176_de_en.pdf (accessed on 11 February 2022).
23. Guo, L.; Li, N.; Jia, F.; Lei, Y.; Lin, J. A Recurrent Neural Network Based Health Indicator for Remaining Useful Life Prediction of Bearings. *Neurocomputing* **2017**, *240*, 98–109. [CrossRef]
24. Tom, K.F. A Primer on Vibrational Ball Bearing Feature Generation for Prognostics and Diagnostics Algorithms. *Sens. Electron Devices ARL.* 2015. Available online: https://apps.dtic.mil/sti/citations/ADA614145 (accessed on 11 February 2022).
25. Javed, K.; Gouriveau, R.; Zerhouni, N.; Nectoux, P. A Feature Extraction Procedure Based on Trigonometric Functions and Cumulative Descriptors to Enhance Prognostics Modeling. In Proceedings of the 2013 IEEE Conference on Prognostics and Health Management (PHM), Gaithersburg, MD, USA, 24–27 June 2013; pp. 1–7. [CrossRef]
26. Backhaus, K.; Erichson, B.; Plinke, W.; Weiber, R. *Multivariate Analysemethoden: Eine Anwendungsorientierte Einführung*; 15. vollständig überarbeitete Auflage; Springer Gabler: Berlin/Heidelberg, Germany, 2018; ISBN 978-3-662-56654-1.
27. Rezaeianjouybari, B.; Shang, Y. Deep Learning for Prognostics and Health Management: State of the Art, Challenges, and Opportunities. *Measurement* **2020**, *163*, 107929. [CrossRef]

28. Breiman, L. Random Forests. *Mach. Learn.* **2001**, *45*, 5–32. [CrossRef]
29. Matzka, S. *Springer Fachmedien Wiesbaden GmbH. Künstliche Intelligenz in den Ingenieurwissenschaften Maschinelles Lernen Verstehen und Bewerten*; Springer Vieweg: Wiesbaden, Germany, 2021; ISBN 978-3-658-34640-9.

 *lubricants*

Review

Current Trends and Applications of Machine Learning in Tribology—A Review

Max Marian [1,*] and **Stephan Tremmel** [2]

[1] Engineering Design, Friedrich-Alexander-University Erlangen-Nuremberg (FAU), Martensstr. 9, 91058 Erlangen, Germany
[2] Engineering Design and CAD, University of Bayreuth, Universitätsstr. 30, 95477 Bayreuth, Germany; stephan.tremmel@uni-bayreuth.de
* Correspondence: marian@mfk.fau.de

Abstract: Machine learning (ML) and artificial intelligence (AI) are rising stars in many scientific disciplines and industries, and high hopes are being pinned upon them. Likewise, ML and AI approaches have also found their way into tribology, where they can support sorting through the complexity of patterns and identifying trends within the multiple interacting features and processes. Published research extends across many fields of tribology from composite materials and drive technology to manufacturing, surface engineering, and lubricants. Accordingly, the intended usages and numerical algorithms are manifold, ranging from artificial neural networks (ANN), decision trees over random forest and rule-based learners to support vector machines. Therefore, this review is aimed to introduce and discuss the current trends and applications of ML and AI in tribology. Thus, researchers and R&D engineers shall be inspired and supported in the identification and selection of suitable and promising ML approaches and strategies.

Keywords: tribology; machine learning; artificial intelligence; triboinformatics; databases; data mining; meta-modeling; artificial neural networks; monitoring; analysis; prediction; optimization

Citation: Marian, M.; Tremmel, S. Current Trends and Applications of Machine Learning in Tribology—A Review. *Lubricants* **2021**, *9*, 86. https://doi.org/10.3390/lubricants9090086

Received: 27 July 2021
Accepted: 25 August 2021
Published: 1 September 2021

Publisher's Note: MDPI stays neutral with regard to jurisdictional claims in published maps and institutional affiliations.

Copyright: © 2021 by the authors. Licensee MDPI, Basel, Switzerland. This article is an open access article distributed under the terms and conditions of the Creative Commons Attribution (CC BY) license (https://creativecommons.org/licenses/by/4.0/).

1. Introduction

Tribology has been and continuous to be one of the most relevant fields in today's society, being present in almost aspects of our lives. The importance of friction, lubrication and wear is also reflected by the significant share of today's world energy consumption [1]. The understanding of tribology can pave the way for novel solutions for future technical challenges. At the root of all advances are multitudes of precise experiments and advanced computer simulations across different scales and multiple physical disciplines [2]. In the context of tribology 4.0 [3] or triboinformatics [4], advanced data handling, analysis, and learning methods can be developed based upon this sound and data-rich foundation and employed to expand existing knowledge. Moreover, tribology is characterized by the fact that it is not yet possible to fully describe underlying processes with mathematical terms, e.g., by differential equations. Therefore, modern Machine Learning (ML) or Artificial Intelligence (AI) methods provide opportunities to explore the complex processes in tribological systems and to classify or quantify their behavior in an efficient or even real-time way [5]. Thus, their potential also goes beyond purely academic aspects into actual industrial applications. The advantages and the potential of ML and AI techniques are seen especially in their ability to handle high dimensional problems and data sets as well as to adapt to changing conditions with reasonable effort and cost [6]. They allow for the identification of relevant relations and/or causality, thus expand the existing knowledge with already available data. Ultimately, through analyses, predictions, and optimizations, transparent and precise recommendations for action could be derived for the engineer, practitioner, or even the potentially smart and adaptive tribological system itself. Nevertheless, compared to other disciplines or domains, e. g. economics and finances [7], health care [8], or manufacturing processes [6], the applicability of ML and AI techniques for tribological issues is still surprisingly underexplored. This is certainly also due to the interdisciplinarity and the quantity of heterogenous data from simulations on different scales or manyfold measurement devices with individual uncertainties. Furthermore, friction and wear characteristics do not represent hard data, but irreversible loss quantities with a dependence on time and test conditions [9].

To help pave the way, a more detailed analysis of the available ML/AI techniques as well as their applicability, strengths and limitations with regard to the requirements of the respective tribological application scenario with its specific, theoretical foundations is essential. Therefore, this contribution aims to introduce the trends and applications of ML algorithms with relevance to the domain of tribology. While other reviews were more generic [10], had a more concise scope [5], or focused on a specific technique (i.e., artificial neural networks [11]), this review article is also intended to cover a wider range of techniques and in particular to shed light on the broad applicability to various fields with tribological issues. Thus, the interested reader shall be provided with a high-level understanding of the capabilities of certain methods with respect to the tribological applications ranging from composite materials over drive technology or manufacturing to surface engineering and lubricant formulations. This article is therefore structured in such a way that first the theoretical background is introduced, and the results of a quantitative meta-literature analysis are presented (Section 2). Thereby, the published work on ML in the field of tribology is clustered according to the level or intention, the scale under consideration, the nature of the database and the area of application. Organized according to the latter, the work and progress reported in literature is then discussed in detail (Sections 3.1–3.6) before the main trends are summarized and concluded (Section 4).

2. Background and a Quantitative Survey on Machine Learning in Tribology

ML is part of AI [12] and thus originally a sub-domain of computer science. AI and ML are formed by logic, probability theory, algorithm theory, and computing [13]. In a first step, ML involves designing computing systems for a special task that can learn from training data over time and develop and refine experience-based models that predict outcomes. The system can thus be used to answer questions in the given field [12]. There are a number of different algorithms that can be used for ML, whereby the suitability is strongly task-dependent. Generally, algorithms can be categorized as "supervised learning" or "unsupervised learning" [12]. For the former, algorithms learn a relation from a given set of input and output data vectors. During learning, a "teacher" (e.g., an expert) provides the correct inputs and outputs. In unsupervised learning, the algorithm generates a statistical model that describes a given data set without the model being evaluated by a "teacher". Furthermore, reinforcement learning features different characteristics, although it is sometimes classified as supervised learning. Instead of induction from pre-classified examples, an "agent" "experiments" with the system and the system responds to the experiments with reward or punishment. The agent thus optimizes the behavior with the goal of maximizing reward and minimizing punishment. While the classification of the three learning types mentioned above is common and widely accepted, there is no consensus on which algorithms should be assigned to which category. One possible allocation following [6] is illustrated in Figure 1.

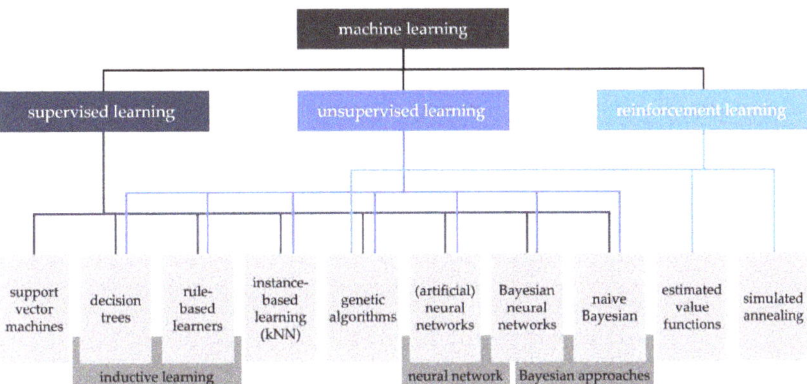

Figure 1. Classification of machine learning techniques. Redrawn from [6] with permission by CC BY 4.0 (Taylor and Francis).

The basic idea of *support vector machines* (SVM) is that a known set of objects is represented by a vector in a vector space. Hyperplanes are introduced into this space to separate the data points. In most cases, only the subset of the training data that lies on the boundaries of two planes is relevant. These vectors are the name-giving support vectors [14]. To account for nonlinear boundaries, kernel functions are an essential part of SVM. By using the kernel trick, the vector space is transformed into an arbitrarily higher-dimensional space, so that arbitrarily nested vector sets are linearly separable [15]. *Decision trees* (DT) are ordered, directed trees that illustrate hierarchically successive decisions [12]. A decision tree always consists of a root node and any number of inner nodes as well as at least two leaves. Each node represents a logical rule, and each leaf represents an answer to the decision problem. The complexity and semantics of the rules are not restricted, although all decision trees can be reduced to binary decision trees. In this case, each rule expression can take only one of two values [16]. A possibility to increase the classification quality of decision trees is the use of sets of decision trees instead of single trees, this is called *decision forests* [17]. If decision trees are uncorrelated, they are called *random forest* (RF) [18]. The idea behind decision forests is that while a single, weak decision tree may not provide optimal classification, a large number of such decision trees are able to do so. A widely used method for generating decision forests is boosting [19]. In *rule-based learners*, the output results from composing individual rules, which are typically expressed in the form "If–Then". Rule-based ML methods typically comprise a set of rules, that collectively make up the prediction model. *K-Nearest-Neighbor algorithms* (kNN) are classification methods in which class assignment is performed considering *k* nearest neighbors, which were classified before. The choice of *k* is crucial for the quality of the classification [16]. In addition, different distance measures can be considered [20]. *Artificial neural networks* (ANN) are essentially modeled on the architecture of natural brains [21]. They are 'a computing system made up of a number of quite simple but highly interconnected processing elements (neurons), which process information by their dynamic state response to external inputs' [12]. The so-called transfer function calculates the neuron's network input based on the weighting of the inputs [22]. Calculating the output value is done by the so-called activation function considering a threshold value [12,22]. Weightings and thresholds for each neuron can be modified in a training process [16]. The overall structure of neurons and interconnections, in particular how many neurons are arranged in a layer and how many neurons are arranged in parallel per layer, is called topology or architecture. The last layer is called the output layer and there can be several hidden layers between the input and the output layer (multilayer ANN) [21]. While single-layer networks can only be used to solve linear problems, multilayer networks also allow the solution of nonlinear problems [12]. Feedforward means, that neuron outputs are routed in processing direction only. Recurrent networks, in contrast, also have feedback loops. Commonly, ANNs are represented in graph theory notation, with nodes representing neurons and edges representing their interconnections.

Already rather early works in the field of tribology from the 1980s can be assigned to the current understanding of ML. For example, Tallian [23,24] introduced computerized databases and expert systems to support tribological design decisions or failure diagnosis. Other initial studies were concerned, for example, with the prediction of tribological properties [25–27] or classification of wear particles [28,29]. Between 1985 and today, almost 130 publications related to ML in tribology were identified within a systematic literature review (see Prisma flow chart in Figure 2a), whereas the number of papers initially increased slowly and more rapidly within the last decade (Figure 2b). During the latter period, the number of publications has more than tripled, which represents a faster growth than the general increase in the number of publications in the field of tribology (the numbers of Scopus-listed publications related to tribology grew by a factor of 2.3 between 2010 and today). It can therefore be highly expected that this trend will continue and that ML techniques will also become increasingly prominent in the field of tribology due to technological advances and decreasing barriers and preconceptions. Therefore, the analysis of the publications with respect to the fields of application is of particular interest, which is illustrated in Figure 2c. Especially in the areas of composite materials, drive technology, and manufacturing, numerous successful implementations of ML algorithms can already be found. Yet, some studies can also be found for surface engineering, lubricant formulation or manufacturing. As depicted in Figure 2d, ML techniques are applied for monitoring tribo-systems or for pure analytical/diagnostic purposes, but especially for predicting and optimizing the tribological behavior with respect to the friction and wear behavior. The scales under consideration are mainly on the macro and/or micro level, see Figure 2e. However, a few works also show the applicability down to the nano scale. Finally, it could be observed that the database for training the ML algorithms can also be generated based on numerical or theoretical fundamentals from simulation models or on information from the literature. However, the vast majority (roughly three quarters) of the published work is based upon experimentally generated data sets (Figure 2f).

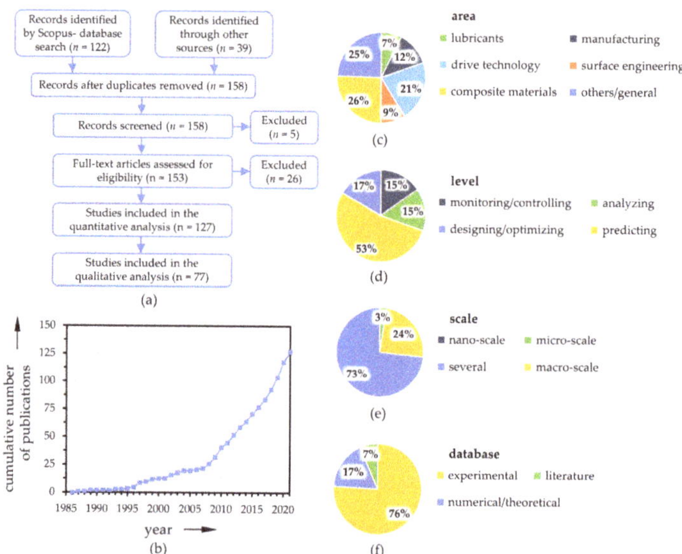

Figure 2. Systematic protocol (Prisma flow chart) for the paper collection/screening (**a**), and number of publications per year (**b**) and clustered by the area of application (**c**), level (**d**), scale (**e**) and database (**f**).

3. Results

As illustrated in the previous section, there is a wide variety of implementations of different ML/AI approaches. In order to give a more detailed overview of applications for ML to solve tribological issues, various cases are presented and discussed in the following. Since the aim is to address interested readers from the field of tribology and to show how ML can be used effectively in their respective fields, this is organized according to the area of application in descending order of the number of published works.

3.1. Composite Materials

ML and AI algorithms are already widely used in the field of composite materials for tribological applications. Generally, there has been a remarkable growth in the large-scale use of materials made from two or more constituent materials with different physical or chemical properties, for example a fiber and/or filler reinforced polymer (PMC), ceramic (CMC) or metal (MMC) matrix composites. The advantages of these materials lie especially in the high strength-to-weight as well as stiffness-to-weight ratios [30]. For a general overview of tribological properties of different composites in dependency of contact and/or environmental conditions, the interested reader is referred to various review articles [31–33]. A major field which already exploited ML approaches to a greater extent have been wear-resistant composites with polymer matrix, for example thermosets such as epoxy or polyester [34] as well as thermoplastics [35], e.g., polyamide (PA), polyphenylene sulfide (PPS), polytetrafluoroethylene (PTFE), polyethylene (PE), polyether ether ketone (PEEK) [36,37], or polypropylene (PP) [38].

3.1.1. Thermoset Matrix Composites

In this way, Padhi and Satapathy [39] applied a Taguchi experimental design of experiments (DoE, 16 data points) in combination with a back propagation ANN to train multi-layered feed-forward networks, predicting the tribological behavior of epoxy composites with short glass fibers (SGF) and/or micro-sized blast furnaces slag (BFS) particles. Based on data obtained from tests in a pin-on-disk setup under dry sliding conditions against a hardened ground steel counter-body and divided into training, test and validation categories and operational and material parameters with significance for the resulting wear rate were thus identified. Thereby, the ANN was able to predict the specific wear rate with low errors between 2.5% and 6.9% for composites without BFS and between 0.9% and 5.1% for composites with BFS. Epoxy composites were also investigated recently by Egala et al. [40] with newly developed natural short castor oil fibers (ricinus communis) as unidirectional reinforcements of different lengths and at a constant volume fraction of 40%. The database consisting of 36 data points was acquired from experiments utilizing a flat pin-on-disk tribometer under dry sliding conditions against a hardened steel disk as a counter-body. Besides fiber lengths, the normal force as well as the sliding distance were varied and the influence on gravimetric wear, interfacial heat, and COF were studied within a full factorial DoE. The experiments were carried out in duplicate and averaged values were used in further data processing. Thereby, the relationships between variation parameters and target values were expressed by linear regression as well as by hidden layer ANNs. For the training of the latter, the data set was randomly split into training (60%), validation (20%), and test (20%) data. To find the best prediction, 73 different ANNs (cascade forward back propagation, feed forward back propagation and layer recurrent) with Levenberg-Marquardt (LM) training function and a varying number of hidden layers (1–4), number of neurons (7–15), and different transfer functions (Logsig, Purelin) were tested stepwise (see Figure 3a–d). It was found that the linear regressions were able to describe the results within errors of ±8%. The best predictions however were provided by a cascade forward back propagation network as well as a feed forward back propagation ANN with architectures as illustrated in Figure 3e,f using Trainlm and Purelin as training and transfer functions. Thereby, the errors were ±5% and ±4.5%, respectively, indicating higher efficiency and reliability in predicting the tribological behavior of studied composites than common regression models.

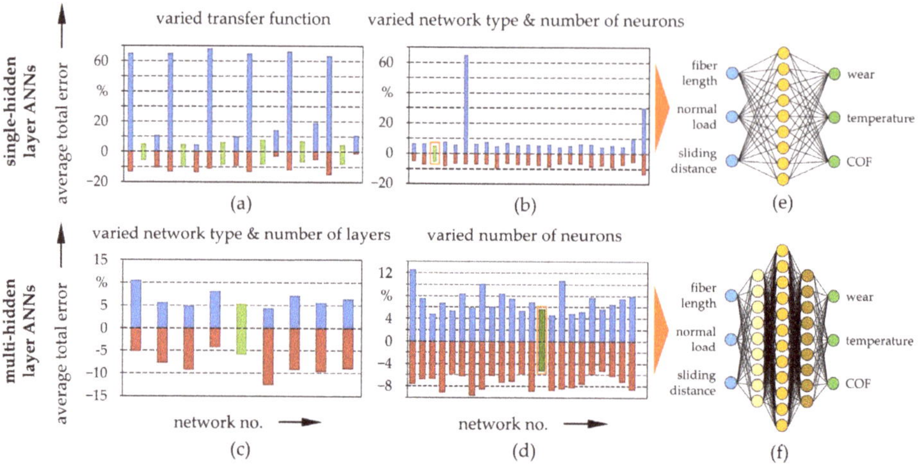

Figure 3. Average total errors for the wear prediction of different ANN architectures when optimizing the transfer function (**a**), the network type and the number of neurons in the single-hidden layer ANN (**b**) as well as the network type and the number of layers (**c**) and the number of neurons in the multi-hidden layer ANN (**d**). Illustration of the architectures of the single- (**e**) and multi-hidden layer ANN (**f**) with least errors. Redrawn from [40] with permission by CC BY 4.0 (Springer).

Nirmal [41] attempted to predict the friction coefficient of treated betelnut fiber reinforced polyester composites by an ANN trained with data from 492 experimental sets of a block-on-disk tribometer against stainless steel under dry sliding conditions with varying normal loads, sliding distances and three different fiber orientations (parallel, anti-parallel and normal). In trial-and-error variations of neuron, layer, and transfer function, an ANN consisting of two hidden layers with 30 and 20 neurons, respectively, trained by LM function and utilizing logsig transfer functions between the hidden layers and a pure linear transfer function to the output layer was found as most capable of predicting the COF based upon the inputs. Albeit other training algorithms (gradient descent back propagation, with momentum and adaptive learning rate, with adaptive learning rate and conjugate gradient back propagation with Powell-Beale restarts) resulted in significantly faster convergence, the LM function featured the lowest errors compared to the test data, especially after repeated training. Thus, sum squared errors (SSE) of less than 10^{-2} were obtained. Similarly, Nasir et al. [42] identified the LM function as most suitable compared to others when training ANNs to predict the COF from 7389 data sets attained in experiments on multi-layered glass fiber reinforced polyester resin rubbing against stainless steel using a disk-on-flat tribometer under different fiber orientations, loads, sliding speeds, and test durations. The prediction model was able to reproduce the trends of the experiments well and accuracies up to 90% were achieved. It was stated, however, that performance was lower compared to other studies due to the large amount of input data as well as larger deviations and fluctuations in the experimental results, especially during running-in periods. Furthermore, it was emphasized that the number of layers as well as neurons have a decisive influence on the results. While multi-hidden layer ANNs mapped partial areas of the input data (e.g., only one fiber orientation) very well, the entire data area was best represented by a single-hidden layer ANN with comparatively many neurons.

3.1.2. Thermoplastic Matrix Composites

Already in the early 2000s, Velten et al. [43,44] evaluated the ability of ANNs to predict tribological properties of short fiber thermoplastic matrix (PA) composites and aid in the material design. Here, the decisive role of the data sets as well as the ANN architecture was emphasized as well. Later, Gyurova et al. [45] modeled the tribological behavior of PPS composites with short carbon fibers (SCF), graphite, PTFE, and titanium dioxide (TiO_2) fillers with over 90 data sets obtained from dry-running pin-on-disk tribometer tests at constant test duration and varied loads and sliding speeds. The data were split into 80% training and 20% testing data and included the material composition (matrix volume fraction, filler, reinforcing agents and lubricants), testing conditions (pressure and sliding speed), as well as characteristic thermo-mechanical properties (tensile and compressive properties) as inputs and the specific wear and the friction coefficient as outputs. For the latter, separate ANNs were trained by a gradient descent back propagation algorithm with momentum and adaptive learning rate to minimize the mean relative error (MRE). These consisted of two hidden layers with 9 and 3 (wear rate) as well as 3 and 1 (COF) neurons, respectively. Thus, most significant inputs could be identified, and it was observed that the MRE for the wear rate (0.60–0.78) was higher than for the sliding friction (0.10–0.12), which was attributed to the rather small database. Furthermore, a so-called optimal brain surgeon (OBS) method was used to prune the ANN through the identification and removal of irrelevant nodes (weight elimination). The architectures as well as exemplary 3D profiles for predicting the wear rate in dependency of the SCF and the TiO_2 content before and after pruning are illustrated in Figure 4. Apparently, both cases matched adequately with the experimental data. Besides higher computational efficiency, the pruned network featured superior prediction accuracy in some areas of the parameter space. Finally, optimal compositions with higher SCF and lower TiO_2 concentrations around 10–15% as well as 3–5%, respectively, could be derived with considerably reduced experimental efforts, which corresponded well to the observations from Jiang et al. [46]. Gyurova and Friedrich [47] evaluated the influence of the data set size on the prediction capabilities of trained ANNs. Utilizing a newly measured database consisting of 124 independent pin-on-disk dry sliding wear tests on PPS matrix composites, the mean relative errors were reduced from above 0.72 to below 0.55 (specific wear rate) and from above 0.11 to beneath 0.10 (COF) compared to previous studies [45,48]. Later, the approach was further enhanced by Busse and Schlarb [49] using the same data, most notably by utilizing a LM training algorithm with mean squared error regularization as performance function, which significantly improved the computational efficiency and, in particular, the accuracy. Independently of the inputs, the wear rate prediction quality was found to be six times higher compared to the comparative studies [45,47].

Zhu et al. [50] also emphasized the crucial role of data set size and reported better agreement of experimental data with the prediction of the friction coefficient than with the volumetric wear losses when applying an ANN to carbon fiber and TiO_2 reinforced PTFE. 12 Different compositions were therefore investigated in block-on-disk dry sliding tests under varying sliding velocities and normal loads. A network trained by gradient search and consisting of three hidden layers (15, 10, and 5 neurons) and tan-sigmoid transfer functions between the input and the hidden layers as well as pure linear transfer functions to the output layer was found to deliver the least mean square errors. Li et al. [51] applied a Monte Carlo-based ANN to predict the tribological behavior of PTFE resin with aramid pulp, potassium titanate whisker (PTW), mica, copper (Cu) as well as silicon dioxide (SiO_2) for ultrasonic motors and compared the performance to a back propagation ANN. The database, an orthogonal table by variation of the composition, was generated from experiments conducted in triplicate on a quasi-static test rig where the specimens were fixed on a dynamic rotor and slid against a phosphor bronze stator at constant speed and load. In combination with a grey relational analysis, it was shown that especially mica and SiO_2 exerted significant roles for friction and wear improvements. The Monte Carlo-based ANN was particularly suitable for predictions with more limited amount of data due to repeated random sampling and the utilization of combinations of different transfer functions (sigmoid, polynomial, tanh, and gauss functions). The authors reported that, in the context of the variation and volatility of the underlying data, the Monte Carlo ANN performed better than the conventional back propagation ANN with root mean squared errors of 0.97 (specific wear rate) and 0.007 (COF) compared to 2.08 and 0.019.

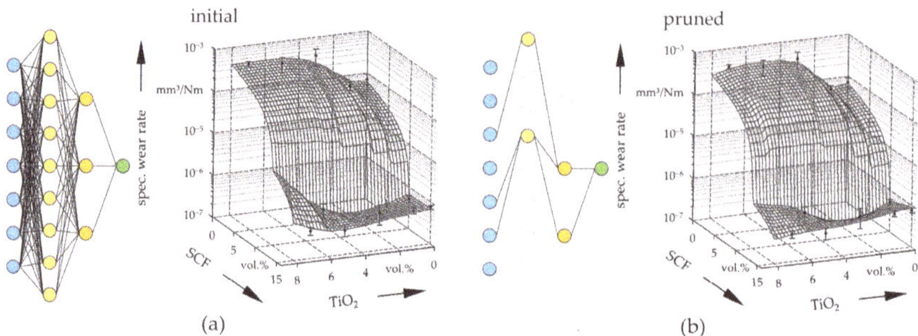

Figure 4. ANN architecture as well as 3D profiles for the specific wear rate in dependency of the SCF and the TiO$_2$ concentration without (**a**) and with (**b**) pruning. Redrawn from [45] with permission (Elsevier).

Kurt and Oduncuoglu [52] utilized 125 data sets extracted from established literature sources to study the effects of normal load and sliding speed in dry sliding experiments as well as the type and weight fraction of various reinforcements in ultrahigh molecular weight PE (UHMWPE) composites by a feed forward back propagation ANN. This involved zinc oxide (ZnO), zeolite, carbon nanotubes (CNT), carbon fibers (CF), graphene oxide (GO), and wollastonite additives, leading to a total number of 11 inputs, whereas the volumetric wear loss was considered as target/output value. In a trial-and-error search, an ANN with a single-hidden layer consisting of 12 neurons and logistic sigmoid transfer functions trained by a LM algorithm was selected. With R^2 values for training and testing above 0.8 as well as mean absolute errors not exceeding 4.1%, it was thus shown that sliding speed and load determined the wear losses more significantly than the particle types and fractions. Recently, Vinoth and Datta [53] also used 153 experimental data sets from literature to predict mechanical properties of UHMWPE composites with multi-walled carbon nanotubes (MWCNT) and graphene reinforcements in dependency of seven input variables comprising composite composition, particle size, and mechanical bulk properties. A feed forward ANN with scaled conjugate gradient back propagation, hyperbolic tangent transfer functions and 3 (for Young's modulus) or 5 (for the ultimate tensile strength) hidden layers were utilized, achieving correlation coefficients for the outputs of 0.93 and 0.97, respectively. Subsequently, a multi-objective (pareto) optimization of the input variables was performed with a non-dominated sorting genetic algorithm. On this basis, samples (pins) of UHMWPE composites with MWCNT and graphene filler ratios considered as optimal were fabricated accordingly and characterized mechanically as well as in tribological tests under dry sliding conditions against cobalt chromium alloy disks. It was actually possible to demonstrate improved properties compared to references and, in particular, excellent wear behavior due to the formation of wear-protecting transfer films on the counter-body.

3.1.3. Metal Matrix Composites

Some successful studies using ML and AI can also be found for composites with soft metals as matrix [54], for example aluminum, copper or zinc and their alloys [55–59]. As such, Stojanović et al. [60] investigated the friction and wear behavior of aluminum hybrid composites with Al-Si alloy matrix and 10 wt.% silicon carbide (SiC) as well as 0, 1, and 3 wt.% graphite. The data sets were generated in lubricated block-on-disk tribometer tests at three sliding speeds, the normal loads and at constant sliding distance with the application of Taguchi's robust orthogonal array design method (27 data points). This was reported to be a simple and efficient methodology. Besides performing ANOVA factor variance analysis and the fitting of a linear regression model, a feed forward back propagation

ANN was developed. Therefore, 70% of the data were used for training, 15% for testing and 15% for validation. The model was trained by LM optimization and consisted of two hidden layers of 20 and 30 neurons, respectively, as well as logarithmic sigmoid and pure linear transfer functions. The values predicted by the ANN provided sufficient agreement with the experiments and were more precise than those provided by the statistical methods used. Similarly, Thankachan et al. [61] compared the performance of a feed forward back propagation ANN with statistical regression analysis when investigating the wear behavior of hybrid copper composites with aluminum nitride and boron nitride particles in dry-running pin-on-disk tribometer tests at different volumetric fractions, loads, sliding speeds and sliding distances by applying Taguchi's orthogonal array. The ANN featured the 4 inputs, one hidden layer with 7 neurons, the specific wear rate as output and was trained by the LM function to optimize the mean absolute error. Thus, the neural network reached higher accuracy than the reference regression model.

Gangwar and Pathak [62] introduced a novel improved bat algorithm (IBA) to train an ANN for predicting the wear behavior of marble dust reinforced zinc-aluminum (Zn-Al) alloy by optimizing the weights, biases and neurons as well as finding minimum mean squared errors, see Figure 5a). The main advantage of the IBA compared to other training algorithms (e.g., back propagation, genetic algorithms or particle swarm optimization) was in the flexibility and stable training through the introduction of a new velocity, position search equation and sugeno inertia weights. This overcame local optima stagnation and enhanced the convergence speed. The evaluation of the specific wear rate was based on data from pin-on-disk experiments with varied filler content, normal load, sliding velocity and distance, as well as ambient temperature (5 levels each) by means of a Taguchi orthogonal array (25 data sets). Thereby, an ANN with 7 neurons in the single-hidden layer was found to be optimal, with a mean squared error of 0.26 and an average prediction accuracy of 97%. Exemplary 3D plots of the wear rate as a function of the variation parameters are shown in Figure 5b). Obtained results and the suggested IBA-ANN approach can thus help to save resources when searching for beneficial stress or material combinations with limited experimental database.

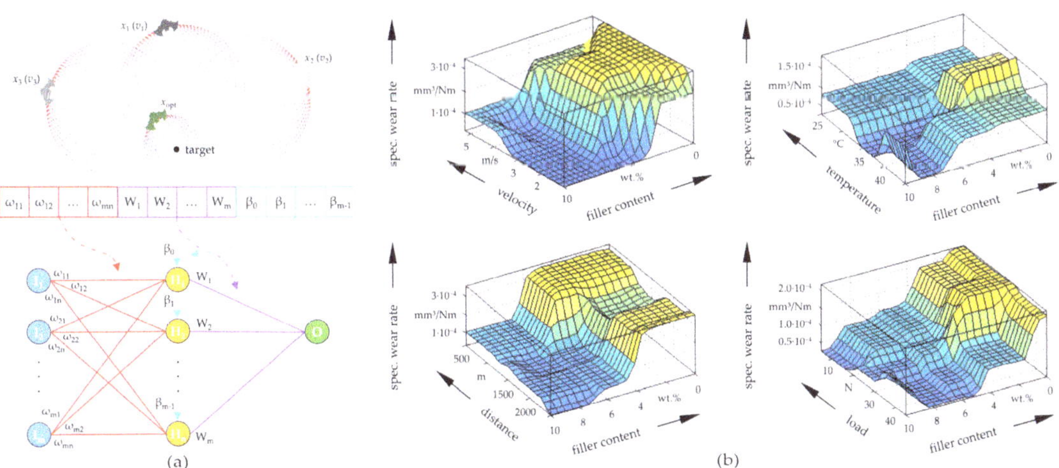

Figure 5. Schematic representation of encoding bat individuals to train the ANN (**a**) as well as 3D response surfaces of influencing factors on the specific wear rate (**b**). Redrawn from [62] with permission (Elsevier).

Very recently, Hasan et al. [63,64] compared five different ML techniques when predicting the friction and wear behavior of aluminum base alloys and graphite composites: ANN, kNN, SVM, gradient boosting machine (GBM), and RF. The 852 data sets were obtained from experimental studies in literature. It was shown that basically all ML approaches were able to adequately describe the tribological behavior from material and tribological test data. Thereby, RF outperformed the other algorithms in predicting the wear behavior, while GBM and KNN had the highest accuracy for the friction behavior for the base alloy and the composite, respectively. This underlines that the right choice of the ML approach is highly dependent on the respective problem formulation.

The works in the area of composite materials are summarized in Table 1 according to the subject, the database, the inputs and outputs, and the ML approach.

3.2. Drive Technology

In the field of drive technology, there are several areas of application for using ML for rolling and sliding bearings, seals, brakes, and clutches, which are involved in systems for motion generation and power transmission.

3.2.1. Rolling Bearings

Rolling bearings are among the most important machine elements, locally transmitting large forces via several rolling contacts. The bearing components are also exposed to complex dynamics and friction occurs in numerous contacts influencing the operation. Bearing failures can be of very different nature. Mostly, they are longer lasting processes between first occurrence of damage and fatal failure. However, damage to rolling bearing components can be quickly observed in the operating behavior of machines and systems, for example in the form of increasing friction, heat, vibration, and noise. Therefore, one possible application for ML is condition monitoring and damage detection [65,66]. Most published work was related to vibration theory rather than tribology [67,68], which is why only some representative examples shall be introduced.

Table 1. Overview of ML approaches successfully applied in the area of composite materials.

Subject	Database, Number of Data Sets (If Applicable Divided in Train/Test/Validation)	Inputs	Outputs	ML Approach	Prediction	Ref.
SGF and BFS reinforced epoxy	experimental (pin-on-disk) Taguchi DoE, 16	BFS content, sliding velocity, normal load, sliding distance	spec. wear rate	back propagation ANN (4:7:1)	Errors < 6.9%	[39]
unidirectional short castor oil fiber reinforced epoxy	experimental (pin-on-disk) full factorial DoE, 36 (60%/20%/20%)	fiber length, normal load, sliding distance	wear, temperature, COF	various ANNs, best results for back propagation ANN (3:9:3 & 3:9:12:9:3)	averaged total errors < 5%	[40]
treated betelnut fiber reinforced polyester	experimental (block-on-disk), 492	fiber orientation, normal load, sliding distance	COF	ANN (3:30:20:1)	SSE < 1%	[41]
glass fiber reinforced polyester	experimental (disk-on-flat), 7389	fiber orientation, rotational speed, normal load, test duration	COF	ANN (4:40:1)	SSE < 15%	[42]
SCF, graphite, PTFE, and TiO$_2$ reinforced PPS	experimental (pin-on-disk), 90 (80%/20%)	matrix vol. fraction, filler, reinforcing agent and lubricant, contact pressure, sliding speed, tensile strength, compressive strength	spec. wear rate, COF	various gradient descent back propagation ANNs (7:9:3:1 for wear, 7:3:1:1 for COF)	MRE < 0.78 (wear), MRE < 0.12 (COF)	[45]
	experimental (pin-on-disk), 124 (80%/20%)				MRE < 0.55 (wear), MRE < 0.10 (COF)	[47]
					MRE < 0.14 (wear), MRE < 0.03 (COF)	[49]

Table 1. *Cont.*

Subject	Database, Number of Data Sets (If Applicable Divided in Train/Test/Validation)	Inputs	Outputs	ML Approach	Prediction	Ref.
CF and TiO$_2$ reinforced PTFE	experimental (block-on-ring), 30–105 (10–98%/2–90%), best results for largest database	PTFE content, carbon fiber content, TiO$_2$ content, sliding speed, normal load, hardness, compressive strength	vol. wear loss, COF	various ANNs, best results for gradient search ANN (7:15:10:5:1)	CoD > 90%	[50]
aramid pulp, PTW, mica, Cu, and SiO$_2$ reinforced PTFE	experimental (rotor/stator test-rig) in orthogonal table DoE, 18 (80%/20%)	aramid pulp content, PTW content, mica content, Cu content, SiO$_2$ content	spec. wear rate, COF	back propagation ANN	RMSE < 2.08 (wear), RMSE < 0.019 (COF)	[51]
				Monte Carlo-based ANN	RMSE < 0.97 (wear), RMSE < 0.007 (COF)	
ZnO, zeolite, CNT, CF, GO, and wollastonite reinforced UHMWPE	experiments from literature, 125	UHMWPE content, ZnO content, Zeolite content, CNT content, CF content, GO content, wollastonite content, normal load, sliding speed	vol. wear loss	back propagation ANN (11:12:1)	R^2 > 0.8, mean total error < 4.1%	[52]
MWCNT and graphene reinforced UHMWPE	experiments from literature, 153	MWCNT fiber diameter, MWCNT fiber length, MWCNT content, graphene sheet length, graphene sheet thickness, graphene content, UHMWPE molecular weight, UHMWPE tensile strength, UHMWPE Young's modulus	Young's modulus, tensile strength	scaled conjugate gradient back propagation ANN (7:3:1 for Young's modulus and 7:5:1 for tensile strength)	R^2 > 0.93 (Young's modulus), R^2 > 0.97 (tensile strength)	[53]
graphite reinforced Al-Si alloy	experimental (block-on-disk) in Taguchi's orthogonal array DoE, 27 (70%/15%/15%)	graphene content, normal load, sliding speed	vol. wear rate, COF	back propagation ANN (3:20:30:2)	R^2 > 0.98	[60]
aluminum nitride and boron nitride reinforced copper	experimental (pin-on-disk) in Taguchi's orthogonal array DoE, 27 (90%/10%)	volume fraction, normal load, sliding velocity, sliding distance	spec. wear rate	back propagation ANN (4:7:1)	errors < 3.4%	[61]
marble dust reinforced Zn-Al alloy	experimental (pin-on-disk) in Taguchi's orthogonal array DoE, 25 (60%/20%/20%)	filler content, normal load, sliding velocity, sliding distance, amb. temperature	spec. wear rate	IBA trained ANN (5:7:1)	MSE < 0.26, accuracy > 97%	[62]

Table 1. *Cont.*

Subject	Database, Number of Data Sets (If Applicable Divided in Train/Test/Validation)	Inputs	Outputs	ML Approach	Prediction	Ref.
Graphite reinforced aluminum alloy	experiments from literature, 852	graphite content, hardness, ductility, processing procedure, heat treatment, SiC content, yield strength, tensile strength, normal load, sliding velocity, sliding distance,	vol. wear rate, COF	back propagation ANN (11:10:10:10:2)	MSE < 0.003 wear) RMSE < 0.06 (wear) R^2 > 0.74 (wear) MSE < 0.004 (COF) RMSE < 0.06 (COF) R^2 > 0.86 (COF)	[63,64]
				kNN	MSE < 0.002 wear) RMSE < 0.04 (wear) R^2 > 0.85 (wear) MSE < 0.007 (COF) RMSE < 0.08 (COF) R^2 > 0.76 (COF)	
				RF	MSE < 0.001 wear) RMSE < 0.04 (wear) R^2 > 0.88 (wear) MSE < 0.004 (COF) RMSE < 0.06 (COF) R^2 > 0.86 (COF)	
				SVM	MSE < 0.006 (COF) RMSE < 0.08 (COF) R^2 > 0.76 (COF)	
				GBM	MSE < 0.002 wear) RMSE < 0.04 (wear) R^2 > 0.86 (wear) MSE < 0.003 (COF) RMSE < 0.05 (COF) R^2 > 0.89 (COF)	

As such, Subrahmanyam and Sujatha [69] investigated the suitability of two different ANNs, namely multilayered feed forward neural network trained with supervised error back propagation (EBP) technique and an unsupervised adaptive resonance theory-2 (ART2) based neural network, for the diagnosis of local defects in deep groove ball bearings. The input vector consisted of eight parameters that were used to describe the vibration signal and the output was a condition rating for the bearing (good/bad) and, if the condition was classified bad, the defect was pinpointed. The authors concluded from their work that the performance of the ANN with EBP was excellent for recognizing ball bearing states. They reported that defective bearings were distinguished from good ones with 100% confidence, while the ANN had a success rate of over 95% in diagnosing localized defects. The results of the ANN with ART2 were ambivalent: The learning process was about 100 times faster than that of the ANN with EBP and defective bearings also were distinguished from good ones with 100% reliability. Yet, the estimation of localized defects was not satisfactory. Furthermore, Kanai et al. [70] presented a condition monitoring method for ball bearings using both, model-based estimation (MBE) and ANN, to guess the vibration velocity and the defect frequency of the rotor-bearing-system. The authors based their study on a three-layered feed forward neural network trained with EBP, where the input vector consisted of 5 parameters (speed, load, defect volume, radial clearance, number of balls) obtained from rig tests on a self-aligning deep groove ball bearing. According to the authors, the ANN shows satisfactory results compared to MBE and experimental tests.

Apart from condition monitoring, ML approaches have recently been utilized for designing rolling bearing components. Schwarz et al. [71] used different ML methods to classify possible cage motion modes of rolling bearings and to predict application-related undesired cage instabilities, see Figure 6. The data set was generated from sophisticated rolling bearing dynamics simulations, which were confirmed by means of experimental investigations on a test rig. Based on the simulations, the authors determined metrics that, in combination, reliably characterize the state of the cage condensed in three classes "stable", "unstable" and "circling". They used these metrics to classify cage motion using quadratic discriminant analysis (QDA). QDA is a method of multivariate statistics to separate different classes on the basis of characteristics [16]. It is interesting to note that we could not discover this method in any other article within our literature survey. To predict the class of cage motion, Schwarz et al. applied decision trees as weak learners within an ensemble classification model based on AdaBoostM1 [72] to achieve good results. Furthermore, Wirsching et al. [73] aimed at tailoring the roller face/rib contact in tapered roller bearings. Geometric parameters were sampled by a Latin hypercube sampling (LHS) and the tribological behavior was predicted by means of elastohydrodynamic lubrication (EHL) contact simulations. Key target variables such as pressure, lubricant gap and friction were approximated by a so-called metamodel of optimal prognosis (MOP) [74] and optimization was carried out using an evolutionary algorithm (EA). The MOP fully automatically filtered non-significant variables and various approaches (polynomial regression, moving least squares, isotropic or anisotropic kriging) were trained to derive the most suitable approximation. The applied ML approach provided very good prediction for most geometries and target values, which was reflected in the high prediction coefficients (CoP) in most cases above 90% and the low errors in mostly below 2% of the optimized pairing between the prediction and verification calculations.

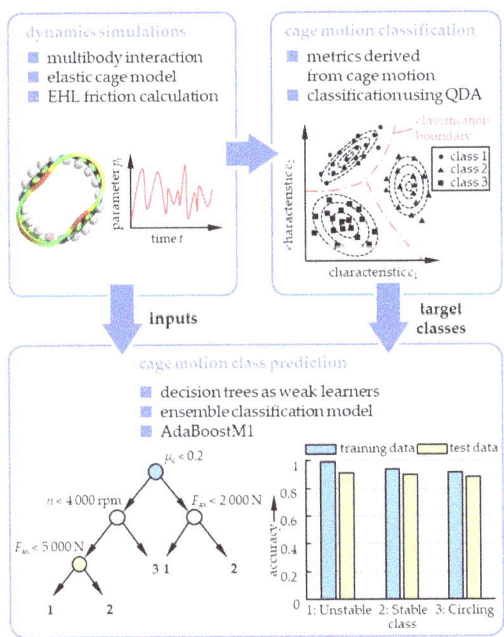

Figure 6. Global scheme for classifying and predicting rolling bearing cage motion modes based on dynamics simulations and ML following [71].

3.2.2. Sliding Bearings

Since the operating behavior of sliding bearings is highly non-linear and depending on numerous parameters, ML methods have been utilized for the analysis and synthesis of the tribosystem. Canbulut et al. [75] analyzed the frictional losses of a hydrostatic slipper bearing using an ANN fed by experimental test data. Input parameters were the average roughness of the rubbing surfaces, relative velocity, supply pressure, hydrostatic pocket ratio, and capillary tube diameter. Three-layered feed forward neural networks containing 10 neurons in the hidden layer trained with EBP were to be found as suitable. The predictive performance of the ANN was evaluated using six operation cases for the bearing, where an exact match of the ANN predictions with the experimental results was reported. Further, using ANNs, Ünlü et al. [76] analyzed the friction and wear behavior of a radial journal bearing (bronze CuSn10/steel SAE 1050 pairing) under dry and lubricated conditions. The ANN with EBP technique was featured a 3:5:5:3 multilayer architecture for the dry case and 3:4:4:3 for the lubricated case. The input vector was described by time, applied load and rotational speed and the outputs were coefficient of friction, journal and bearing weight loss. Input data were collected from previously published experiments. The ANN predictions show high agreement to the experimental data and the authors stated that such ANNs can effectively reduce the number of future experiments. Furthermore, Moder et al. [77] showed that supervised ML algorithms can be used to predict the lubrication regime of hydrodynamic radial journal bearings based on given torques. Therefore, the torque time series were first analyzed using Fast Fourier Transformation (FFT) and manually assigned to lubrication regimes. Two ML algorithms were used for the classification task: Logistic regression and deep neural networks. Based on their results, the authors concluded that even shallow neural networks as well as logistic regressions are able to reach high accuracy for the given problem. It was indicated that data scaling was essential, while feature scaling, which is often applied in data analysis, was not suitable for the FFT classification. Prost et al. [78] investigated the feasibility of classifying the operating condition (running-in, steady, pre-critical, critical) of a translationally oscillating self-lubricating journal bearing using an ensemble learning algorithm. To this end, the authors applied a semi-supervised random forest classifier (RFC), which was based on the aggregation of a large number of independent decision trees. The RFC was trained with high-resolution force signals from experiments and showed a very high classification accuracy in validation experiments. The authors pointed out, that labeling the data is essential and requires expert knowledge. As this step is very tedious and time-consuming, they suggested a semi-automated process based on principal components analysis and k-means clustering algorithms. Francisco et al. [79] studied how far ML can be used to optimize connecting rod big-end bearings. They combined sophisticated finite element (FE) simulations with a nondominated sorting genetic algorithm, which allowed them to minimize the frictional losses and functioning severity of the bearing by optimizing 10 parameters. The authors concluded that metamodels based on previous simulations and including all relevant parameters allow the optimization of a tribological system in a very time and resource saving way.

3.2.3. Seals

Seals play an important role in mechanical drive technology as they separate lubricants or operating fluids and the environment of the drive train from each other. Contact seals frequently affect the friction behavior in the whole drive train, and they are exposed to wear. Increasing requirements demand more precise descriptions of the tribological behavior of contact seals in design phases as well as condition monitoring [80,81]. Logozzo and Valigi [82] suggested ANNs as an alternative for analytical models to predict friction instabilities and critical angular speeds of face seals during shaft decelerations. The authors studied different feed forward neural networks with 2:x:1 architecture ($x = 6$, 8, 10, 12, 15, 16), trained with supervised EBP technique. Thereby, 10 neurons in the hidden layer showed the best training convergence. Input data were collected from experimentally validated tribo-dynamics simulations based on a lumped parameter model with 2 degrees of freedom. The input vector of the ANNs consisted of two parameters (axial and torsional stiffness). The authors pointed out that unlike deterministic models, the ANNs were not able to explain the phenomena of frictional instability but provided a smart way to define parameters in the design phase for the avoidance of frictional instabilities. Yin et al. [83] used a SVM regression to monitor the status of a gas face seal based on acoustic emissions (AE). Input data as well as validation data were collected from rig tests. To generate the representative vectors with satisfactory experimental agreement, the AE power concentration in several key bands within certain durations were used.

3.2.4. Brakes and Clutches

Brakes and clutches are safety-relevant components and have to work reliably even under extreme conditions. They are usually integrated in closed loop control systems, which makes it necessary to describe and optimize the braking and coupling behavior, involving complex squealing and wear phenomena. Accordingly, ML approaches have been applied in this area as well [84,85]. For example, Aleksendrić et al. [86,87] applied an ANN to model the speed-dependent cold performance of brakes They considered 18 material composition parameters, five manufacturing parameters and three operating condition parameters as inputs. Friction data was collected from test rig experiments. Since it is not known a priori which model provides the best prediction quality, the authors investigated 18 different architectures with five different learning algorithms (LM, Bayesian regulation, resilient back propagation, scaled conjugate gradient and gradient decent). The best prediction results were provided by a 26:8:4:1 double-hidden-layer architecture trained by a Bayesian regulation algorithm. The authors stated that their ANN has shown sufficient flexibility to generalize the influences of unknown types of friction material on their cold performance. The methodology was later extended to predict materials recovery performance [88] and brake wear [89] by the same authors. Basically, the procedure was similar to the work described above and the best prediction results were attained from a single hidden layer ANN (25:5:1) trained by a Bayesian regulation algorithm. Timur and Aydin [90] investigated whether the friction coefficient of brakes can be predicted by means of ML based upon experimental training data (1050 points). Comparing different regression methods (linear, least median squared linear, Gaussian processes, pace, simple linear, isotonic, SVM) and 10-fold cross-validation, they noted that all algorithms showed a correlation coefficient larger than 0.99 and a root mean squared error below 0.01. However, isotonic regression allowed the fastest model building.

The prediction of friction coefficient for automobile brake as well as clutch materials against steel using ML algorithms was also addressed by Senatore et al. [91], who showed how to obtain a comprehensive view on the influence of the main sliding parameters. Based upon experimental data from pin-on-disk tests with varying sliding speed, acceleration and contact pressure (200 data sets), the authors trained two different supervised feed-forward double-hidden-layer EPB ANNs with a 3:6:3:1 architecture for braking and 3:6:7:1 for the clutch material, respectively. The authors concluded that ANNs have confirmed suitability for valid prediction of friction coefficients, with utility being enhanced by significance as well as sensitivity analysis of input parameters. A possible application could be in more accurate friction maps for electronic control purpose. However, the authors also discussed the limitations of the approach, in particular pointing out extrapolation errors. Comparable findings were obtained by Grzegorzek and Scieszka [92], who used a similar methodology (in this case feed-forward EPB ANN with 6:12:1 architecture) to investigate the friction behavior of industrial emergency brakes from 408 data sets. The authors self-described their work as being at a preliminary stage, yet they were able to demonstrate the performance of ANN against various models of multiply regression analysis.

The works in the area of drive technology are summarized in Table 2 according to the subject, the database, the inputs and outputs, and the ML approach.

Table 2. Overview of ML approaches successfully applied in the area of drive technology.

Subject	Database, Number of Data Sets (If Applicable Divided in Train/Test/Validation)	Inputs	Outputs	ML Approach	Prediction	Ref.
groove ball bearing defect diagnosis	experimental (bearing test-rig), 108 (90%/10%)	peak value of amplitude, average of top five peak values of amplitude, peak value of auto-correlation function, standard deviation, kurtosis	bearing state	EBP ANN	success rate > 95%	[69]
				ART2 ANN	success rate = 100%	
ball bearing condition monitoring	experimental (bearing test-rig), 145 (75%/25%)	speed, load, defect volume, radial clearance, number of balls	vibration velocity	back propagation ANN (5:12:1)	errors < 14%	[70]
cage motion mode classification in rolling bearings	numerical (dynamics simulation) LHS, 4000	cage mass, cage bending stiffness, pocket clearance, guidance clearance, bearing type, COF, axial force, radial force, bending moment, rotational speed	CDI	QDA and DT	accuracy > 91%	[71]
TRB roller/face rib contact geometry design	numerical (EHL simulation) in LHS, 370 (70%/30%)	roller face radius, eccentricity, rib radius	max. pressure, min. film height, COF	MOP	CoP > 90%, errors < 2%	[73]
frictional power losses of hydrostatic slipper bearings	experimental (hydrostatic slipper test-rig)	average roughness, relative velocity, supply pressure, hydrostatic pocket ratio, capillary tube diameter	frictional power loss	back propagation ANN	errors < 1.9%	[75]
dry and lubricated journal bearing behavior	experimental (journal bearing test-rig), 4	time, load, rotational speed	COF, bearing weight loss, journal weight loss	EBP ANN (3:5:5:3 for dry and 3:4:4:3 for lubricated case)	mean errors < 4% (dry), mean errors < 5.3% (lubricated),	[76]
journal bearing lubrication regime prediction	experimental (journal bearing test-rig), 888 (80%/20%)	frictional torque	lubrication regime	FFT+ back propagation ANN (1:256:128:64:32:16:8:1)	accuracy > 99%	[77]
journal bearing operating condition classification	experimental (journal bearing test-rig), 9 (75%/25%)	time, lateral force	operating state	RFC (DT)	accuracy > 94%	[78]

Table 2. *Cont.*

Subject	Database, Number of Data Sets (If Applicable Divided in Train/Test/Validation)	Inputs	Outputs	ML Approach	Prediction	Ref.
connecting rod big-end bearing design	numerical (elastic HL simulation) in CCF DoE, 9	oil viscosity at ref. temperature, oil viscosity at ref. pressure, oil thermo-viscosity coefficient, oil piezo-viscosity coefficient, oil piezo-viscosity index, oil supply pressure, lemon shape, shell bore relief depth, shell bore relief length, barrel shape, radial clearance	pressure times velocity product, power loss	nondominated sorting genetic algorithm	$R^2 > 0.99$	[79]
face seal friction instability prediction	numerical (dynamics simulation), 40 (90%/10%)	axial stiffness torsional stiffness	critical speed	various ANNs, best results for EBP ANN (2:10:1)	$R^2 > 0.97$	[82]
disk brake performance	experimental (inertial dynamometer), 275 (70%/10%/20%)	applied pressure, initial speed, number of braking events, phenolic resin, iron oxide, barites, calcium carbonate, brass chips, aramid, mineral fiber, vermiculite, steel fiber, glass fiber, brass powder, copper powder, graphite, friction dust, molybdenum disulphide, aluminum oxide, silica, magnesium oxide, spec. molding pressure, molding temperature, molding time, heat treatment temperature, heat treatment time	brake factor	various ANNs, best results for Bayesian ANN (26:8:4:1)	sufficient (not quantified)	[86, 87]
brake materials	experimental (inertial dynamometer), 408 (34%/33%/33%)	sliding speed, contact pressure, temperature, binder resin, premix masterbatch, residuum	COF	EPB ANN (6:12:1)	errors < 4%	[92]
clutch materials	experimental (pin-on-disk), 200 (50%/25%/25%)	sliding speed, sliding acceleration, contact pressure	COF	EPB ANN (3:6:3:1) / EPB ANN (3:6:7:1)	sufficient within the data range (not quantified)	[91]

3.3. Manufacturing

ML approaches were also employed in the area of manufacturing technology, for example, for process monitoring or in quality control/image recognition [6]. There are also some studies related to tribology, particularly regarding friction stir welding [93–95], but also for forming or machining [96]. Sathiya et al. [97] modeled the relationship between friction welding process parameters as heating pressure, heating time, upsetting pressure as well as upsetting time and output parameters (tensile strength and metal loss) when joining similar stainless steel by means of a back propagation ANN with 9 neurons in the single-hidden layers. The database (14 data points) was generated from corresponding experiments. Subsequently, different optimization strategizes based upon the ANN's prediction were compared: Genetic algorithm, simulated annealing algorithm, and particle swarm optimization. Among them, the genetic algorithm was reported to be most suitable and good agreement was found between the prediction of tensile strength and metal loss for optimized process parameters with respective validation experiments. Similarly, Tansel et al. [98] and Atharifar [99] applied and confirmed the suitability of ANNs for optimizing friction stir welding processes. However, the latter further introduced an optimization of the back propagation ANNs using a genetic algorithm (genetically optimized neural network systems) to maximize the prediction quality. Anand et al. [100] compared the performance of an ANN (4:9:2) with a response surface methodology approach (quadratic polynomial models) when optimizing friction welding with respect to tensile strength and burn-off length. The data (30 data sets) were generated with experiments within a five-level, four variable centrale composite DoE (CCD). It was observed that the ANN featured higher accuracy by a factor of two compared to the response surface. In turn, Dewan et al. [101] compared back propagation neural networks with adaptive neuro-fuzzy interference systems (ANFIS) [102] when predicting tensile properties in dependency of spindle speed, plunge force and welding speed from a rather small database (73 data points). Here, 1200 different ANFIS models were developed with varying number and type of membership functions as well as input combinations. It was reported the optimized ANFIS provided lower prediction errors than the ANN.

In addition to process optimization, ML approaches have also been used for monitoring friction stir welding. Baraka et al. [103] made use of process signals (traverse and downward tool force) to predict the weld quality. This was based upon frequency analysis by FFT, and an interval type 2 radial base function (RBF) neural network trained by an adaptive error propagation algorithm that effectively provided continuous feedback to the operator with an accuracy above 80%. Das et al. [104] also used real-time process signals (torque) for internal defect identification in friction welding. The experimentally obtained signals were analyzed by discrete wavelet transformation, statistical features (dispersion, asymmetry, excess) as well as general regression models and ML methods, namely SVM and back propagation ANN (3:5:1, log-sigmoid transfer functions) trained by the gradient descent method to predict tensile strength. It was reported the prediction performance of the SVM (0.5% error) was superior to regression (13.6%) and the ANN (3.1%).

Regarding other manufacturing processes, Fereshteh-Saniee et al. [105] trained a feedforward back propagation ANN with 21 neurons in the single hidden-layer (tan-sigmoid transfer function) from over 700 FE simulations to determine material flow and friction factors in one-step ring forming. Thereby, obtained load curves showed good agreement with experimental validation tests, featuring an accuracy of 99% and 97% for grease lubricated and dry conditions, respectively. The difference was traced back to higher variations of friction for unlubricated forming. Furthermore, Bustillo et al. [106] attempted to predict surface roughness and mass loss during turning, grinding, or electric discharge machining based upon surface isotropy levels and different ML approaches: Artificial regression trees, multilayer perceptions (MLP), RBF networks, and random forest. The most accurate approach for predicting the loss of mass was found to be RBF, while the MLP most precisely predicted the arithmetic mean roughness. However, the model parameters of both approaches had to be tuned very carefully and even small changes led to a substantial increase of errors. In contrast, satisfactory accuracy without any tuning stage could be obtained using the random forest ensembles. It was also reported that the prediction quality was comparatively sound even outside the training record as well as for smaller data sets.

The works in the area of manufacturing technology are summarized in Table 3 according to the subject, the database, the inputs and outputs, and the ML approach.

Table 3. Overview of ML approaches successfully applied in the area of manufacturing technology.

Subject	Database, Number of Data Sets (If Applicable Divided in Train/Test/Validation)	Inputs	Outputs	ML Approach	Prediction	Ref.
friction stir welding process optimization	experimental (friction stir welding), 14	heating pressure, heating time, upsetting pressure, upsetting time	tensile strength, metal loss		MSE < 0,01%	[97]
	experimental (friction stir welding), 30			back propagation ANN (4:9:2)	RMSE < 0.98 (tensile strength), RMSE < 0.05 (tensile strength),	[100]
	experimental (friction stir welding), 73 (60%/20%/20%)	rotational speed, welding speed, plunge force, empirical force index	tensile strength	various ANNs, best results for back propagation ANN (3:5:1)	mean absolute error < 7.7%	[101]
				ANFIS	mean absolute error < 10.1%	
friction stir welding process monitoring	experimental (friction stir welding), 25 (80%/20%)	rotational speed, welding speed	weld threshold for downward force, weld threshold for traverse force	RBF trained ANN	accuracy > 80%	[103]
	experimental (friction stir welding), 64 (60%/25%/15%)	rotational speed, welding speed, shoulder diameter	tensile strength	SVM	error < 0.5%	[104]
				back propagation ANN	error < 3%	
ring forming	numerical (FE simulation), 700	polynomial regression factors to fit load-displacement curves	strain hardening exponent, strength coefficient, COF	ANN (8:21:3:3)	accuracy > 97%	[105]

3.4. Surface Engineering

Approaches to enhance the tribological behavior of components by modifying their surfaces can be subsumed under the term surface engineering [107]. This involves adjusting the surface topography with and without compositional changes through as well as the application of coatings. Examples include, among others, tailoring the roughness and/or statistically distributed or discrete micro-textures, carburizing, nitriding, anodizing, electroplating, weld hardfacing, thermal spraying, chemical, or physical vapor deposition (CVD, PVD) [107]. Some studies have also applied ML approaches to better understand or design the surface modifications.

3.4.1. Coatings

Cetinel [108] used a single-hidden layer feed forward ANN to predict the COF and wear loss of thermally sprayed aluminum titanium oxide (Al_2O_3-TiO_2) coatings. The database was created by reciprocal pin-on-block tribometer tests under dry as well as acid conditions different loads. In the ANN, the test conditions were the inputs and—after trial-and-error testing of different configurations—the hidden layer consisted of 80 neurons. Furthermore, the ANN provided 63 outputs in the form of the COF and linear wear progress at different times of the experiments. Thus, the tribological behavior over the test period could be mapped very well. Sahraoui et al. [109] analyzed the friction and wear behavior of high-velocity oxy-fuel (HVOF) sprayed Cr-C-Ni-Cr and WC-Co coatings as well as electroplated hard chromium by means of a feed forward ANN. The database consisted of 180 training and 180 test data sets from dry-running pin-on-disk tribometer tests of the coated test specimens against brass disks at various normal loads and sliding speeds. An ANN with sigmoid transfer functions as well as two hidden layers (6 and 4 neurons) was found to be suitable for predicting the COF within variabilities between 5.8% and 10.8%. The main advantage of the model in this study was that the friction coefficients could be predicted comparatively well for a range of parameters up to 7 times larger than those contained in the training data. Upadhyay and Kumaraswamidhas [110,111] applied a back propagation ANN to optimize multilayer nitride coatings on tool steel deposited by unbalanced reactive PVD magnetron sputtering. The input parameters comprised bias voltage and gas flow rate as well as time, velocity and load within pin-on-disk sliding tests. The data was split into 70% training, 15% validation, and 15% test data. Training was based on the LM function and the most favorable ANN consisted of 20 neurons in the hidden layer. Thus, the wear rate as well as the COF could be predicted within errors of less than 10%.

3.4.2. Surface Texturing

Otero et al. [112] attempted to optimize surface micro-textures fabricated by photolithography and chemical etching processes in order to reduce the COF of EHL contacts by means of an ANN. The data was obtained from tests on a mini-traction machine (steel ball-on-micro-textured copper disk) at various loads, total speeds and slip conditions. The ANN consisted of 7 inputs (average velocity, SRR, load, minor and major axis dimensions, depth and texturing density), 20 neurons in the hidden layer, and the COF as output. Thus, load case-dependent ranges for beneficial texture parameters could be derived. Additionally, referring to tests on samples with pores or micro-textures on a lubricated mini traction machine at different test conditions, Boidi et al. [113] applied an RBF to predict the wear behavior of sintered components. The database included 1704 experimental sets with different sum velocities and slip, as well as geometric or statistical characteristics of the dimples, grooves and pores, respectively. A Hardy multiquadric RBF was found to provide an excellent fit with an overall correlation of 0.93, especially with regard to the standard deviations of the tribological experiments. Mo et al. [114] utilized statistical methods as well as a back propagation ANN with 60 neurons in the single-hidden layer to investigate the role of micro-texture shape deviations and dimensional uncertainties on the tribological performance. The database was founded on physical modeling approaches in the form of simulations of parallel, hydrodynamically lubricated (HL) contacts and randomly split into 70% training and 30% validation data. The trained ANN was able to predict the relationships between geometric micro-texture parameters (e.g., dimple diameter, depth, area density etc.) and the frictional force as well as the load carrying capacity with an accuracy of 99.7% and 97.5%, respectively. Thus, the influences of statistical deviations (e.g., roundness errors, standard deviations of the dimensional parameters, etc.) could be estimated and optimal, robust optima could be retrieved by means of a genetic algorithm. Similarly, Marian et al. [115,116] utilized a MOP [74] to model the influence of micro-textures in EHL contacts as well as an EA to optimize the micro-texture geometry and distribution. Based upon a LHS (70 data sets) and contact simulations, the contact pressure, lubricant film height, and frictional force were predicted with CoPs larger than 82%, allowing subsequent optimization with an EA. Zambrano et al. [117] used reduced order modeling (ROM) to predict and optimize the frictional behavior of surface textures in dynamic rubber applications under different operating conditions. It is noteworthy that this was based on a limited number of experimental measurements and the ROM was fed with microscope-based texture measurements. In this sense, besides nominal texture parameters, the real geometries as well as their deviations and uncertainties have been evaluated with good accuracy.

The works in the area of surface engineering are summarized in Table 4 according to the subject, the database, the inputs and outputs, and the ML approach.

Table 4. Overview of ML approaches successfully applied in the area of surface engineering.

Subject	Database, Number of Data Sets (If Applicable Divided in Train/Test/Validation)	Inputs	Outputs	ML Approach	Prediction	Ref.
thermally sprayed Al_2O_3-TiO_2 coatings	experimental (pin-on-disk), 8	load, environment (dry or acid)	linear wear, COF at different time steps	back propagation ANN (2:80:63)	sufficient (not quantified)	[108]
HVOF sprayed Cr-C-Ni-Cr and WC-Co coatings and electroplated hard chromium	experimental (pin-on-disk), 360 (50%/50%)	material type, normal load, sliding velocity, sliding distance	COF	back propagation ANN (4:6:4:1)	errors < 11%	[109]
multilayer nitride PVD coatings	experimental (pin-on-disk), 246 (70%/15%/15%)	time, normal load, sliding velocity, lap, bias voltage, gas flow rate	spec. wear rate, COF	back propagation ANN (6:5:5:2)	errors < 1%	[110, 111]
surface texture design for EHL contacts	experimental (mini traction machine), 2000 (90%/5%/5%)	average velocity, slide-to-roll ratio, normal load, minor axis, major axis, texture depth, texture density	COF	various ANNs, best results for back propagation ANN (7:20:1)	MSE < 0,1%, R^2 > 0.99	[112]
	experimental (mini traction machine), 1704	entrainment speed, slide-to-roll ratio, surface feature ball, surface feature disk	COF	Hardy multiquadric RBF	R^2 > 0.935	[113]
	numerical (EHL simulation) in LHS, 70 (70%/30%)	texture diameter, texture depth, texture distance	max. pressure, min. film height, COF	MOP	CoP > 83%	[115, 116]
surface texture design for HL contacts	numerical (HL simulation)	dimple diameter, depth, area density, and various statistical deviations	COF, load carrying capacity	various ANNs, best results for back propagation ANN (41:20:2)	accuracy > 99.7% (COF), accuracy > 97.5% (load carrying capacity)	[114]

3.5. Lubricants

ML/AI approaches have also been used in the development and formulation of lubricants [118] and their additives [119] intended for the use in tribological systems. As such, Durak et al. [120] analyzed the effects of PTFE-based additives in mineral oil onto the frictional behavior of hydrodynamic journal bearings (252 data sets) by the aid of a feed forward back propagation ANN. An architecture with three inputs as studied in respective experiments (load, velocity, additive concentration), two hidden layers of 5 and 3 neurons, and the COF as output resulted in an accuracy of 98%. Therefore, optimal concentrations depending on the load case could be identified with rather little experimental effort. Humelnicu et al. [121] applied an ANN to investigate the tribological behavior of vegetable oil-diesel fuel mixtures. The data were generated in pin-on-disk tests at constant conditions, whereas the concentration of rapeseed and sunflower oil was varied and the averaged COF values of five repetitions of each combination was used for further processing. The neural network was trained with a back propagation algorithm and tangential transfer functions and the architecture considered as most suitable with relative deviations between 0.2% and 2.3% was built of three hidden layers with 2, 6, and 9 neurons, respectively. Bhaumik et al. [122,123] also applied a multi-hidden layer feed forward ANN to design lubricant formulations with vegetable oil blends (coconut, castor and palm oil) and various friction modifiers (MWCNT and graphene) based upon 80 data sets obtained from four-ball-tests as well as 120 data sets from pin-on-disk tests as reported in various literature. The respective material and test conditions were also included as influencing factors. For building the ANN, hyperbolic tangent transfer functions and a scaled conjugate gradient back propagation algorithm were used. Good prediction quality was thus achieved for the 11 and 13 inputs in the four-ball- and pin-on-disk tests, respectively, with accuracies over 92%. In addition to the influences of the lubricant and material properties, significant differences were also revealed due to the test setup. In an optimization based on the ANN using a genetic algorithm, it was also possible to derive ideal lubricant formulations, the suitability of which was actually demonstrated by subsequent preparation and corresponding experimental validation. Lately, Mujtaba et al. [124] utilized a Cuckoo search algorithm to optimize an extreme learning machine (ELM) and a response surface methodology (RSM) in predicting the tribological behavior of biodiesel from palm-sesame oil in dependency of ultrasound-assisted transesterification process variables. Based on a Box-Behnken experimental design, the biodiesel yield was predicted, whereby the ELM featured a better performance than RSM, and optimized. In tribological experiments on a four-ball-tester, improved friction and wear behavior compared to reference lubricants was also demonstrated with the derived blend.

In addition to these more macro-tribological approaches, some studies can also be found that tend to target even smaller scales [125]. For example, Sattari Baboukani et al. [126] employed a Bayesian modeling and transfer learning approach to predict maximum energy barriers of the potential surface energy, which corresponds to intrinsic friction, of various 2D materials from the graphene and the transition metal dichalcogenide (TMDC) families when sliding against a similar material with the aim of application as lubricant additives. The input variables for the model in the form of different descriptors (structural, electronic, thermal, electron-phonon coupling, mechanical and chemical effects) were extracted from density function theory (DFT) and molecular dynamics (MD) simulation studies in literature. The applied Bayesian model accommodated the sparse and noisy data set and estimated the maximum energy barrier as target variable as well as its uncertainty and potentially missing data. The predictions were validated against MD simulations, whereas excellent agreement with mean squared errors mostly below 0.25 were found. Thus, the application of the ML approach not only allowed for the prediction estimation of the applicability for tribological purposes of ten previously underexplored 2D materials, but also initiated discussion on novel empirical correlations and physical mechanisms.

The works in the area of lubricant formulation are summarized in Table 5 according to the subject, the database, the inputs and outputs, and the ML approach.

Table 5. Overview of ML approaches successfully applied in the area of lubricant formulation.

Subject	Database, Number of Data Sets (If Applicable Divided in Train/Test/Validation)	Inputs	Outputs	ML Approach	Prediction	Ref.
PTFE-based additives in mineral oil	experimental (journal bearing test-rig), 252 (80%/20%)	load, velocity, additive concentration	COF	back propagation ANN (3:5:3:1)	accuracy > 98%	[120]
vegetable oil-diesel fuel mixtures	experimental (pin-on-disk), 135	sunflower concentration, rapeseed concentration	COF	back propagation ANN (2:2:6:9:1)	RMSE < 0,1%	[121]
lubricant formulations with vegetable oil blends and friction modifiers (MWCNT, graphene)	literature (pin-on-disk, four-ball-tests), 200	speed, normal load, temperature, ball/pin/disk hardness, coconut oil content, castor oil content, palm oil content, MWCNT content, MWCNT size, graphene content, graphene dimensions	COF	scaled conjugate gradient back propagation ANN	accuracy > 92%	[122, 123]
biodiesel formulation	experimental (transesterification), 30	time, catalyst concentration, methanol-to-oil ratio, duty cycle	biodiesel yield	RSM	$R^2 > 0.994$, MSE < 0.023, RMSE < 0.151	[124]
				Cuckoo ELM	$R^2 > 0.996$, MSE < 0.024, RMSE < 0.117	
lubricant additives	literature and numerical (DFT and MD simulation)	lattice constant, c/a ratio, bond angle, interlayer space, M-X length, X-X length, M-radii, hexagonal width, in-plane stiffness, cohesive energy, binding energy, bandgap energy, thermal conductivity, average mass	maximum energy barrier	Bayesian model	MSE < 0.25	[126]

3.6. Others/General

Apart from aforementioned areas, a wide variety of studies can be found in fields which also related to tribology, but that were not assigned to the traditional core and are therefore not included in more detail in this review. The tribology of driven piles in clay [127], plate tectonics and earthquakes [128], or motion control [129,130] can be mentioned as examples. Nevertheless, some selected research shall be presented that did not necessarily fit into one of the upper categories but had a rather general scope. As such, already in 2002, Ao et al. [131] introduced an ANN to predict the evolution of surface topography during the wear process. The proposed approach utilized surface measurements at a finite number of time intervals during tribological experiments in a conformal block-on-ring configuration. The back-propagation ANN with sigmoid transfer functions was trained with the LM algorithm and statistical surface parameters (RMS roughness, skewness, kurtosis, and autocorrelation). Together with initial surface parameters, the corresponding 3D topography in worn conditions could be estimated by surface synthesis. Thereby, good prediction quality could be achieved, especially if the autocorrelation function did not experience stronger changes. So, this was not just about predicting and optimizing target variables but was rather already a step towards semi-physical modeling. Thereby, the usage of ML/AI in the field of tribology may not be limited to forward approaches, which predict the tribological behavior based on some input data sets in the context of an experimental design. Accordingly, Haviez et al. [132] later developed a modified ANN model, which was used to solve actual physical equations describing the phenomena of fretting wear. Interestingly, this eliminated the necessity for iterative learning, e.g., by back propagation, or other regularization techniques. Thus, the fretting wear damage could be predicted with higher efficiency and accuracy than by a conventional back propagation ANN trained with experimental data, highlighting the ability of generalization albeit the rather low level of complexity. Similarly, Argatov and Chai [133] suggested an ANN-based modeling framework for analyzing the dry sliding wear during running-in from pin-on-disk tribometer tests. The authors attempted to derive the true wear coefficient instead of the specific wear rate at given conditions, contact pressures and sliding velocities. This was based upon the integral and differential forms of the Archard's wear equation as well as single-hidden layer ANN with sigmoid transfer functions. They applied their approach to various data from the literature ranging from cermet coatings, zirconia reinforced aluminum hybrid composites to nickel–chromium alloys and reported good efficiency and agreement. Very recently, Almqvist [134] derived a physics informed neural network (PINN) to solve the initial and boundary value problems described by linear ordinary differential equations and to solve the second order Reynolds differential equation. Thereby, comparable results to analytical solutions were obtained. The advantage of the present approach is not in accuracy or efficiency, but in the fact that it is a mesh-free method that is not data-driven. The author hypothesized that this concept could be generalized in the future and lead to a more accurate and efficient solution of related but nonlinear problems than the currently available routines.

Finally, two papers shall be highlighted that addressed other approaches than ANNs and/or other scales as well. Bucholz et al. [135] used a dataset from dry sliding pin-on-disk tests with different ceramic pairings having different intrinsic properties and inorganic minerals to develop a predictive model. The latter was generated by the recursive partitioning method, resulting in a graphical expression of the classification of observations according to similarities determined by variable importance in projection and the error some of squares. The obtained regression tree as illustrated in Figure 7a) demonstrated a satisfactory coefficient of determination above 0.89 when comparing prediction and experiment (Figure 7b). Finally, Perčic et al. [136] recently trained various ML/AI approaches to predict the nanoscale friction of alumina (Al_2O_3), titanium dioxide (TiO_2), molybdenum disulphide (MoS2), and aluminum (Al) thin films in dependency of several process parameters, including normal forces, sliding velocities, and temperature. The data were acquired by lateral force microscopy (LFM) within a centroidal Voronoi tessellation (CVT) design of experiments, whereas 2/3 of the data were generally used for training and 1/3 for validation. The study employed MLP ANN, random DT and RF, support vector regression (SVR), age-layered population structure (ALPS), grammatical evolution (GE), and symbolic regression multi-gene programming (SRMG). The suitability for predicting the frictional force for these approaches was further evaluated with respect to the mean absolute error, the root mean squared error and the coefficient of determination. Thereby, the SRMG model showed the best performance with prediction accuracies (determination coefficient) between 72% and 91%, depending on the sample type. This allowed to derive simple functional descriptions of the nanoscale friction for studied variable process parameters.

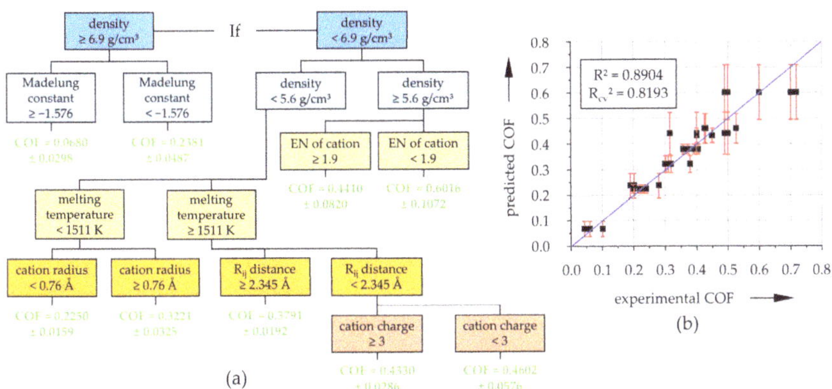

Figure 7. Dendrogram for the COF estimation from recursive partitioning (**a**) and comparison between experimental and predicted values (**b**). Redrawn from [135] with permission (Springer).

4. Summary and Concluding Remarks

Tribology naturally involves multiple interacting features and processes, where machine learning and artificial intelligence approaches are feasible to support sorting through the complexity of patterns and identifying trends on a much larger scale than the human brain is capable of. Computers are able to fit thousands of properties, which enables for a much wider search of the available solution space and allows quantitative fits to a broad range of properties. Predictions do not have to be limited to averaged or global values/outputs but could also cover locally and timely resolved evolutions and bridge the gap between different scales. Therefore, ML and AI might change the landscape of what is possible going beyond the mere understanding of mechanisms towards designing novel and/or potentially smart tribological systems. As is also evident from the quantified survey, ML has hence already been employed in many fields of tribology, from composite materials and drive technology to manufacturing, surface engineering, and lubricants. The intent of ML might not necessarily be to create conclusive predictive models but can be seen as complementary tool to efficiently achieve optimum designs for problems, which elude other physically motivated mathematical and numerical formulations. We assume that, besides the availability of larger amounts of experimental data, this is the reason for the comparatively large number of investigations on composite materials.

The challenge is that a ML approach does not necessarily guide towards the specific problem solution and the selection as well as optimization of a qualified algorithm is of decisive importance. Accordingly, there is a wide variety of approaches that have already been successfully applied to answer tribological research questions. A summary is provided in Table 6, which is—together with Tables 1–5—intended to support researchers in identifying initial selections.

Table 6. Overview of ML approaches successfully applied in various areas of tribology.

ML Approach	Application Area													
	Composite Materials			Drive Technology				Manufacturing			Surface Engineering			
	Thermoset Matrix	Thermoplastic Matrix	Metal Matrix	Rolling Bearings	Sliding Bearings	Seals	Brakes And Clutches	Friction Stir Welding	Forming	Machining	Coating	Texturing	Lubricants	Others
ANN	[39–42]	[43–53]	[60–64]	[69,70]	[75–77]	[82]	[86–89, 91,92]	[97–101,103,104]	[105]	[106]	[108–111]	[112, 114]	[120–123]	[131–134, 136]
ANFIS								[101]						
Bayesian													[126]	
DT			[63,64]	[60]						[106]				[135, 136]
KNN			[63,64]											
MOP				[73]								[115, 116]		
QDA				[71]										
RF			[63,64]		[78]					[106]				[136]
RBF										[106]		[113]		
SVM			[63,64]			[83]	[90]	[104]						[136]

Apparently, a large share of the research discussed in this article (roughly three quarters) was based on ANNs. However, even still, there are manifold possibilities concerning architecture, training algorithms, or transfer functions. Other ML approaches are still less commonly used for tribological issues but are justifiably coming more into focus and can be more effective for some problems. The reproducibility and comparability of the prediction quality from the various approaches and studies is frequently hampered by the sometimes ambiguous underlying database and the lack of information on the implementation of ML approaches withing publications as well as the use of different error/accuracy measures. Most of the works also comprised forward ML models, which were developed to predict the tribological behavior as output based on various input parameters such as material or test conditions. In principle, however, inverse models to characterize the materials and surfaces [54] or physics-informed ML approaches [134] can also be applied. With a closer assessment of the intentions and objectives of the studies, as well as the overrepresentation of ANNs, one might get the impression that ML is in many cases being used to serve its own ends. The added value compared to physical modeling or statistical evaluation based on more classical regressions is not always evident. A few studies, however, manage to extract real insights and thus additional knowledge from a large and broad database. The comprehensive works in the field of composite materials from Kurt and Oduncuoglu [52], Vinoth and Datta [53], and Hasan et al. [63,64] utilizing literature-extracted databases may be highlighted here and can serve as excellent examples. The current showstopper is still the availability of sufficient and comparable datasets as well as the handling of uncertainties regarding test conditions and deviations. In this respect, we would like to encourage authors to also publish the underlying databases and the corresponding models in appendices or data repositories. Moreover, there is great potential to automatize and optimize the data acquisition and processing, which is presently still very manual in the field of tribology, in order to unfold the knowledge already available in institutes, enterprises or in the literature by means of machine learning.

Author Contributions: Conceptualization, M.M.; methodology, M.M. and S.T.; formal analysis and investigation, M.M. and S.T.; data curation, M.M.; visualization, M.M.; writing—original draft preparation, review and editing, M.M. and S.T. Both authors have read and agreed to the published version of the manuscript.

Funding: This research received no external funding.

Institutional Review Board Statement: Not applicable.

Informed Consent Statement: Not applicable.

Data Availability Statement: The data on the quantitative evaluation are available upon request from the corresponding author.

Acknowledgments: The authors kindly acknowledge the continuous support of Friedrich-Alexander-University Erlangen-Nuremberg (FAU) and University of Bayreuth, Germany.

Conflicts of Interest: The authors declare no conflict of interest.

Abbreviations

AE	acoustic emission
AI	artificial intelligence
ALPS	age-layered population structure
ANFIS	adaptive neuro-fuzzy interference system
ANN	artificial neural network
ART	adaptive resonance theory
BFS	blast furnaces slag
CCD	centrale composite design
CDI	cage dynamics indicator
CF	carbon fiber
CMC	ceramic matrix composite
CNT	carbon nanotube
CoD	coefficient of determination
COF	coefficient of friction
CoP	coefficient of prognosis
CVT	centroidal voronoi tessellation
DFT	density function theory
DoE	design of experiments
DT	decision tree
EA	evolutionary algorithm
EBP	error back propagation
EHL	elastohydrodynamic lubrication
ELM	extreme learning machine
FE	finite element
FFT	fast fourier transformation
GBM	gradient boosting machine
GE	grammatical evolution
GO	graphene oxide
HL	hydrodynamic lubrication
HVOF	high-velocity oxy-fuel
IBA	improved bat algorithm
kNN	k-nearest neighbor
LFM	lateral force microscopy
LHS	latin hypercube sampling
LM	levenberg-marquardt
MBE	model-based estimation
MD	molecular dynamics
ML	machine learning
MLP	multilayer perception

References

1. Holmberg, K.; Erdemir, A. Influence of tribology on global energy consumption, costs and emissions. *Friction* **2017**, *5*, 263–284. [CrossRef]
2. Vakis, A.; Yastrebov, V.; Scheibert, J.; Nicola, L.; Dini, D.; Minfray, C.; Almqvist, A.; Paggi, M.; Lee, S.; Limbert, G.; et al. Modeling and simulation in tribology across scales: An overview. *Tribol. Int.* **2018**, *125*, 169–199. [CrossRef]

3. Ciulli, E. Tribology and industry: From the origins to 4.0. *Front. Mech. Eng.* **2019**, *5*, 103. [CrossRef]
4. Zhang, Z.; Yin, N.; Chen, S.; Liu, C. Tribo-informatics: Concept, architecture, and case study. *Friction* **2021**, *9*, 642–655. [CrossRef]
5. Rosenkranz, A.; Marian, M.; Profito, F.J.; Aragon, N.; Shah, R. The use of artificial intelligence in tribology—A perspective. *Lubricants* **2021**, *9*, 2. [CrossRef]
6. Wuest, T.; Weimer, D.; Irgens, C.; Thoben, K.-D. Machine learning in manufacturing: Advantages, challenges, and applications. *Prod. Manuf. Res.* **2016**, *4*, 23–45. [CrossRef]
7. Ghoddusi, H.; Creamer, G.G.; Rafizadeh, N. Machine learning in energy economics and finance: A review. *Energy Econ.* **2019**, *81*, 709–727. [CrossRef]
8. Kaieski, N.; da Costa, C.A.; Righi, R.D.R.; Lora, P.S.; Eskofier, B. Application of artificial intelligence methods in vital signs analysis of hospitalized patients: A systematic literature review. *Appl. Soft Comput.* **2020**, *96*, 106612. [CrossRef]
9. Kügler, P.; Marian, M.; Schleich, B.; Tremmel, S.; Wartzack, S. tribAIn—Towards an explicit specification of shared tribological understanding. *Appl. Sci.* **2020**, *10*, 4421. [CrossRef]
10. Ji, Y.; Bao, J.; Yin, Y.; Ma, C. Applications of artificial intelligence in tribology. *Recent Patents Mech. Eng.* **2016**, *9*, 193–205. [CrossRef]
11. Argatov, I. Artificial Neural Networks (ANNs) as a novel modeling technique in tribology. *Front. Mech. Eng.* **2019**, *5*, 1074. [CrossRef]
12. Bell, J. *Machine Learning: Hands-On for Developers and Technical Professionals*; Wiley: Hoboken, NJ, USA, 2014; ISBN 978-1-118-88906-0.
13. Wittpahl, V. *Künstliche Intelligenz*; Springer: Berlin, Germany, 2019; ISBN 978-3-662-58041-7.
14. Müller, A.C.; Guido, S. *Introduction to Machine Learning with Python: A Guide for Data Scientist*, 1st ed.; O'Reilly: Beijing, China, 2016; ISBN 978-1-4493-6941-5.
15. Schölkopf, B.; Smola, A.J. *Learning with Kernels: Support Vector Machines, Regularization, Optimization, and Beyond*; MIT Press: Cambridge, MA, USA, 2002; ISBN 978-0-262-19475-4.
16. Bishop, C.M. Pattern recognition and machine learning. In *Information Science and Statistics*; Springer: New York, NY, USA, 2006; pp. 21–24, ISBN 9780387310732.
17. Freund, Y. Boosting a weak learning algorithm by majority. *Inf. Comput.* **1995**, *121*, 256–285. [CrossRef]
18. Breiman, L. Random forests. *Mach. Learn.* **2001**, *45*, 5–32. [CrossRef]
19. Schapire, R.E. The strength of weak learnability. *Mach. Learn.* **1990**, *5*, 197–227. [CrossRef]
20. Forsyth, D. *Applied Machine Learning*; Springer: Cham, Switzerland, 2019; ISBN 978-3-030-18113-0.
21. Sarkar, D.; Bali, R.; Sharma, T. *Practical Machine Learning with Python: A Problem-Solver's Guide to Building Real-World Intelligent Systems*; Apress L.P.: Berkeley, CA, USA, 2017; ISBN 978-1-4842-3206-4.
22. Kinnebrock, W. *Neuronale Netze: Grundlagen, Anwendungen, Beispiele*; 2., verb. Aufl.; Oldenbourg: München, Germany, 1994; ISBN 3-486-22947-8.
23. Tallian, T.E. A computerized expert system for tribological failure diagnosis. *J. Tribol.* **1989**, *111*, 238–244. [CrossRef]
24. Tallian, T.E. Tribological design decisions using computerized databases. *J. Tribol.* **1987**, *109*, 381–386. [CrossRef]
25. Jones, S.P.; Jansen, R.; Fusaro, R.L. Preliminary investigation of neural network techniques to predict tribological properties. *Tribol. Trans.* **1997**, *40*, 312–320. [CrossRef]
26. Karkoub, M.; Elkamel, A. Modelling pressure distribution in a rectangular gas bearing using neural networks. *Tribol. Int.* **1997**, *30*, 139–150. [CrossRef]
27. Santner, E. Computer support in tribology—Experiments and database. *Tribotest* **1996**, *2*, 267–280. [CrossRef]
28. Myshkin, N.; Kwon, O.; Grigoriev, A.; Ahn, H.-S.; Kong, H. Classification of wear debris using a neural network. *Wear* **1997**, *203–204*, 658–662. [CrossRef]
29. Umeda, A.; Sugimura, J.; Yamamoto, Y. Characterization of wear particles and their relations with sliding conditions. *Wear* **1998**, *216*, 220–228. [CrossRef]
30. Chowdhury, M.A.; Nuruzzaman, D.M.; Rahaman, M.L. Tribological behavior of composite materials—A review. *Recent Pat. Mech. Eng.* **2010**, *1*, 123–128. [CrossRef]
31. Tyagi, R.; Davim, J.P. *Processing Techniques and Tribological Behavior of Composite Materials*; Engineering Science Reference, an Imprint of IGI Global: Hershey, PA, USA, 2015; ISBN 9781466675308.
32. Omrani, E.; Menezes, P.L.; Rohatgi, P.K. State of the art on tribological behavior of polymer matrix composites reinforced with natural fibers in the green materials world. *Eng. Sci. Technol. Int. J.* **2016**, *19*, 717–736. [CrossRef]
33. Milosevic, M.; Valášek, P.; Ruggiero, A. Tribology of natural fibers composite materials: An overview. *Lubricants* **2020**, *8*, 42. [CrossRef]
34. Nayak, S.K.; Satapathy, A. Wear analysis of waste marble dust-filled polymer composites with an integrated approach based on design of experiments and neural computation. *Proc. Inst. Mech. Eng. Part J J. Eng. Tribol.* **2019**, *234*, 1846–1856. [CrossRef]
35. Friedrich, K.; Reinicke, R.; Zhang, Z. Wear of polymer composites. *Proc. Inst. Mech. Eng. Part J J. Eng. Tribol.* **2002**, *216*, 415–426. [CrossRef]
36. Liujie, X.; Davim, J.P.; Cardoso, R. Prediction on tribological behaviour of composite PEEK-CF30 using artificial neural networks. *J. Mater. Process. Technol.* **2007**, *189*, 374–378. [CrossRef]
37. Zhang, G.; Guessasma, S.; Liao, H.; Coddet, C.; Bordes, J.-M. Investigation of friction and wear behaviour of SiC-filled PEEK coating using artificial neural network. *Surf. Coat. Technol.* **2006**, *200*, 2610–2617. [CrossRef]

38. Padhi, P.K.; Satapathy, A.; Nakka, A.M. Processing, characterization, and wear analysis of short glass fiber-reinforced polypropylene composites filled with blast furnace slag. *J. Thermoplast. Compos. Mater.* **2013**, *28*, 656–671. [CrossRef]
39. Padhi, P.K.; Satapathy, A. Analysis of sliding wear characteristics of BFS filled composites using an experimental design approach integrated with ANN. *Tribol. Trans.* **2013**, *56*, 789–796. [CrossRef]
40. Egala, R.; Jagadeesh, G.V.; Setti, S.G. Experimental investigation and prediction of tribological behavior of unidirectional short castor oil fiber reinforced epoxy composites. *Friction* **2021**, *9*, 250–272. [CrossRef]
41. Nirmal, U. Prediction of friction coefficient of treated betelnut fibre reinforced polyester (T-BFRP) composite using artificial neural networks. *Tribol. Int.* **2010**, *43*, 1417–1429. [CrossRef]
42. Nasir, T.; Yousif, B.F.; McWilliam, S.; Salih, N.D.; Hui, L.T. An artificial neural network for prediction of the friction coefficient of multi-layer polymeric composites in three different orientations. *Proc. Inst. Mech. Eng. Part C J. Mech. Eng. Sci.* **2010**, *224*, 419–429. [CrossRef]
43. Zhang, Z.; Friedrich, K.; Velten, K. Prediction on tribological properties of short fibre composites using artificial neural networks. *Wear* **2002**, *252*, 668–675. [CrossRef]
44. Velten, K.; Reinicke, R.; Friedrich, K. Wear volume prediction with artificial neural networks. *Tribol. Int.* **2000**, *33*, 731–736. [CrossRef]
45. Gyurova, L.A.; Miniño-Justel, P.; Schlarb, A.K. Modeling the sliding wear and friction properties of polyphenylene sulfide composites using artificial neural networks. *Wear* **2010**, *268*, 708–714. [CrossRef]
46. Jiang, Z.; Gyurova, L.A.; Schlarb, A.K.; Friedrich, K.; Zhang, Z. Study on friction and wear behavior of polyphenylene sulfide composites reinforced by short carbon fibers and sub-micro TiO2 particles. *Compos. Sci. Technol.* **2008**, *68*, 734–742. [CrossRef]
47. Gyurova, L.A.; Friedrich, K. Artificial neural networks for predicting sliding friction and wear properties of polyphenylene sulfide composites. *Tribol. Int.* **2011**, *44*, 603–609. [CrossRef]
48. Jiang, Z.; Gyurova, L.; Zhang, Z.; Friedrich, K.; Schlarb, A.K. Neural network based prediction on mechanical and wear properties of short fibers reinforced polyamide composites. *Mater. Des.* **2008**, *29*, 628–637. [CrossRef]
49. Busse, M.; Schlarb, A.K. A novel neural network approach for modeling tribological properties of polyphenylene sulfide reinforced on different scales. In *Tribology of Polymeric Nanocomposites*; Elsevier: Cham, Switzerland, 2013; pp. 779–793. [CrossRef]
50. Zhu, J.; Shi, Y.; Feng, X.; Wang, H.; Lü, X. Prediction on tribological properties of carbon fiber and TiO2 synergistic reinforced polytetrafluoroethylene composites with artificial neural networks. *Mater. Des.* **2009**, *30*, 1042–1049. [CrossRef]
51. Li, S.; Shao, M.; Duan, C.; Yan, Y.; Wang, Q.; Wang, T.; Zhang, X. Tribological behavior prediction of friction materials for ultrasonic motors using Monte Carlo-based artificial neural network. *J. Appl. Polym. Sci.* **2019**, *136*, 47157. [CrossRef]
52. Kurt, H.I.; Oduncuoglu, M. Application of a neural network model for prediction of wear properties of ultrahigh molecular weight polyethylene composites. *Int. J. Polym. Sci.* **2015**, *2015*, 315710. [CrossRef]
53. Vinoth, A.; Datta, S. Design of the ultrahigh molecular weight polyethylene composites with multiple nanoparticles: An artificial intelligence approach. *J. Compos. Mater.* **2020**, *54*, 179–192. [CrossRef]
54. Kordijazi, A.; Zhao, T.; Zhang, J.; Alrfou, K.; Rohatgi, P. A review of application of machine learning in design, synthesis, and characterization of metal matrix composites: Current status and emerging applications. *JOM* **2021**, *73*, 2060–2074. [CrossRef]
55. Genel, K.; Kurnaz, S.; Durman, M. Modeling of tribological properties of alumina fiber reinforced zinc–aluminum composites using artificial neural network. *Mater. Sci. Eng. A* **2003**, *363*, 203–210. [CrossRef]
56. Kumar, K.R.; Mohanasundaram, K.M.; Arumaikkannu, G.; Subramanian, R. Artificial neural networks based prediction of wear and frictional behaviour of aluminium (A380)–fly ash composites *Tribol. Mater. Surf. Interfaces* **2012**, *6*, 15–19. [CrossRef]
57. Saravanan, S.D.; Senthilkumar, M. Prediction of tribological behaviour of rice husk ash reinforced aluminum alloy matrix composites using artificial neural network. *Russ. J. Non-Ferrous Metals* **2015**, *56*, 97–106. [CrossRef]
58. Arif, S.; Alam, T.; Ansari, A.H.; Shaikh, M.B.N.; Siddiqui, M.A. Analysis of tribological behaviour of zirconia reinforced Al-SiC hybrid composites using statistical and artificial neural network technique. *Mater. Res. Express* **2018**, *5*, 056506. [CrossRef]
59. Vettivel, S.; Selvakumar, N.; Narayanasamy, R.; Leema, N. Numerical modelling, prediction of Cu–W nano powder composite in dry sliding wear condition using response surface methodology. *Mater. Des.* **2013**, *50*, 977–996. [CrossRef]
60. Stojanović, B.; Vencl, A.; Bobić, I.; Miladinović, S.; Skerlić, J. Experimental optimisation of the tribological behaviour of Al/SiC/Gr hybrid composites based on Taguchi's method and artificial neural network. *J. Braz. Soc. Mech. Sci. Eng.* **2018**, *40*, 311. [CrossRef]
61. Thankachan, T.; Prakash, K.S.; Kamarthin, M. Optimizing the tribological behavior of hybrid copper surface composites using statistical and machine learning techniques. *J. Tribol.* **2018**, *140*, 031610. [CrossRef]
62. Gangwar, S.; Pathak, V.K. Dry sliding wear characteristics evaluation and prediction of vacuum casted marble dust (MD) reinforced ZA-27 alloy composites using hybrid improved bat algorithm and ANN. *Mater. Today Commun.* **2020**, *25*, 101615. [CrossRef]
63. Hasan, S.; Kordijazi, A.; Rohatgi, P.K.; Nosonovsky, M. Triboinformatic modeling of dry friction and wear of aluminum base alloys using machine learning algorithms. *Tribol. Int.* **2021**, *161*, 107065. [CrossRef]
64. Hasan, S.; Kordijazi, A.; Rohatgi, P.K.; Nosonovsky, M. Triboinformatics Approach for friction and wear prediction of al-graphite composites using machine learning methods. *J. Tribol.* **2021**, *144*, 011701. [CrossRef]
65. Kankar, P.; Sharma, S.C.; Harsha, S. Fault diagnosis of ball bearings using machine learning methods. *Expert Syst. Appl.* **2011**, *38*, 1876–1886. [CrossRef]

66. Vargas-Machuca, J.; García, F.; Coronado, A.M. Detailed comparison of methods for classifying bearing failures using noisy measurements. *J. Fail. Anal. Prev.* **2020**, *20*, 744–754. [CrossRef]
67. Desavale, R.G.; Venkatachalam, R.; Chavan, S.P. Antifriction bearings damage analysis using experimental data based models. *J. Tribol.* **2013**, *135*, 041105. [CrossRef]
68. Gomes, G.F.; Mendez, Y.A.D.; Alexandrino, P.D.S.L.; Da Cunha, S.S.; Ancelotti, A.C. A Review of vibration based inverse methods for damage detection and identification in mechanical structures using optimization algorithms and ANN. *Arch. Comput. Methods Eng.* **2019**, *26*, 883–897. [CrossRef]
69. Subrahmanyam, M.; Sujatha, C. Using neural networks for the diagnosis of localized defects in ball bearings. *Tribol. Int.* **1997**, *30*, 739–752. [CrossRef]
70. Kanai, R.A.; Desavale, R.G.; Chavan, S.P. Experimental-based fault diagnosis of rolling bearings using artificial neural network. *J. Tribol.* **2016**, *138*, 031103. [CrossRef]
71. Schwarz, S.; Grillenberger, H.; Tremmel, S.; Wartzack, S. Prediction of rolling bearing cage dynamics using dynamics simulations and machine learning algorithms. *Tribol. Trans.* **2021**, 1–23. [CrossRef]
72. Freund, Y.; Schapire, R.E. A Decision-theoretic generalization of on-line learning and an application to boosting. *J. Comput. Syst. Sci.* **1997**, *55*, 119–139. [CrossRef]
73. Wirsching, S.; Marian, M.; Bartz, M.; Stahl, T.; Wartzack, S. Geometrical optimization of the EHL roller face/rib contact for energy efficiency in tapered roller bearings. *Lubricants* **2021**, *9*, 67. [CrossRef]
74. Most, T.; Will, J. Metamodel of Optimal Prognosis—An Automatic Approach for Variable Reduction and Optimal Meta-Model Selection. In Proceedings of the 2008 Weimarer Optimierungs und Stochastiktage 5.0, Weimar, Germany, 20–21 November 2008. [CrossRef]
75. Canbulut, F.; Yildirim, S.; Sinanoglu, C. Design of an artificial neural network for analysis of frictional power loss of hydrostatic slipper bearings. *Tribol. Lett.* **2004**, *17*, 887–899. [CrossRef]
76. Ünlü, B.S.; Durmuş, H.; Meriç, C. Determination of tribological properties at CuSn10 alloy journal bearings by experimental and means of artificial neural networks method. *Ind. Lubr. Tribol.* **2012**, *64*, 258–264. [CrossRef]
77. Moder, J.; Bergmann, P.; Grün, F. Lubrication regime classification of hydrodynamic journal bearings by machine learning using torque data. *Lubricants* **2018**, *6*, 108. [CrossRef]
78. Prost, J.; Cihak-Bayr, U.; Neacşu, I.; Grundtner, R.; Pirker, F.; Vorlaufer, G. Semi-supervised classification of the state of operation in self-lubricating journal bearings using a random forest classifier. *Lubricants* **2021**, *9*, 50. [CrossRef]
79. Francisco, A.; Lavie, T.; Fatu, A.; Villechaise, B. Metamodel-assisted optimization of connecting rod big-end bearings. *J. Tribol.* **2013**, *135*, 041704. [CrossRef]
80. Li, X.; Fu, P.; Chen, K.; Lin, Z.; Zhang, E. The contact state monitoring for seal end faces based on acoustic emission detection. *Shock Vib.* **2015**, *2016*, 1–8. [CrossRef]
81. Zhang, Z.; Li, X. Acoustic Emission Monitoring for Film Thickness of Mechanical Seals Based on Feature Dimension Reduction and Cascaded Decision. In Proceedings of the 2014 Sixth International Conference on Measuring Technology and Mechatronics Automation, Zhangjiajie, China, 10–11 January 2014; pp. 64–70. [CrossRef]
82. Logozzo, S.; Valigi, M.C. Investigation of instabilities in mechanical face seals: Prediction of critical speed values. In *Advances in Mechanism and Machine Science*; Uhl, T., Ed.; Springer International Publishing: Cham, Switzerland, 2019; pp. 3865–3872. [CrossRef]
83. Yin, Y.; Liu, X.; Huang, W.; Liu, Y.; Hu, S. Gas face seal status estimation based on acoustic emission monitoring and support vector machine regression. *Adv. Mech. Eng.* **2020**, *12*, 168781402092132. [CrossRef]
84. Yin, Y.; Bao, J.; Yang, L. Wear performance and its online monitoring of the semimetal brake lining for automobiles. *Ind. Lubr. Tribol.* **2014**, *66*, 100–105. [CrossRef]
85. Xiao, G.; Zhu, Z. Friction materials development by using DOE/RSM and artificial neural network. *Tribol. Int.* **2010**, *43*, 218–227. [CrossRef]
86. Aleksendrić, D.; Duboka, Č.; Mariotti, G.V. Neural modelling of friction material cold performance. *Proc. Inst. Mech. Eng. Part D J. Automob. Eng.* **2008**, *222*, 1201–1209. [CrossRef]
87. Aleksendrić, D.; Barton, D.C. Neural network prediction of disc brake performance. *Tribol. Int.* **2009**, *42*, 1074–1080. [CrossRef]
88. Aleksendrić, D.; Barton, D.C.; Vasić, B. Prediction of brake friction materials recovery performance using artificial neural networks. *Tribol. Int.* **2010**, *43*, 2092–2099. [CrossRef]
89. Aleksendric, D. Neural network prediction of brake friction materials wear. *Wear* **2010**, *268*, 117–125. [CrossRef]
90. Timur, M.; Aydin, F. Anticipating the friction coefficient of friction materials used in automobiles by means of machine learning without using a test instrument. *Turk. J. Electr. Eng. Comput. Sci.* **2013**, *21*, 1440–1454. [CrossRef]
91. Senatore, A.; D'Agostino, V.; Di Giuda, R.; Petrone, V. Experimental investigation and neural network prediction of brakes and clutch material frictional behaviour considering the sliding acceleration influence. *Tribol. Int.* **2011**, *44*, 1199–1207. [CrossRef]
92. Grzegorzek, W.; Scieszka, S.F. Prediction on friction characteristics of industrial brakes using artificial neural networks. *Proc. Inst. Mech. Eng. Part J J. Eng. Tribol.* **2013**, *228*, 1025–1035. [CrossRef]
93. Gupta, S.K.; Pandey, K.; Kumar, R. Artificial intelligence-based modelling and multi-objective optimization of friction stir welding of dissimilar AA5083-O and AA6063-T6 aluminium alloys. *Proc. Inst. Mech. Eng. Part L J. Mater. Des. Appl.* **2016**, *232*, 333–342. [CrossRef]

94. D'Orazio, A.; Forcellese, A.; Simoncini, M. Prediction of the vertical force during FSW of AZ31 magnesium alloy sheets using an artificial neural network-based model. *Neural Comput. Appl.* **2018**, *31*, 7211–7226. [CrossRef]
95. Huggett, D.J.; Liao, T.W.; Wahab, M.A.; Okeil, A. Prediction of friction stir weld quality without and with signal features. *Int. J. Adv. Manuf. Technol.* **2018**, *95*, 1989–2003. [CrossRef]
96. Das, S.; Roy, R.; Chattopadhyay, A. Evaluation of wear of turning carbide inserts using neural networks. *Int. J. Mach. Tools Manuf.* **1996**, *36*, 789–797. [CrossRef]
97. Sathiya, P.; Aravindan, S.; Haq, A.N.; Paneerselvam, K. Optimization of friction welding parameters using evolutionary computational techniques. *J. Mater. Process. Technol.* **2009**, *209*, 2576–2584. [CrossRef]
98. Tansel, I.N.; Demetgul, M.; Okuyucu, H.; Yapici, A. Optimizations of friction stir welding of aluminum alloy by using genetically optimized neural network. *Int. J. Adv. Manuf. Technol.* **2010**, *48*, 95–101. [CrossRef]
99. Atharifar, H. Optimum parameters design for friction stir spot welding using a genetically optimized neural network system. *Proc. Inst. Mech. Eng. Part B J. Eng. Manuf.* **2009**, *224*, 403–418. [CrossRef]
100. Anand, K.; Shrivastava, R.; Tamilmannan, K.; Sathiya, P. A Comparative study of artificial neural network and response surface methodology for optimization of friction welding of incoloy 800 H. *Acta Metall. Sin. Engl. Lett.* **2015**, *28*, 892–902. [CrossRef]
101. Dewan, M.W.; Huggett, D.J.; Liao, T.W.; Wahab, M.A.; Okeil, A. Prediction of tensile strength of friction stir weld joints with adaptive neuro-fuzzy inference system (ANFIS) and neural network. *Mater. Des.* **2016**, *92*, 288–299. [CrossRef]
102. Jang, J.-S. ANFIS: Adaptive-network-based fuzzy inference system. *IEEE Trans. Syst. Man Cybern.* **1993**, *23*, 665–685. [CrossRef]
103. Baraka, A.; Panoutsos, G.; Cater, S. A real-time quality monitoring framework for steel friction stir welding using computational intelligence. *J. Manuf. Proc.* **2015**, *20*, 137–148. [CrossRef]
104. Das, B.; Pal, S.; Bag, S. Torque based defect detection and weld quality modelling in friction stir welding process. *J. Manuf. Proc.* **2017**, *27*, 8–17. [CrossRef]
105. Fereshteh-Saniee, F.; Nourbakhsh, S.H.; Pezeshki, S.M. Estimation of flow curve and friction coefficient by means of a one-step ring test using a neural network coupled with FE simulations. *J. Mech. Sci. Technol.* **2012**, *26*, 153–160. [CrossRef]
106. Bustillo, A.; Pimenov, D.; Matuszewski, M.; Mikolajczyk, T. Using artificial intelligence models for the prediction of surface wear based on surface isotropy levels. *Robot. Comput. Manuf.* **2018**, *53*, 215–227. [CrossRef]
107. Hutchings, I.; Shipway, P. Surface engineering. In *Tribology*; Elsevier: Amsterdam, The Netherlands, 2017; pp. 237–281, ISBN 9780081009109.
108. Cetinel, H. The artificial neural network based prediction of friction properties of Al2O3-TiO2 coatings. *Ind. Lubr. Tribol.* **2012**, *64*, 288–293. [CrossRef]
109. Sahraoui, T.; Guessasma, S.; Fenineche, N.; Montavon, G.; Coddet, C. Friction and wear behaviour prediction of HVOF coatings and electroplated hard chromium using neural computation. *Mater. Lett.* **2004**, *58*, 654–660. [CrossRef]
110. Upadhyay, R.; Kumaraswamidhas, L. Multilayer nitride coating performance optimized by an artificial neural network approach. *Ciência Tecnol. Mater.* **2016**, *28*, 47–54. [CrossRef]
111. Upadhyay, R.K.; Kumaraswamidhas, L.A. Friction and wear response of nitride coating deposited through PVD magnetron sputtering. *Tribol. Mater. Surf. Interfaces* **2016**, *10*, 196–205. [CrossRef]
112. Otero, J.E.; Ochoa, E.D.L.G.; Vallinot, I.B.; Tanarro, E.C. Optimising the design of textured surfaces for reducing lubricated friction coefficient. *Lubr. Sci.* **2016**, *29*, 183–199. [CrossRef]
113. Boidi, G.; da Silva, M.R.; Profito, F.J.; Machado, I.F. Using machine learning Radial Basis Function (RBF) method for predicting lubricated friction on textured and porous surfaces. *Surf. Topogr. Metrol. Prop.* **2020**, *8*, 044002. [CrossRef]
114. Mo, F.; Shen, C.; Zhou, J.; Khonsari, M.M. Statistical analysis of the influence of imperfect texture shape and dimensional uncertainty on surface texture performance. *IEEE Access* **2017**, *5*, 27023–27035. [CrossRef]
115. Marian, M.; Grützmacher, P.; Rosenkranz, A.; Tremmel, S.; Mücklich, F.; Wartzack, S. Designing surface textures for EHL point-contacts—Transient 3D simulations, meta-modeling and experimental validation. *Tribol. Int.* **2019**, *137*, 152–163. [CrossRef]
116. Tremmel, S.; Marian, M.; Zahner, M.; Wartzack, S.; Merklein, M. Friction reduction in EHL contacts by surface microtexturing—Tribological performance, manufacturing and tailored design. *Ind. Lubr. Tribol.* **2019**, *71*, 986–990. [CrossRef]
117. Zambrano, V.; Brase, M.; Hernández-Gascón, B.; Wangenheim, M.; Gracia, L.; Viejo, I.; Izquierdo, S.; Valdés, J. a digital twin for friction prediction in dynamic rubber applications with surface textures. *Lubricants* **2021**, *9*, 57. [CrossRef]
118. Shea, T.M.; Gunsel, S. Modeling base oil properties using NMR spectroscopy and neural networks. *Tribol. Trans.* **2003**, *46*, 296–302. [CrossRef]
119. Dai, K.; Gao, X. Estimating antiwear properties of lubricant additives using a quantitative structure tribo-ability relationship model with back propagation neural network. *Wear* **2013**, *306*, 242–247. [CrossRef]
120. Durak, E.; Salman, O.; Kurbanoglu, C. Analysis of effects of oil additive into friction coefficient variations on journal bearing using artificial neural network. *Ind. Lubr. Tribol.* **2008**, *60*, 309–316. [CrossRef]
121. Humelnicu, C.; Ciortan, S.; Amortila, V. Artificial neural network-based analysis of the tribological behavior of vegetable oil–diesel fuel mixtures. *Lubricants* **2019**, *7*, 32. [CrossRef]
122. Bhaumik, S.; Mathew, B.R.; Datta, S. Computational intelligence-based design of lubricant with vegetable oil blend and various nano friction modifiers. *Fuel* **2019**, *241*, 733–743. [CrossRef]
123. Bhaumik, S.; Pathak, S.; Dey, S.; Datta, S. Artificial intelligence based design of multiple friction modifiers dispersed castor oil and evaluating its tribological properties. *Tribol. Int.* **2019**, *140*, 105813. [CrossRef]

124. Mujtaba, M.; Masjuki, H.; Kalam, M.; Ong, H.C.; Gul, M.; Farooq, M.; Soudagar, M.E.M.; Ahmed, W.; Harith, M.; Yusoff, M. Ultrasound-assisted process optimization and tribological characteristics of biodiesel from palm-sesame oil via response surface methodology and extreme learning machine—Cuckoo search. *Renew. Energy* **2020**, *158*, 202–214. [CrossRef]
125. Summers, A.Z.; Gilmer, J.B.; Iacovella, C.R.; Cummings, P.T.; McCabe, C. MoSDeF, a Python framework enabling large-scale computational screening of soft matter: Application to chemistry-property relationships in lubricating monolayer films. *J. Chem. Theory Comput.* **2020**, *16*, 1779–1793. [CrossRef] [PubMed]
126. Baboukani, B.S.; Ye, Z.; Reyes, K.G.; Nalam, P.C. Prediction of nanoscale friction for two-dimensional materials using a machine learning approach. *Tribol. Lett.* **2020**, *68*, 1–14. [CrossRef]
127. Moayedi, H.; Hayati, S. Artificial intelligence design charts for predicting friction capacity of driven pile in clay. *Neural Comput. Appl.* **2019**, *31*, 7429–7445. [CrossRef]
128. Rouet-Leduc, B.; Hulbert, C.; Bolton, D.C.; Ren, C.X.; Riviere, J.; Marone, C.; Guyer, R.A.; Johnson, P.A. Estimating fault friction from seismic signals in the laboratory. *Geophys. Res. Lett.* **2018**, *45*, 1321–1329. [CrossRef]
129. Tijani, I.; Akmeliawati, R. Support vector regression based friction modeling and compensation in motion control system. *Eng. Appl. Artif. Intell.* **2012**, *25*, 1043–1052. [CrossRef]
130. Tijani, I.B.; Wahyudi, M.; Talib, H. Adaptive neuro-fuzzy inference system (ANFIS) for friction modelling and compensation in motion control system. *Int. J. Model. Simul.* **2011**, *31*, 32–41. [CrossRef]
131. Ao, Y.; Wang, Q.; Chen, P. Simulating the worn surface in a wear process. *Wear* **2002**, *252*, 37–47. [CrossRef]
132. Haviez, L.; Toscano, R.; El Youssef, M.; Fouvry, S.; Yantio, G.; Moreau, G. Semi-physical neural network model for fretting wear estimation. *J. Intell. Fuzzy Syst.* **2015**, *28*, 1745–1753. [CrossRef]
133. Argatov, I.I.; Chai, Y.S. Artificial neural network modeling of sliding wear. *Proc. Inst. Mech. Eng. Part J J. Eng. Tribol.* **2020**, *235*, 748–757. [CrossRef]
134. Almqvist, A. Fundamentals of physics-informed neural networks applied to solve the Reynolds boundary value problem. *Lubricants* **2021**, *9*, 82. [CrossRef]
135. Bucholz, E.W.; Kong, C.S.; Marchman, K.R.; Sawyer, W.G.; Phillpot, S.; Sinnott, S.B.; Rajan, K. Data-driven model for estimation of friction coefficient via informatics methods. *Tribol. Lett.* **2012**, *47*, 211–221. [CrossRef]
136. Perčić, M.; Zelenika, S.; Mezić, I. Artificial intelligence-based predictive model of nanoscale friction using experimental data. *Friction* **2021**, *9*, 1726–1748. [CrossRef]

MDPI
St. Alban-Anlage 66
4052 Basel
Switzerland
Tel. +41 61 683 77 34
Fax +41 61 302 89 18
www.mdpi.com

Lubricants Editorial Office
E-mail: lubricants@mdpi.com
www.mdpi.com/journal/lubricants

www.ingramcontent.com/pod-product-compliance
Lightning Source LLC
LaVergne TN
LVHW071015170726
843515LV00006B/1139